Chemistry

Study and Revision Guide

Christopher Talbot
Richard Harwood

Acknowledgements

We thank the following for their invaluable advice on the content of selected chapters: Professor Norman Billingham, University of Sussex, Professor Jon Owen, University of Southampton, Professor David Jenkins, University of York, Gordon Wood, Dr. David Fairley, Overseas Family School, Singapore, Professor Stanley Furrow, Pennsylvania State University, Professor Philip Walker, University of Surrey, Dr. David Cooper, University of Liverpool, Professor Laurence Harwood, University of Reading, Professor Mike Williamson, University of Sheffield and Dr. Jon Cooper, University College, London.

Photo credits:

p.130 all © Richard Harwood

The Options and answers are free online at www.hoddereducation.com/IBextras

Although every effort has been made to ensure that website addresses are correct at time of going to press, Hodder Education cannot be held responsible for the content of any website mentioned in this book. It is sometimes possible to find a relocated web page by typing in the address of the home page for a website in the URL window of your browser.

Hachette UK's policy is to use papers that are natural, renewable and recyclable products and made from wood grown in well-managed forests and other controlled sources. The logging and manufacturing processes are expected to conform to the environmental regulations of the country of origin.

Orders: please contact Hachette UK Distribution, Hely Hutchinson Centre, Milton Road, Didcot, Oxfordshire, OX11 7HH. Telephone: +44 (0)1235 827827. Email education@hachette. co.uk. Lines are open from 9 a.m. to 5 p.m., Monday to Saturday, with a 24-hour message answering service. You can also order through our website: www.hoddereducation.com

First published in 2017 by
Hodder Education,
An Hachette UK Company
Carmelite House
50 Victoria Embankment
London EC4Y 0DZ
www.hoddereducation.com

Impression number   10 9 8 7 6 5 4 3

Year   2021

Cover photo © ESA/Herschel/PACS/MESS Key Programme Supernova Remnant Team; NASA, ESA and Allison Loll/Jeff Hester (Arizona State University)

Illustrations by Aptara Inc.

Typeset in Goudy Oldstyle 10/12 pts by Aptara Inc.

Printed and bound by CPI Group (UK) Ltd, Croydon, CR0 4YY

A catalogue record for this title is available from the British Library.

ISBN: 9781471899713

Contents

How to use this revision guide   vi

Getting to know Papers 1, 2 and 3   vi

Assessment objectives   vii

Countdown to the exams   viii

Core

Topic 1 Stoichiometric relationships   1

■ 1.1 Introduction to the particulate nature of matter and chemical change   1

■ 1.2 The mole concept   6

■ 1.3 Reacting masses and volumes   10

■ 1.4 Gases   14

Topic 2 Atomic structure   19

■ 2.1 The nuclear atom   19

■ 2.2 Electron configuration   23

Topic 3 Periodicity   32

■ 3.1 Periodic table   32

■ 3.2 Periodic trends   34

Topic 4 Chemical bonding and structure   44

■ 4.1 Ionic bonding and structure   44

■ 4.2 Covalent bonding   49

■ 4.3 Covalent structures   54

■ 4.4 Intermolecular forces   60

■ 4.5 Metallic bonding   62

Topic 5 Energetics/thermochemistry   66

■ 5.1 Measuring energy changes   66

■ 5.2 Hess's law   71

■ 5.3 Bond enthalpies   76

Topic 6 Chemical kinetics   79

■ 6.1 Collision theory and rates of reaction   79

Topic 7 Equilibrium   90

■ 7.1 Equilibrium   90

Topic 8 Acids and bases 99

- 8.1 Theories of acids and bases 99
- 8.2 Properties of acids and bases 102
- 8.3 The pH scale 103
- 8.4 Strong and weak acids and bases 105
- 8.5 Acid deposition 108

Topic 9 Redox processes 112

- 9.1 Oxidation and reduction 112
- 9.2 Electrochemical cells 122

Topic 10 Organic chemistry 127

- 10.1 Fundamentals of organic chemistry 127
- 10.2 Functional group chemistry 134

Topic 11 Measurement, data processing and analysis 144

- 11.1 Uncertainties and errors in measurements and results 144
- 11.2 Graphical techniques 151
- 11.3 Spectroscopic identification of organic compounds 154

Additional higher level (AHL)

Topic 12 Electrons in atoms 164

- 12.1 Electrons in atoms 164

Topic 13 The periodic table – the transition metals 171

- 13.1 First-row d-block elements 171
- 13.2 Coloured complexes 176

Topic 14 Chemical bonding and structure 186

- 14.1 Further aspects of covalent bonding and structure 186
- 14.2 Hybridization 190

Topic 15 Energetics/thermochemistry 198

- 15.1 Energy cycles 198
- 15.2 Entropy and spontaneity 202

Topic 16 Chemical kinetics 214

- 16.1 Rate expression and reaction mechanism 214
- 16.2 Activation energy 225

Topic 17 Equilibrium 227

- 17.1 The equilibrium law 227

Topic 18 Acids and bases   236

■ 18.1 Lewis acids and bases   236

■ 18.2 Calculations involving acids and bases   238

■ 18.3 pH curves   244

Topic 19 Redox processes   252

■ 19.1 Electrochemical cells   252

Topic 20 Organic chemistry   264

■ 20.1 Types of organic reactions   264

■ 20.2 Synthetic routes   270

■ 20.3 Stereoisomerism   273

Topic 21 Measurement and analysis   281

■ 21.1 Spectroscopic identification of organic compounds   281

Option chapters and answers

Option chapters and answers appear on the website accompanying this book:
www.hoddereducation.com/IBextras

How to use this revision guide

Welcome to the Chemistry for the IB Diploma Revision Guide!

This book will help you plan your revision and work through it in a methodological way. The guide follows the Chemistry syllabus topic by topic, with revision and practice questions to help you check your understanding.

■ Features to help you succeed

Expert tip

These tips give advice that will help you boost your final grade.

■ QUICK CHECK QUESTIONS

Use these questions provided throughout each section to make sure you have understood a topic. They are short knowledge-based questions that use information directly from the text.

Common mistake

These identify typical mistakes that candidates make and explain how you can avoid them.

Worked example

Some parts of the course require you to carry out mathematical calculations, plot graphs, and so on: these examples show you how.

Key definitions

The definitions of essential key terms are provided on the page where they appear. These are words that will help you have a clear understanding of important ideas. A comprehensive **glossary** of chemical terms should be downloaded from the Hodder Plus website: https://www.hoddereducation.co.uk/ibextras/chemistryfortheib

You can keep track of your revision by ticking off each topic heading in the book. Tick each box when you have:
■ revised and understood a topic
■ tested yourself using the **Quick check questions**.

Online material can be found on the website accompanying this book www.hoddereducation.com/IBextras/

Online material included:
■ option chapters
■ answers to Quick check questions
■ glossary

Use this book as the cornerstone of your revision. Don't hesitate to write in it and personalize your notes. Use a highlighter to identify areas that need further work. You may find it helpful to add your own notes as you work through each topic. Good luck!

Getting to know Papers 1, 2 and 3

At the end of your two year IB Chemistry course you will sit three papers – Papers 1, 2 and 3. Paper 1 is worth 20% of the final marks, Paper 2 is worth 36% of the final marks and Paper 3 is worth 20% of the final marks.

The other assessed part of the course (24%) is made up of the Internal Assessment (practical work), which is marked by your teacher and then moderated by the IBO.

Here is some general advice for the exams:
■ Make sure you have learnt the command terms (e.g. evaluate, explain and outline): there is a tendency to focus on the content in the question rather than the command term, but if you do not address what the command term is asking of you then you will not be awarded marks.
■ Answer all questions and do not leave gaps.
■ Do not write outside the answer boxes provided – if you do so this work will not be marked.
■ If you run out of room on the page, use continuation sheets and indicate clearly that you have done this on the cover sheet. The fact that the question continues on another sheet of paper needs to be clearly indicated in the text box provided.
■ Plan your time carefully before the exams – this is especially important for Papers 2 and 3.

■ Paper 1

Paper 1 consists of multiple choice questions with four responses. There is no reading time for these papers. HL Paper 1 has 40 questions (15 in common with SL) to be answered in one hour. The SL Paper 1 has 30 questions (15 in common with HL) to be answered in 45 minutes. There is no data booklet provided but a periodic table is printed at the front of the papers. A calculator is not allowed for Paper 1.

■ Paper 2

The HL paper 2 has a maximum of 95 marks and lasts 2 hours and 15 minutes. The SL paper 2 has a maximum of 50 marks and lasts 1 hour and 15 minutes. A data booklet is provided and a calculator is allowed. There is five minutes reading time for the papers. There are a variety of types of questions in three sections.

■ Paper 3

The SL and HL Paper 3 both have a maximum of 45 marks and each lasts 1 hour and 15 minutes. A calculator is needed and a data booklet is provided. There is five minutes reading time for the papers. Paper 3 has a compulsory Section A and Section B, where one Option question is answered.

Assessment objectives

To successfully complete the course, you need to have achieved the following objectives:

1 Demonstrate knowledge and understanding of:
   - facts, concepts and terminology
   - methodologies and techniques
   - methods of communicating scientific information.

2 Formulate, analyse and evaluate:
   - hypotheses, research questions and predictions
   - methodologies and techniques
   - primary and secondary data
   - scientific explanations.

3 Demonstrate the appropriate research, experimental and personal skills necessary to carry out insightful and ethical investigations.

Countdown to the exams

4-8 weeks to go

- Start by looking at the syllabus and make sure you know exactly what you need to revise.
- Look carefully at the checklist in this book and use it to help organize your class notes and to make sure you have covered everything.
- Work out a realistic revision plan that breaks down the material you need to revise into manageable pieces. Each session should be around 25–40 minutes with breaks in between. The plan should include time for relaxation.
- Read through the relevant sections of this book and refer to the expert tips, common mistakes, key definitions and worked examples.
- Tick off the topics that you feel confident about, and highlight the ones that need further work.
- Look at past papers. They are one of the best ways to check knowledge and practise exam skills. They will help you identify areas that need further work.
- Try different revision methods, e.g. summary notes, mind maps, flash cards.
- Test your understanding of each topic by working through the 'Quick check' questions provided in each chapter. Also work through past exam papers.
- Make notes of any problem areas as you revise, and ask a teacher to go over them in class.

My exams

Chemistry Paper 1

Date:...................................

Time:...................................

Location:..............................

Chemistry Paper 2

Date:...................................

Time:...................................

Location:..............................

Chemistry Paper 3

Date:...................................

Time:...................................

Location:..............................

One week to go

- Aim to fit in at least one more timed practice of entire past papers, comparing your work closely with the mark scheme.
- Examine the checklist carefully to make sure you haven't missed any of the topics.
- Tackle any final problems by getting help from your teacher or talking them over with a friend.

The day before the examination

- Look through this book one final time. Look carefully through the information about Papers 1, 2 and 3 to remind yourself what to expect in the different papers.
- Check the time and place of the exams.
- Make sure you have all the equipment you need (e.g. extra pens, a watch, tissues). Make sure you have a calculator – this is needed in both papers.
- Allow some time to relax and have an early night so you are refreshed and ready for the exams.

Topic 1 Stoichiometric relationships

1.1 Introduction to the particulate nature of matter and chemical change

Essential idea: Physical and chemical properties depend on the ways in which different atoms combine.

Elements and compounds

■ Atoms of different elements combine in fixed ratios to form compounds, which have different properties from their component elements.

Elements

There are 92 naturally occurring **elements**. An element consists of one type of particle. These particles may be atoms or molecules (atoms of the same type bonded together covalently). Most elements consist of single atoms, and their formulas are simply the symbol for the element. However, some elements exist as diatomic molecules, having the formula X_2. Seven elements behave like this under normal conditions; hydrogen, oxygen, nitrogen and the halogens (group 17 – fluorine, chlorine, bromine and iodine).

> **Key definition**
> **Element** – a substance that cannot be broken down by any chemical reaction into simpler substances.

One possible way to remember them is by learning the mnemonic involving their symbols:

I Have **N**o **Br**ight **O**r **Cl**ever **F**riends!

Atoms of different elements have different sizes and different masses (Figure 1.1). Elements can be classified into metals, non-metals or metalloids.

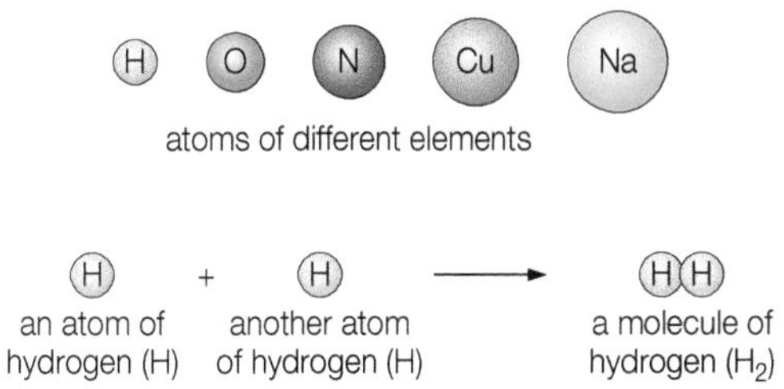

atoms of different elements

 + ⟶
an atom of   another atom   a molecule of
hydrogen (H)   of hydrogen (H)   hydrogen (H_2)

Figure 1.1 Representation of atoms

Compounds

Some pure substances are made up of a single element, although there may be more than one atom of the element in a particle of the substance. For example, as we have seen above, hydrogen is diatomic: molecules of hydrogen contain two hydrogen atoms and have the formula H_2 (H–H).

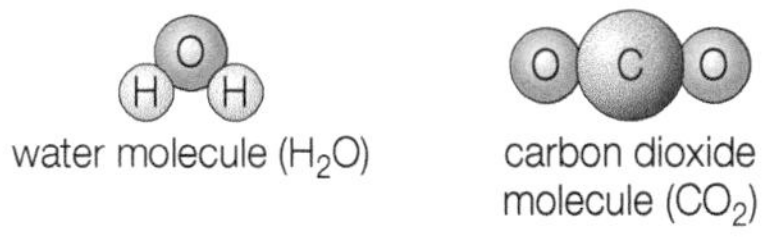

water molecule (H_2O)

carbon dioxide molecule (CO_2)

Figure 1.2 Representation of the molecular compounds water and carbon dioxide

Compounds are made up of two or more different atoms (or ions) that have chemically bonded in fixed ratios; but not always in molecules – we cannot use the term molecule when considering ionic compounds. For example, molecules of carbon dioxide (CO_2) contain one carbon atom and two oxygen atoms. However, zinc iodide is also a compound and contains zinc and iodide ions. It has the chemical formula, ZnI_2 [Zn^{2+} $2I^-$].

Zinc iodide is a compound, and can be broken down chemically into its constituent elements: iodine and zinc. This can be carried out by heating (thermal decomposition) or electrolysis of molten zinc iodide (electrolytic decomposition).

Laws of chemical combination

The law of conservation of mass states that during any physical or chemical change, the total mass of the products remains equal to the total mass of the reactants. The law of definite proportions states that a compound always contains the same elements combined together in the same proportion by mass. The law of multiple proportions states that when two elements combine with each other to form one or more compounds, the masses of one of the elements which combine with a fixed mass of the other, are in an integer ratio to one another.

NATURE OF SCIENCE

Making quantitative measurements with replicates to ensure reliability – definite and multiple proportions. The various laws of chemical combination were formulated after repeated accurate measurements of mass and gas volumes were performed on a variety of chemical reactions.

Mixtures

Revised

- **Mixtures** contain more than one element and/or compound that are not chemically bonded together and so retain their individual properties.
- Mixtures are either homogeneous or heterogeneous.

The components of a mixture (Figure 1.3) may be elements or compounds. These components are not chemically bonded together and therefore the components of a mixture keep their individual properties.

If all the components of a mixture are in the same phase the mixture is *homogeneous*. In a homogeneous mixture, the components remain mixed with each other and the composition is uniform. If the components of a mixture are in different phases the mixture is *heterogeneous*. There is a phase boundary between two phases.

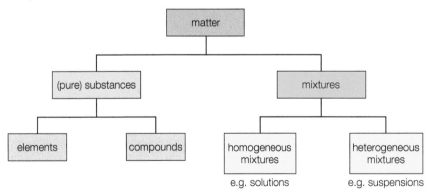

Figure 1.3 Classification of matter

Homogeneous mixtures can be separated by certain physical methods, e.g. distillation/crystallization for a sodium chloride solution, fractional distillation for liquid air, petroleum, or an ethanol/water mixture.

Heterogeneous mixtures can be separated by a different set of physical (mechanical) methods, e.g. filtration for a suspension of fine sand in water.

States of matter and changes of state

Revised ☐

One of the most fundamental ideas in chemistry is that all matter consists of
particles since we use this idea to explain many of the chemical and physical
behaviours that are observed (the *kinetic model* of matter).

Figure 1.4 shows the interconversions between the three states of matter and the
arrangement of their particles (ions, atoms or molecules). These changes of state
are physical changes and occur at constant pressure. Table 1.1 summarizes the
properties of the three states of matter and their description according to kinetic
molecular theory.

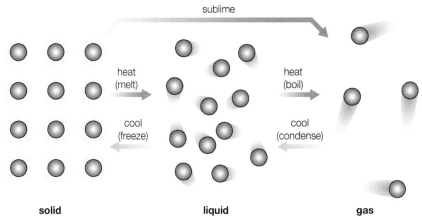

Figure 1.4 The three states of matter and their interconversion

Table 1.1 Properties of the three states of matter and their description by kinetic molecular theory

Solids	Liquids	Gases
● Particles are close together.	● Particles are usually slightly further apart than in the solid.	● Particles are far apart from one another.
● Particles have lower kinetic energy than in the other two states.	● Particles have greater values of kinetic energy than those in the solid state.	● Particles have much more kinetic energy than the other two states.
● Particles can only vibrate about fixed positions.	● Particles can move (translation) so diffusion may occur.	● Particles move at high speeds in straight lines between collisions.
● There are strong forces of attraction between particles.		● There is very little attraction between the particles (no intermolecular forces operating between the particles of an ideal gas).
Properties		
Fixed shape	Variable shape	Variable shape (occupies whole container due to diffusion)
Fixed volume	Fixed volume	Variable volume (affected by temperature and pressure)
Incompressible	Incompressible	Highly compressible
High density	Intermediate density	Very low density

Substances can change their state of matter as a result of temperature changes (or changes in pressure). Table 1.2 shows the changes in energy and movement that occur during a change in state (the shaded cells) as a result of changes in temperature. Remember that the temperature of a substance is directly related to the average kinetic of all its particles.

When the temperature of a crystalline solid is raised, the particles vibrate increasingly vigorously until they can no longer be held in an ordered arrangement (lattice) by the forces of attraction and the solid melts. Raising the temperature of a liquid increases the average speed of the particles until their kinetic energy is sufficient to overcome the forces of attraction between them: the liquids boils. Evaporation is the conversion of the liquid to the gaseous state, below the boiling point. It occurs at the surface of a liquid, but boiling involves the entire liquid and bubble formation.

Table 1.2 The physical changes of state taking place as a substance is heated (left hand column) or cooled down (right hand column). These transitions are often represented by a heating or cooling curve.

			Cooling (exothermic)
Gas	Particles gain kinetic energy and vibrate, rotate and move faster, move randomly, and move further apart (gas expands). *Temperature of the gas rises.*	Particles lose kinetic energy and vibrate, rotate and translate less, come closer together (gas contracts). *The temperature decreases to condensation point.*	Gas
Evaporation, boiling	Particles use the thermal energy supplied to overcome attractive forces between particles and escape the liquid. *Temperature remains the same.*	Particles come closer together and release energy (latent heat). *Temperature remains the same.*	Condensation
Liquid	Particles gain kinetic energy and vibrate, rotate and move faster. *The temperature rises to boiling point.*	Particles lose kinetic energy and vibrate, rotate and translate less. *The temperature decreases to freezing point.*	Liquid
Melting	Particles use the thermal energy supplied to overcome attractive forces between them. *Temperature remains the same.*	Particles lose energy (latent heat) as attractive forces are formed and the particles return to the lattice or regular arrangement. *Temperature remains the same.*	Freezing
Solid	Particles gain kinetic energy and vibrate more and move further apart. Substance expands but maintains its shape. *Temperature increases to melting point.*	Particles lose kinetic energy – *temperature decreases* – and its particles vibrate less and move closer together.	Solid

Heating (endothermic)

On cooling the reverse changes occur. The particles of the gas gradually slow down as the temperature falls until the forces of attraction are able to condense them together and form a liquid. Cooling the liquid causes further loss of kinetic energy until eventually the particles form a crystalline solid where they vibrate about fixed points.

> **Expert tip**
>
> Sublimation is the conversion of a solid to gas at constant temperature; vapour deposition is the reverse process.

Chemical equations

Revised

During a chemical reaction new substances are formed. There is an enthalpy change between the reacting chemicals (system) and its surroundings: heat will be released or absorbed. This is due to bonds in the reactants being broken and bonds in the products being formed. There is a fixed relationship between the masses of the *reactants* and *products*, known as the stoichiometry of the reaction.

Chemical reactions can be described by balanced equations showing the formulas of the substances and their physical states (under standard conditions). Equations need to be balanced with (stoichiometric) *coefficients* because atoms cannot be created or destroyed during a chemical reaction.

Consider the following balanced equation:

$2H_2(g) + O_2(g) \rightarrow 2H_2O(l)$

Qualitatively, an equation states the names of the reactants and products (and gives their physical states using state symbols).

Quantitatively, it expresses the following relationships:
- the *relative number of molecules* of the reactants and products – here 2 molecules of hydrogen react with 1 molecule of oxygen to form 2 molecules of water
- the (*amounts*) in moles of the particles that form the reactants and products – 2 moles of hydrogen molecules react with 1 mole of oxygen molecules to form 2 moles of water molecules (one mole is 6.02×10^{23} particles)
- the *relative masses* of reactants and products – 4.04 g of molecular hydrogen reacts with 32.00 g of molecular oxygen to form 36.04 g of water molecules; the law of conservation of mass is obeyed
- the *relative volumes* of gaseous reactants and products (Avogadro's law) – 2 volumes of hydrogen react with 1 volume of oxygen to form liquid water. For example, $2 \, dm^3$ of hydrogen would react with $1 \, dm^3$ of oxygen.

Simple equations are balanced by a process of 'trial and error' (inspection). However, it is helpful to balance the formula which contains the maximum number of elements first.

■ **QUICK CHECK QUESTION**

3   Balance the following equations.

$Cl_2(g) + NaOH(aq) \rightarrow NaCl(aq) + NaClO_3(aq) + H_2O(l)$

$MnO_2(s) + HCl(aq) \rightarrow MnCl_2(aq) + Cl_2(g) + H_2O(l)$

$Cs(s) + H_2O(l) \rightarrow CsOH(aq) + H_2(g)$

$CuO(s) + NH_3(g) \rightarrow Cu(s) + N_2(g) + H_2O(l)$

$Na(s) + O_2(g) \rightarrow Na_2O(s)$

▨ Ionic equations

An ionic equation may be written when a reaction occurs in solution in which some of the ions originally present are removed from solution, or when ions not originally present are formed. Usually, ions are removed from solution by one or more of the following processes:
- formation of an insoluble ionic compound (a precipitate)
- formation of molecules containing only covalent bonds (water, for instance)
- formation of new ionic species and
- formation of a gas (e.g. carbon dioxide).

To deduce an ionic equation follow these steps:
1   Write down the balanced molecular equation.

2   Convert those chemicals that are ions in solution into their ions.

3   Cross out spectator ions that appear on both sides of the equation in solution as they have not changed during the reaction.

4   Check that the atoms and charges balance in the ionic equation.

■ **QUICK CHECK QUESTION**

4   Write ionic equations for the following reactions.

$AgNO_3(aq) + NaBr(aq) \rightarrow AgBr(s) + NaNO_3(aq)$

$CH_3COOH(aq) + LiOH(aq) \rightarrow CH_3COOLi(aq) + H_2O(l)$

$Zn(s) + FeSO_4(aq) \rightarrow ZnSO_4(aq) + Fe(s)$

$Na_2CO_3(s) + 2HCl(aq) \rightarrow CO_2(g) + H_2O(l) + 2NaCl(aq)$

$CuSO_4(aq) + 2NaOH(aq) \rightarrow Na_2SO_4(aq) + Cu(OH)_2(s)$

Common mistake

It should also be noted that some reactions do *not* occur, even though balanced equations can be written, for example, $Cu(s) + H_2SO_4(aq) \rightarrow CuSO_4(aq) + H_2(g)$. Hence, the activity series (Topic 9, Redox processes) should be consulted before equations for replacement reactions are written.

Expert tip

If you balance an equation in a multiple choice question and deduce the sum of the coefficients, do not forget the 1's in your calculation. So for the equation

$2Pb(NO_3)_2 \rightarrow 2PbO + 4NO_2 + O_2$

the answer is **9**. Note that the *stoichiometric coefficients in a balanced equation are whole numbers*.

Expert tip

It is good practice to include state symbols in equations (and equilibrium expressions) and they are essential in thermochemical equations.

Expert tip

An ionic equation should be balanced with respect to atoms (elements) and charge. The total charges on both sides of the ionic equation should be the same.

Expert tip

Insoluble salts include halides of lead(II), barium sulfate, most silver salts (except silver nitrate) and most carbonates (except group 1). For straightforward precipitation reactions the simplest way to construct the ionic equation is to write down the precipitate on the product side, then fill in the ions that would make that precipitate as the reactants.

■ Types of reaction

A direct synthesis reaction involves the combination of two or more elements to produce a single product. Single replacement (displacement) reactions occur when one element replaces another in a compound. These types of reaction are both redox reactions as they involve a transfer of electrons.

Decomposition reactions involve a single reactant being broken down into two or more products. Precipitation (or double displacement) reactions occur between ions in aqueous solution to form an insoluble precipitate. Acid–base (Brønsted–Lowry) reactions involve a transfer of protons (H^+).

■ **QUICK CHECK QUESTION**

5   Write ionic equations for the following reactions and state the type of reaction.

calcium reacting with dilute nitric acid

barium chloride solution reacting with sodium sulfate solution

copper(II) oxide reacting with hydrochloric acid

Expert tip

Pure water is a liquid and represented as $H_2O(l)$ and not an aqueous solution, $H_2O(aq)$.

1.2 The mole concept

Revised ☐

Essential idea: The mole makes it possible to correlate the number of particles with the mass that can be measured.

Mole concept and the Avogadro constant

Revised ☐

■ The mole is a fixed number of particles and refers to the amount, n, of a substance.

As particles such as atoms, ions and molecules are extremely small, chemists measure amounts of substance using a fixed amount called the **mole**. This fixed amount contains a certain number of particles referred to as the **Avogadro constant**, which has a value of $6.02 \times 10^{23}\,mol^{-1}$.

The mass of one mole of a substance has units of gram per mole ($g\,mol^{-1}$). Multiplying an amount of substance (in number of moles) by Avogadro's constant converts it to the number of particles present. Dividing the number of particles present by the Avogadro constant gives the amount (in mol) (Table 1.3).

Table 1.3 Amounts of selected substances and the number of particles in multiples of Avogadro's constant

Amount of substance/mol	Mass/g	Number of particles (in terms of N_A or L)
1 mole of aluminium, Al	26.98	N_A or L atoms
1 mole of oxygen molecules, O_2	32.00	$2N_A$ or $2L$ atoms or N_A or L molecules
1 mole of ethene, C_2H_4	26.04	$6N_A$ or $6L$ atoms or N_A or L molecules
1 mole of calcium nitrate, $Ca(NO_3)_2$	164.10	N_A or L formula units of $Ca(NO_3)_2$ N_A or L ions of Ca^{2+}, $2N_A$ or $2L$ ions of NO_3^- Total: $3N_A$ or $3L$ ions

Key definitions

Mole – that amount (n) of substance that contains the same number of particles (atoms, ions, molecules or electrons) as there are atoms in exactly 12 g of isotope carbon-12.

Avogadro constant – the number of particles in a mole; found (experimentally) to be 6.02×10^{23} (to 3 s.f.) and has the symbol L or N_A.

Molar mass has the units of $g\,mol^{-1}$

Revised ☐

M refers to the *molar mass*. It is the mass of substance in grams that contains one mole of particles (Figure 1.5). It has units of gram per moles, $g\,mol^{-1}$, and the same numerical value as the relative atomic mass, A_r, or relative formula mass, M_r. The molar mass of a compound is the sum of the molar masses of all the atoms in the molecule or formula unit.

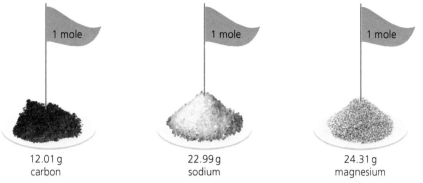

Figure 1.5 The mole concept for selected elements

In each pile there is one mole of atoms (6.02×10^{23} atoms). The relative atomic masses of carbon, sodium and magnesium are 12.01, 22.99 and 24.31. The molar masses are 12.01 g mol⁻¹, 22.99 g mol⁻¹ and 24.31 g mol⁻¹.

Masses of atoms

- Masses of atoms are compared on a scale relative to ^{12}C and are expressed as relative atomic mass (A_r) and relative formula/molecular mass (M_r).

Relative atomic mass and **relative formula** (or **molecular**) **mass** are pure numbers without units. Do not write them with units of grams. Molar mass has units of gram per mole, not gram.

The term relative molecular mass only really applies to simple molecules (a group of covalently bonded atoms); the concept of relative formula mass is used for giant structures like ionic substances (e.g. sodium chloride, copper(II) sulfate) and macromolecules (e.g. silicon dioxide).

> **Key definitions**
>
> **Relative atomic mass, A_r**
> (Figure 1.6) – the weighted average (according to their percentage abundances) of the atomic masses of its isotopes compared to 1/12th of (the mass of) of ^{12}C.
>
> **Relative formula (molecular) mass, M_r** – the sum of the relative atomic masses of the elements present in the formula unit or molecule.

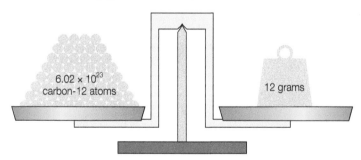

Figure 1.6 Illustration of the Avogadro constant

To calculate the mass of one molecule (or one ion or atom) in grams, calculate the molar mass of the substance and divide by Avogadro's constant.

> **Expert tip**
>
> Remember that the mass (in grams or kilograms) of a single atom or molecule is going to be extremely small.

■ QUICK CHECK QUESTIONS

6 Deduce molar masses of the following species: uranium atom, U, sulfur hexafluoride molecule, SF_6, lead(IV) ion, Pb^{4+}, and ammonium sulfate, $(NH_4)_2SO_4$ $[2NH_4^+ SO_4^{2-}]$.

7 Calculate the number of atoms in 0.5 moles of atoms of nitrogen (N) and 0.2 moles of molecules of nitrogen (N_2).

8 Calculate the mass (in milligram) of 0.020 moles of calcium sulfate, $CaSO_4$.

9 Calculate the mass (in gram) of one molecule of nitrogen dioxide, NO_2.

10 Calculate the amount, mass and number of chloride ions in 0.400 mol of magnesium chloride, $MgCl_2$.

11 0.10 mol of a substance has a mass of 4.00 g. Calculate its molar mass.

12 One argon atom has a mass of 6.64×10^{-26} kg. Calculate the relative atomic mass of argon.

> **Common mistake**
>
> Relative atomic and molecular masses are not referenced to hydrogen atoms, ^1H, the lightest atoms. Carbon-12 has been chosen (for practical reasons) as the reference nuclide for comparing the relative masses of atoms and molecules. Remember too that all *relative* masses do not have units, they are simply a number.

Interconversion between amount and other properties

The amount of a pure substance (solid, liquid or gas) can be found from the mass of the substance and the molar mass:

$$\text{amount (mol)} = \frac{\text{mass of substance (g)}}{\text{molar mass (g mol}^{-1})}$$

In the laboratory substances are most conveniently measured by weighing for solids and by volume for liquids and gases. However, gases and liquids can both be weighed and masses and volumes can be converted via density (density = $\frac{\text{mass}}{\text{volume}}$).

Solution of problems involving the relationships between the number of particles, the amount of substance in moles and the mass in grams

Revised

The relationships between the amount of substance, number of particles (atoms, ions and molecules), mass of pure solid, liquid or gas (and volume of pure gas) are very important (Figure 1.7). Many stoichiometry calculations involve converting from one part of this relationship to another.

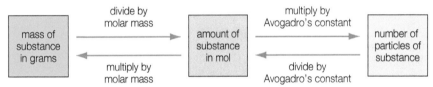

Figure 1.7 Summary of interconversions between amount, mass and number of particles

■ QUICK CHECK QUESTIONS

13 Calculate the mass of carbon dioxide produced from heating 5.50 g of sodium hydrogen carbonate which on heating will completely decompose and form sodium carbonate, water and carbon dioxide.

14 Calculate the mass of titanium(IV) oxide formed when 10.00 g of titanium(IV) chloride reacts with water:

$TiCl_4(s) + 2H_2O (l) \rightarrow TiO_2(s) + 4HCl(aq)$

15 Determine the mass of fluorine (F_2) required to produce 150.00 g of C_4F_{10}.

$C_4H_{10}(g) + 10F_2(g) \rightarrow C_4F_{10}(g) + 10HF(g)$

Formulas of compounds

Revised

■ The empirical formula and molecular formula of a compound give the simplest ratio and the actual number of atoms present in a molecule respectively.

Empirical formulas

■ Interconversion of the percentage composition by mass and the empirical formula

An empirical formula (obtained by experiments) shows the simplest whole number ratio of atoms of each element. It can be deduced from the percentage composition by mass of the compound. The percentage composition values can be converted directly into mass by assuming 100 g of the pure compound. The amounts of each atom can then be calculated using the molar masses. The numbers may need to be multiplied to obtain integer values.

■ Obtaining and using experimental data for deriving empirical formulas from reactions involving mass changes

The empirical formula can also be calculated when given mass data. The mass of a missing element can be calculated by difference. The masses of each element are converted to amounts (mol) via the molar masses. The masses are divided by the smallest value. The numbers may need to be multiplied to obtain integer values.

■ Molecular formulas

■ Determination of the molecular formula of a compound from its empirical formula and molar mass

The molecular formula shows the actual number of atoms of each element in a molecule of the substance. It can be obtained from the empirical formula if the molar mass of the compound is known. The empirical and molecular formulas are connected by the following relationship:

$$\frac{\text{molecular formula}}{\text{empirical formula}} = n \text{ (where } n = 1, 2, 3, \text{ etc.)}$$

The empirical formula of organic compounds can be obtained from combustion data. The masses of carbon and hydrogen can be obtained from the masses of water and carbon dioxide (via their percentage by masses of oxygen and carbon).

■ Structural formulas

Structural formulas show the arrangement of atoms and bonds within a molecule and are used in organic chemistry. Figure 1.8 shows the structural formulas of phosphorus(III) and phosphorus(V) oxides. The molecular formulas are P_4O_6 and P_4O_{10}, and the empirical formulas are P_2O_3 and P_2O_5, respectively.

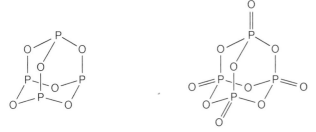

Figure 1.8 The structures of phosphorus(III) oxide and phosphorus(V) oxide

> **Expert tip**
>
> The empirical formula of a compound often may not correspond to a stable species. For example, the empirical formula of H_2O_2 is OH, which does not exist as a stable species.

■ QUICK CHECK QUESTIONS

16 Deduce the empirical formulas of the molecular formulas of the following compounds:

C_6H_6, C_6H_{12}, PH_3, B_2H_6, N_2O_4, Fe_2O_3, C_2H_2

17 The composition by mass of a compound is 40.2% potassium, 32.9% oxygen and 26.9% chromium. Determine the empirical formula of the compound.

18 When 2.67 g of copper reacts with excess sulfur (in the form of atoms), the mass of the compound formed is 4.01 g. Determine the empirical formula of the compound.

19 A compound has an empirical formula of CH_2O. Its molar mass is approximately 180 g mol⁻¹. Determine its molecular formula.

20 1.615 g of an anhydrous salt was placed in moist air. After complete hydration its mass increased to 2.875 g. The composition by mass of the salt was Zn = 40.6%, S = 19.8% and O = 39.6%. Determine the empirical formula of the compound.

21 0.5000 g of an organic compound containing carbon, hydrogen and oxygen formed 0.6875 g of carbon dioxide and 0.5625 g of water on complete combustion. Determine the empirical formula of the compound.

1.3 Reacting masses and volumes

Essential idea: Mole ratios in chemical equations can be used to calculate reacting ratios by mass and gas volume.

Reacting masses

A balanced equation shows the reacting ratios of the amounts (in moles) of the reactants and products (stoichiometry). To find the mass of products formed in a reaction the mass of the reactants, the molar masses of the reactants and the balanced equation are needed. The stoichiometry of the reaction can also be found if the amounts (or masses) of each reactant that exactly react together and the amounts (or masses) of each product formed are known.

The example below shows how to calculate the approximate mass of iron(III) oxide needed to produce 28 g of iron by reduction with carbon monoxide.

■ **Step 1:** write the balanced equation

$Fe_2O_3(s) + 3CO(g) \rightarrow 2Fe(s) + 3CO_2(g)$

■ **Step 2:** convert the given mass data into moles

mass of iron required = 28 g

$$\text{number of moles of iron} = \frac{28\,g}{56\,g\,mol^{-1}} = 0.5\,mol\,\text{of iron}$$

0.5 mol of iron needs 0.25 mol of Fe_2O_3

■ **Step 3:** use molar ratio from equation
The equation tells us that to get 2 mol of Fe we need 1 mol of Fe_2O_3

■ **Step 4:** convert answer into grams

mass = number of moles × molar mass

$= 0.25\,mol \times 160\,g\,mol^{-1}$

$= 40\,g$

■ Limiting and excess reactants

Reactants can be either limiting or excess.

Chemicals react in stoichiometric ratios according to the balanced chemical equation. However, reagents are often not present in stoichiometric amounts; one of the reagents would be present as a *limiting reagent*, while other(s) would be present *in excess* (Figure 1.9). A limiting reactant is the first reagent to be consumed during the reaction, which causes the reaction to stop.

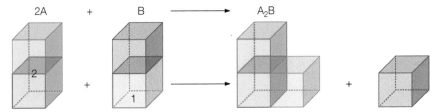

Figure 1.9 Illustrating how excess reactant B (in purple) remains unused after the reaction has taken place

To identify the limiting reactant write the balanced equation. Calculate the amount of each reagent that is involved in the reaction. Convert the reactant quantities to the corresponding amounts (mol). Taking into account the stoichiometric ratio, compare which of the reactants produces the smallest amount of product to determine the limiting reagent.

Expert tip

When finding which reactant is limiting it is useful to do the following simple calculation. If the number of moles of each reactant is divided by its coefficient in the balanced equation, then the smallest number indicates the limiting reactant.

Percentage yield calculations

The experimental yield can be different from the theoretical yield.

Chemists use percentage yield calculations to assess the efficiency of a chemical reaction or series of chemical reactions. Few reactions are completely efficient and most reactions, especially organic reactions, give relatively low yields.

The reaction may be incomplete so that some of the reactants do not react. There may be side reactions producing by-products instead of the required chemical product. Recovery of all the product from the reaction mixture is usually impossible. Some of the product is usually lost during transfer of the chemicals from one container to another.

The *theoretical yield* is the mass of the product assuming that the reaction goes to completion according to the stoichiometric equation and the synthesis is 100% efficient. The *actual yield* is the mass of the product obtained.

$$\text{percentage yield} = \frac{\text{actual mass of product}}{\text{theoretical yield}} \times 100$$

A percentage yield must be less than 100%.

Often one of the chemicals in a reaction mixture is present in an amount which limits the theoretical yield. The other reactants are added in excess to make sure that the most valuable chemical is converted to the required product.

> ■ **QUICK CHECK QUESTIONS**
>
> 22 Ammonia gas reacts with heated copper(II) oxide to form copper, water and nitrogen gas. Formulate a balanced equation. Determine the limiting reagent if 20.00 g of ammonia reacts with 85.00 g of copper(II) oxide. Calculate the percentage yield if 68.50 g of copper is produced.
> 23 State two reasons why the experimental yield from a reaction is less than the theoretical yield.

Solutions

Revised ■

The molar concentration of a solution is determined by the amount of solute and the volume of solution.

Concentration is the amount of solute (dissolved substance) in a known volume of solution (solute and solvent). It is expressed either in grams per cubic decimetre ($g\,dm^{-3}$), or more usually in moles per cubic decimetre ($mol\,dm^{-3}$). For very dilute solutions (or gases) it can be expressed in parts per million (ppm).

The amount of a solute can be calculated from the following expression:

amount (mol) = volume of solution (dm^3) × molar concentration ($mol\,dm^{-3}$)

A molar concentration can be converted to a mass concentration by the following expression:

mass concentration ($g\,dm^{-3}$) = molar mass ($g\,mol^{-1}$) × molar concentration ($mol\,dm^{-3}$).

Expert tip

Read questions carefully to establish whether the question is concerned with the solute or one or more of the ions released when the solute dissolves.

> ■ **QUICK CHECK QUESTIONS**
>
> 24 Determine the concentration in $g\,dm^{-3}$ and $mol\,dm^{-3}$ of the solution formed by dissolving 4.00 g of solid sodium hydroxide in 125.00 cm^3 of aqueous solution.
> 25 Calculate the concentrations (in $mol\,dm^{-3}$) of all the ions in 100.00 cm^3 of 0.020 $mol\,dm^{-3}$ aluminium sulfate solution, $Al_2(SO_4)_3$(aq).
> 26 Calculate the mass of potassium manganate(VII) needed to prepare 250.00 cm^3 of 0.02 $mol\,dm^{-3}$ $KMnO_4$(aq).

Titration

Titration can be used to find the reacting volumes of solutions. From the volumes and concentrations the equation can be determined. Titration of a solution of unknown concentration against a standard solution, with the equations for the reaction, allows an unknown concentration to be determined. If the solution to be titrated is concentrated then it may need to be diluted before titration. A dilution factor then needs to be applied in the calculation.

A standard solution is one of known concentration.

■ **QUICK CHECK QUESTIONS**

27 $1.00\,cm^3$ of concentrated hydrochloric acid was transferred with a graduated pipette to a $100.00\,cm^3$ volumetric flask. The volume was made up to $100.00\,cm^3$ with distilled water. A $10.00\,cm^3$ portion of the diluted solution from the volumetric flask was titrated with $KOH(aq)$ and was neutralized by $24.35\,cm^3$ of potassium hydroxide of concentration $0.0500\,mol\,dm^{-3}$. Calculate the concentration of the original concentrated hydrochloric acid in $mol\,dm^{-3}$.

28 An unknown group 1 metal carbonate reacts with hydrochloric acid:

$M_2CO_3(aq) + 2HCl(aq) \rightarrow 2MCl(aq) + CO_2(g) + H_2O(l)$

A $3.960\,g$ sample of M_2CO_3 was dissolved in distilled water to make $250.00\,cm^3$ of solution. A $25.00\,cm^3$ portion of this solution required $32.85\,cm^3$ of $0.175\,mol\,dm^{-3}$ hydrochloric acid for complete reaction. Calculate the molar mass of M_2CO_3 and determine the identify of the metal M.

29 A $20.00\,g$ sample of a cleaning solution containing aqueous ammonia ($NH_3(aq)$) was dissolved in water and the solution was made up to $500.00\,cm^3$ in a volumetric flask.

A $25.00\,cm^3$ portion of this solution was then neutralized with $26.85\,cm^3$ of $0.200\,mol\,dm^{-3}$ sulfuric acid. Calculate the percentage by mass of ammonia in the cleaning solution.

■ Back titration

Not all substances can be quantitatively determined by a direct titration method. Back titration (Figure 1.10) is an indirect titration method usually used when a compound is an insoluble solid in which the end point is difficult to detect, or a reaction is too slow.

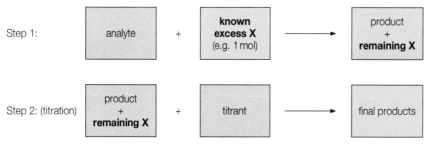

Figure 1.10 Concept of a back titration

Similarly, compounds which contain impurities that may interfere with direct titration or contain volatile substances (ammonia, iodine, etc.) that may result in inaccuracy arising due to loss of substance during titration require a back titration method.

■ Redox and precipitation titrations

Calculations for redox titrations are identical to that of simple acid–base titrations. Completion of a redox reaction may be shown by the final disappearance of a reactant. Titrations involving precipitation often involve determining the chloride ion concentration by titration with silver nitrate. As the silver nitrate solution is slowly added, a precipitate of silver chloride forms:

$Ag^+(aq) + Cl^-(aq) \rightarrow AgCl(s)$

The end point of the titration occurs when all the chloride ions are precipitated.

■ Consecutive reactions

It often takes more than one step to get the desired product. Reactions that are carried out one after another in sequence to yield a final product are called consecutive reactions. Any substance that is produced in one step and is consumed in another step of a multistep process is called an intermediate. The overall reaction is the chemical equation that expresses all the reactions occurring in a single overall equation.

■ Parts per million

When dealing with the levels of pollution in water, the concentrations are often very low, even though they can be dangerous. An example would be the levels of heavy metals in river water, for instance. In this case, the statutory allowed levels are often quoted in terms of the unit, parts per million (ppm).

The number of milligrams of solute per kg of solution (mg/kg) is equivalent to one ppm, since 1 mg = 10^{-3} g and 1 kg = 10^3 g. Assuming the density of water is 1.00 g cm⁻³, 1 dm³ of solution has a mass of 1 kg and hence, 1 mg dm⁻³ = 1 ppm. This is approximately true for dilute aqueous solutions. Parts per million concentrations are essentially mass ratios (solute to solution) × a million (10^6).

To convert concentrations in mg dm⁻³ (or ppm in dilute solution) to molarity (mol dm⁻³), divide the mass in grams by the molar mass of the analyte (chemical under analysis) to convert mass in milligram (mg) into a corresponding number of moles. To convert from molarity to mg dm⁻³ (or ppm in dilute solution), multiply by the molar mass of the analyte.

■ Atom economy

This term refers to the efficiency of a chemical process in terms of the atoms that are lost as by-products to the product. Industrial processes with poor atom economy are inefficient in terms of resources and often produce undesirable waste. The atom economy is calculated as a percentage of desired product divided by the total mass of products and scaled up by multiplying by 100 to make a percentage.

$$\text{Atom economy} = \frac{\text{molar mass of desired products}}{\text{total molar mass of all reactants}} \times 100$$

1.4 Gases

Kinetic theory of gases

The kinetic theory of gases assumes that a gas (Figure 1.11) is composed of widely spaced tiny particles (atoms or molecules); the particles are in rapid and random motion, moving in straight lines (between collisions); the particles have negligible volume; there are no intermolecular forces; the particles frequently collide with each other and with the walls of the container. Gas pressure arises from the force caused by collisions of gas molecules with the walls of the container. All collisions are perfectly elastic, that is, no momentum is lost on collision; the average kinetic energy of gas particles is directly proportional to the absolute temperature and remains constant at any temperature.

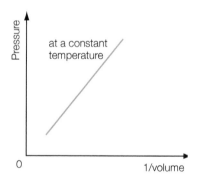

Figure 1.11 The kinetic theory description of a gas

Gas laws

The volume of a gas is changed by changes in temperature and pressure. The changes in the volume of a gas do not depend on the chemical nature of the gas. The behaviour of gases can be described by the gas laws. The gas laws were formulated from early experiments with gases, but can also be derived from the ideal gas equation.

■ Boyle's law

Boyle's law (Figure 1.12) states that, at a constant temperature, the volume of a fixed mass of an ideal gas is inversely proportional to its pressure. The mathematical expressions of Boyle's law are: $V \propto \dfrac{1}{P}$ (at constant temperature) or $PV = k$ (a constant) or $P_1V_1 = P_2V_2$, where P_1V_1 = original pressure and volume, respectively, and P_2V_2 = final pressure and volume, respectively.

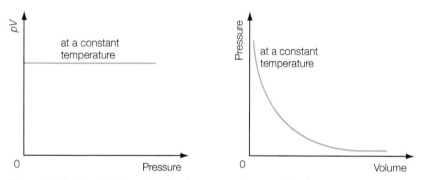

Figure 1.12 Graphical methods of representing Boyle's law

■ Charles' law

Charles' law states that, at a constant pressure, the volume of a fixed mass of an ideal gas is directly proportional to its absolute temperature. The mathematical expressions of Charles' law are: $V \propto T$ $(T = 273 + t\,°C)$, or $\dfrac{V}{T}$ = constant, or $\dfrac{V_1}{T_1} = \dfrac{V_2}{T_2}$.

Charles' law can be represented graphically (Figure 1.13) by plotting the volume of the gas against the temperature. On extrapolation, the graph meets the temperature axis at approximately −273 °C (to 3 s.f.). This temperature is called absolute zero and is adopted as the zero on the absolute or thermodynamic temperature scale (units of kelvin).

Expert tip

Any consistent units for pressure and volume may be used in gas law calculations, but the temperature must always be expressed in kelvin, because the gas laws are derived from thermodynamics.

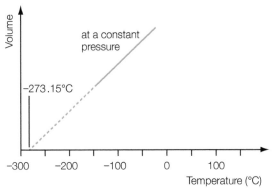

Figure 1.13 Charles' law

Temperature scales

To demonstrate this proportionality, you must use temperatures in kelvin, K (see Figure 1.13). If you use degrees Celsius (°C), then the graph will still be a straight line, but it does not go through the absolute zero.

When using these relationships for gases, it is important to use temperature values on the absolute temperature scale – where the values are in kelvin. This scale is the SI scale for temperature and is the true thermodynamic scale of temperature.

The Celsius scale is also commonly used in scientific and everyday work, but its basis is related to the physical properties of water (the 'fixed points' of the freezing point and the boiling point). The size of the 'degree' on both scales is the same, so it is easy to convert between the two (Figure 1.14).

The Fahrenheit temperature scale can also be used in everyday terms, but is no longer used in scientific work. This scale also uses the fixed points for water, with the values being 32°F and 312°F for the freezing point and boiling point respectively.

temperature (K) = temperature (°C) + 273.15

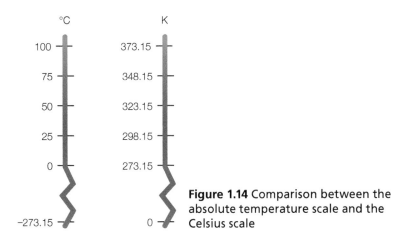

Figure 1.14 Comparison between the absolute temperature scale and the Celsius scale

> **Common mistake**
>
> Note that a doubling of the temperature in degrees Celsius is *not* a doubling of the absolute temperature. For example a doubling of the temperature from 200 °C to 400 °C is only a rise from (200 + 273) = 473 K to (400 + 273) = 673 K, that is, a ratio of 673/473 or 1.42.

Boyle's law and Charles' law can be combined to give the general gas equation:

$$\frac{P_1V_1}{T_1} = \frac{P_2V_2}{T_2}$$

This can be used to solve any gas law problem.

Pressure law

The pressure law states that for a fixed mass of gas (at constant volume) its absolute temperature is directly proportional to pressure. The mathematical expressions of the pressure law are: $P \propto T$, or $\frac{P}{T}$ = constant or $\frac{P_1}{T_1} = \frac{P_2}{T_2}$, where P_1 represents the initial pressure, T_1 represents the initial absolute temperature, P_2 represents the final pressure and T_2 represents the final absolute temperature. Table 1.4 summarizes the various gas laws.

> **Expert tip**
>
> When using this general gas equation $\left(\frac{P_1V_1}{T_1} = \frac{P_2V_2}{T_2}\right)$ you will need to be careful with the units you use:
>
> - Most importantly, *T* must be in kelvin!
> - The units of pressure and volume can be any, so long as they are the same for both the values in the equation.

Table 1.4 Summary of the gas laws

	Constant factor	Relationship	Effect
Boyle's law	Temperature	$V \propto \dfrac{1}{P}$	$P_1 \times V_1 = P_2 \times V_2$
Charles' law	Pressure	$V \propto T$	$\dfrac{V_1}{T_1} = \dfrac{V_2}{T_2}$
Pressure law	Volume	$P \propto T$	$\dfrac{P_1}{T_1} = \dfrac{P_2}{T_2}$
Avogadro's law	Volume and amount	$V \propto n$	$\dfrac{V_1}{n_1} = \dfrac{V_2}{n_2}$

■ **QUICK CHECK QUESTIONS**

36 Indicate whether gas pressure increases or decreases with each of the following changes in a sealed container: increasing the temperature, increasing the volume and increasing the number of gas molecules.

37 $50.00\,cm^3$ of a gas at $100.00\,kPa$ is compressed (at constant temperature) to a volume of $20.00\,cm^3$. Determine the final pressure of the gas.

38 $50.0\,cm^3$ of a sample of a gas is heated (at constant pressure) from 25.0 to 200.0 °C. Calculate the final volume of the gas if the pressure remains constant.

39 a A container is filled with gas at a pressure of $15\,kPa$ and is cooled from 2 °C to –40 °C. Calculate the final pressure, assuming the volume remains constant.

   b The density of a gas at STP is $1.78\,g\,dm^{-3}$. Calculate the molar mass of the gas.

Avogadro's law

Avogadro's law states that equal volumes of all gases at the same temperature and pressure contain equal numbers of atoms or molecules (Figure 1.15). The volume occupied by one mole of any gas (molar volume) depends on the temperature and pressure of the gas. The molar volume of any gas at standard temperature and pressure (STP) is $22.7\,dm^3$ (Figure 1.15). The conditions of STP are: temperature 0 °C (273 K); pressure: $1.00 \times 10^5\,Pa\,N\,m^{-2}$ (1 bar).

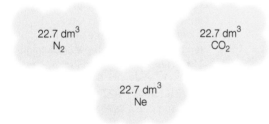

Figure 1.15 An illustration of the molar gas volume (at STP). All samples of the gases contain the same number of particles (atoms or molecules)

The molar gas volume allows interconversion (Figure 1.16) between the volume of a pure gas (behaving ideally) and its amount (in moles). The ratios of reacting volumes of gases can be used to deduce the stoichiometry of a reaction.

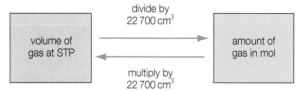

Figure 1.16 Summary of interconversions between the amount of gas and volume (at STP)

Common mistake

The molar gas volume of $22.7\,dm^3\,mol^{-1}$ only applies to gases (behaving ideally). It can be applied to all gases for IB calculations. It does not apply to pure liquids.

■ **QUICK CHECK QUESTIONS**

40 $100\,cm^3$ of carbon monoxide react with $50\,cm^3$ of oxygen to form $100\,cm^3$ of carbon dioxide. Deduce the equation for the reaction.

41 $50\,cm^3$ of hydrogen were exploded with $80\,cm^3$ of air (80% by volume nitrogen and 20% by volume oxygen). Calculate the composition of the resulting gas, all measurements being made at STP.

42 Lead(II) nitrate decomposes to form lead(II) oxide, nitrogen dioxide and oxygen. Determine the total volume of gas (at STP) produced when $33.10\,g$ of lead(II) nitrate is completely decomposed by strong heating.

43 Determine the volume of sulfur trioxide produced by the complete reaction of $1000\,cm^3$ of sulfur dioxide with oxygen. Determine the volume of oxygen required to react with the sulfur dioxide.

44 Calculate the volume occupied (at STP) by the following gases: $0.250\,mol$ of chlorine and $106.35\,g$ of chlorine gas.

The ideal gas equation

Revised ■

The behaviour of an ideal gas can be described by the ideal gas equation:

$pV = nRT$

The ideal gas equation allows the conversion from volumes of gases to amounts (in moles) at any particular temperature and pressure. When feeding values into the ideal gas equation you should be careful to use the correct units:

■ pressure, p, must be in Pa ($N\,m^{-2}$); if kPa are given, multiply by 10^3

■ volume, V, must be in m^3; if dm^3 are given, divide by 10^3, or if cm^3 are given, divide by 10^6

■ amount of gas (mol), n: this is often calculated using $n = \dfrac{m}{M}$

■ absolute temperature, T must be in kelvin; if °C is given, add 273 (or, very precisely, 273.15)

■ gas constant, $R = 8.31\,J\,K^{-1}\,mol^{-1}$.

The ideal gas equation can be used to determine the relative molecular mass, M_r, of compounds in the gaseous state, and to determine the density, ρ, of a gas.

By recording the mass (m) of the gas, the pressure (p), temperature (T) and volume of the gas (V) and using this modified form of the ideal gas equation shown below, the relative molecular mass of the compound in its gaseous state can be determined.

If the relative molecular mass (M_r), pressure (p) and the absolute temperature (T) of the gas are known, then an alternative form of the ideal gas equation can be used to find the density (ρ) of the compound in its gaseous state.

$\rho = \dfrac{pM_r}{RT}$

Expert tip

Suitable units for a calculation involving any of these forms of the ideal gas equation are SI base units for pressure (Pa) and volume (m^3). Temperature must be expressed in kelvin.

■ **QUICK CHECK QUESTIONS**

45 A sample of a gas has a mass of $0.112\,g$ and it occupies a volume of $81.80\,cm^3$ at a temperature of 127 °C and a pressure of $100\,kPa$. Calculate the relative molecular mass of the gas (to the nearest integer).

46 A syringe containing $0.15\,g$ of a volatile liquid was injected into a graduated gas syringe at a temperature of 90.0 °C and a pressure of $101\,kPa$. The vapour occupied a volume of $62.2\,cm^3$. Calculate the relative molecular mass of the gas.

47 The density of a gas is $2.615\,g\,cm^{-3}$ at $298\,K$ and $101\,kPa$. Calculate the molar mass of the gas.

■ ## Comparison between a real gas and an ideal gas

Chemists make predictions about gas behaviour by assuming that the gases are ideal and the gas laws are exactly obeyed. This assumption allows chemists to make very accurate estimations of measurements.

The molecules of an ideal gas are assumed to be points with no attractive or repulsive forces operating between the particles (atoms or molecules). It is a hypothetical state but gases (especially the noble gases) approach ideal behaviour under certain conditions. Table 1.5 summarizes the difference between the behaviour of real and ideal gases.

Table 1.5 The behaviour of real and ideal gases

	Real gas	Ideal gas
1.	Each gas particle has a definite volume and hence, occupies space. So the volume of the gas is the volume of the particles and the spaces between them.	Each gas particle has a negligible (almost zero) volume and hence, the volume of the gas is the volume of the space between the particles.
2.	Attractive forces are present between the particles.	There are negligible attractive forces operating between the particles.
3.	Collisions between the particles are non-elastic, and hence, there is a change (loss or gain) of kinetic energy from the particles after collision.	Collisions between the particles are elastic, and hence, there is no loss of kinetic energy from the particles after collision.

- Real gases can be compressed into liquids but molecular kinetic theory predicts that ideal gases cannot be liquefied.
- Ideal gases obey the ideal gas equation: $pV = nRT$.
- A gas behaves most like an ideal gas at high temperatures and low pressures when the particles move fast and are far apart, which minimizes attractive forces between particles.

NATURE OF SCIENCE

Chemists and physicists use simplified models – the three 'gas laws' – based on ideal behaviour to describe gases. Deviation of the experimental values from the predicted values led to a refinement of models of gas to include intermolecular forces and the actual volume of gas particles. The gas laws were developed from empirical data (observations and measurements) and were all later explained in terms of molecular kinetic theory.

■ Deviation from ideal gas behaviour

The deviation from ideal gas behaviour is shown by plotting pV against p (Figure 1.17) or $\frac{pV}{RT}$ against p. For an ideal gas, pV must be constant at all pressures if the temperature is constant. When pV is plotted against p, or when $\frac{pV}{RT}$ is plotted against p, a straight line is obtained which runs parallel to the x-axis of pressure. The extent of deviation from ideal behaviour depends on pressure, temperature, the size of the gas particles and the strength of the intermolecular forces.

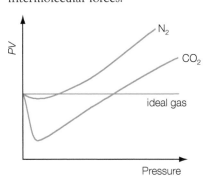

Figure 1.17 Deviation from ideal behaviour at high pressure

■ **QUICK CHECK QUESTIONS**

48 State three assumptions underlying the kinetic theory of gases.

49 Explain Boyle's law using this theory.

50 State and explain the conditions at which gases would show almost ideal behaviour.

■ **QUICK CHECK QUESTION**

51 State which of the following gases would behave most ideally (under the same conditions of temperature and pressure: hydrogen, fluorine and hydrogen fluoride.

State and explain which one would show the greatest deviation from ideal behaviour.

Topic 2 Atomic structure

2.1 The nuclear atom

Essential idea: The mass of an atom is concentrated in its minute, positively charged nucleus.

Atoms

- Atoms contain a positively charged dense nucleus composed of protons and neutrons (nucleons).
- Negatively charged electrons occupy the space outside the nucleus.

An atom is the smallest particle of an element. The atoms of a given element are chemically identical to each other, but chemically different from the atoms of every other element.

■ Nuclear model of the atom

The sub-atomic particles in the atom are the proton, neutron and the electron. The properties of the sub-atomic particles are summarized in Table 2.1.

Table 2.1 Properties of sub-atomic particles

Sub-atomic particle	Proton (p)	Neutron (n)	Electron (e)
Mass ($\times 10^{-27}$ kg)	1.673	1.675	9.109×10^{-4}
Relative mass (amu)	1	1	$\frac{1}{1840} = 5 \times 10^{-4}$
Charge ($\times 10^{-19}$ C)	+ 1.6022	0	− 1.6022
Relative charge	+ 1	0	− 1
Effect of electric field (potential difference)	Deflected to the negatively charged plate, proving that the proton is positively charged. The small deflection angle is due to its heavy mass.	Not deflected, proving that the neutron is not charged.	Deflected to the positively charged plate, proving that the electron is negatively charged. The large deflection angle is due to its light mass.

The proton and neutron (nucleons) have similar masses and are tightly bound to form the nucleus at the centre of the atom. The nucleus of the atom contains all the positive charge and most of the mass of the atom.

This simple model of the atom (Figure 2.1) was proposed by Rutherford following his gold foil experiment, which involved the scattering of alpha particles by gold nuclei. The electrons occupy shells around the nucleus. Electrons in different shells have different energies.

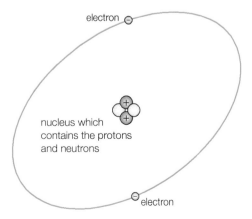

Figure 2.1 Simple nuclear model of a helium-4 atom

Common mistake

The electron and proton have charges of −1 and +1. These are relative charges. The actual charges are very small and expressed in coulombs. It is the relative masses and charges of these sub-atomic particles that you need to remember.

Expert tip

There is repulsion between the protons in the nucleus but many atoms are stable because of the presence of the strong nuclear force that binds the protons and neutrons together to form a nucleus of very high density.

Atoms of different elements have different numbers of protons. An element contains atoms which have the same number of protons in their nuclei. The number of protons (Z) is also known as the atomic number. The sum of the protons and neutrons in the nucleus is known as the mass or nucleon number (A).

The atomic number and mass number of the nuclide of an element may be indicated by a subscript and superscript number. The difference between the two numbers is the number of neutrons (Figure 2.2).

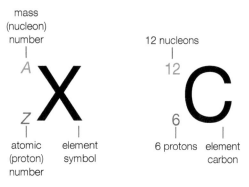

Figure 2.2 Standard notation for specifying the nuclide of an atom

NATURE OF SCIENCE

Rutherford found a method of probing inside atoms using alpha particles from a radioactive substance as small and fast-moving 'bullets'. Alpha particles are helium-4 nuclei, $^4He^{2+}$. When alpha particles were fired at a very thin sheet of gold, the majority of alpha particles passed through. However, some of the alpha particles were deflected by the gold foil and few of them bounced back. Rutherford explained these results by proposing that atoms have a positively charged dense nucleus (with nearly all the mass) surrounded by a much larger region of empty space containing electrons. This nuclear model was a paradigm shift – a totally new way of describing atoms.

Atoms are electrically neutral because the number of protons is equal to the number of electrons. Atoms can gain or lose electrons to form ions (Table 2.2), which have an electrical charge because the numbers of protons and electrons are different. You should be able to identify the number subatomic particles in any atom, positive ion, or negative ion given data for the species concerned (Table 2.2).

Table 2.2 Nuclide notation for atoms and ions

	Atom	**Anion**	**Cation**
Nuclide representation	$^A_Z W$	$^P_Q X^{R-}$	$^P_Q Y^{R+}$
Number of protons	Z	Q	Q
Number of neutrons	$A - Z$	$P - Q$	$P - Q$
Number of electrons	Z	$Q + R$	$Q - R$

▨ Isotopes

Isotopes (Figure 2.3) are atoms of the same element that differ in the number of neutrons. They have the same atomic number but different mass numbers. Isotopes of elements have different nucleon numbers, relative isotopic masses and numbers of neutrons in the nucleus.

Isotopes have the same chemical properties but different physical properties (for example, density, diffusion rate, mass and other properties that depend on mass).

■ QUICK CHECK QUESTION

1 a State the relative charges, relative masses and positions of protons, neutrons and electrons within an atom.

   b Which sub-atomic particles are nucleons?

Expert tip

You need to be able to recognize that Z stands for the atomic number, and A for the mass number of an atom (nuclide).

■ QUICK CHECK QUESTIONS

2 Deduce the numbers of protons, electrons and neutrons in the following oxygen-containing species:

^{16}O, ^{17}O, ^{18}O, $^{16}O^{2-}$, $^{17}O^{2+}$, $^{16}O_2$, $^{18}O_2^+$.

3 Identify the species with the following composition: 9 protons, 10 neutrons and 10 electrons.

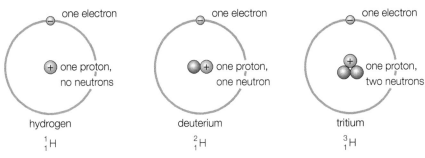

Figure 2.3 The three isotopes of hydrogen: hydrogen-1 (protium), hydrogen-2 (deuterium) and hydrogen-3 (tritium)

Relative atomic mass

- The mass spectrometer is used to determine the relative atomic mass of an element from its isotopic composition

The **relative atomic mass** (A_r) of an element is the weighted average atomic mass of its naturally occurring isotopes, e.g., bromine contains 50.52% of ^{79}Br and 49.48% of ^{81}Br. Hence its relative atomic mass, A_r, is $\dfrac{(50.52 \times 79) + (49.48 \times 81)}{100} = 79.99$.

$$\text{relative atomic mass} = \frac{\text{average mass of one atom of element}}{\frac{1}{12} \times \text{mass of one atom of carbon-12}}$$

Mass spectrometer

The relative atomic mass of an element can be determined in a mass spectrometer (Figure 2.4). The instrument can also be used to help determine the structure of organic molecules (Topic 21 Measurement and analysis).

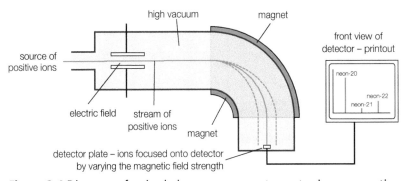

Figure 2.4 Diagram of a single beam mass spectrometer in cross-section

The vaporization chamber is used to vaporize solid or liquid samples. The ionization chamber ionizes (via electron bombardment) the gaseous atoms or molecules to form unipositive ions. The ions are accelerated in an electric field and deflected in a magnetic field. The angle of deflection is inversely proportional to the mass to charge ratio.

■ **QUICK CHECK QUESTION**

6 Antimony (Sb) consists of two isotopes: Sb-121 with an isotopic mass of 120.904 and Sb-123 with an isotopic mass of 122.904. The relative atomic mass of antimony is 121.80.

   a Calculate the percentage abundance of each isotope.

   b State the nuclide used to compare relative atomic and molecular masses.

The ions are focused on an ion detector, which is connected to a recorder. In a mass spectrometer the magnetic field strength is steadily changed so that particles of increasing mass arrive one after another at the detector.

A mass spectrum (Figure 2.5) is produced and the number of peaks shows the number of isotopes of the element. The height of the peaks shows the relative abundance of the isotopes.

Expert tip

The most common isotope of hydrogen, ^1_1H, does not have any neutrons.

■ **QUICK CHECK QUESTIONS**

4 a Define the term isotope.

   b Draw simple labelled diagrams showing the structure and composition of the atoms ^6_3Li and ^7_3Li.

5 State two differences in the physical properties of $^{235}UF_6(g)$ and $^{238}UF_6(g)$.

Revised

Expert tip

In principle any stable nuclide of any element can be used as the basis for the relative atomic mass scale. ^{12}C has been agreed as the standard because it is abundant, can be easily identified in a mass spectrometer and is easy to handle since it is a solid under standard conditions.

Common mistake

The relative atomic mass of an ion of an element is the same as that of its atom. The mass of the electrons can be neglected.

It is the relative atomic mass value of an element that is given in the Periodic Table, not the mass of individual atoms of the element.

Expert tip

Knowledge of the operation of the mass spectrometer will not be tested but is based on simple physical principles.

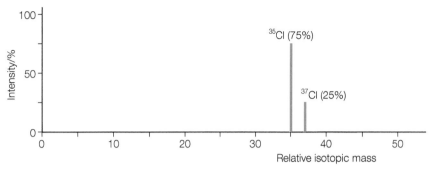

Figure 2.5 Mass spectrum of a sample of naturally occurring chlorine atoms

---

■ QUICK CHECK QUESTIONS

7 Silicon exists as three naturally occurring isotopes. The three isotopes have relative isotopic mass numbers of 28, 29 and 30 respectively.

 a Write down the formulas of the three gaseous unipositive ions that would form in a mass spectrometer after ionization.

 The silicon atoms in the sample have the following isotopic composition by mass.

Isotope	$^{28}_{14}Si$	$^{29}_{14}Si$	$^{30}_{14}Si$
Relative abundance	92.20	4.70	3.10

 b Calculate the relative atomic mass of silicon to two decimal places.

8 The mass spectrum of gaseous copper atoms is shown below. Calculate the relative atomic mass of copper (to one decimal place).

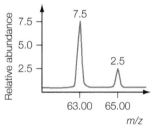

Figure 2.6 Mass spectrum of gaseous copper atoms

9 Explain why chlorine can form molecules with three different relative atomic masses.

The deflection or path of an ion in the mass spectrometer depends on:
■ the absolute mass of the ion
■ the charge of the ion
■ the strength of the magnetic field
■ the velocity (or speed) of the ion which is controlled by the strength of the electric field.

The mass and charge of an ion are combined into its $\frac{mass}{charge}$ ratio or $\frac{m}{z}$ (e.g. $\frac{m}{z}$ for $^{32}S^{2+} = 16$). For each isotope its path or $\frac{m}{z}$ ratio is compared with the path or $\frac{m}{z}$ ratio of the $^{12}C^+$ ion. *The lower this ratio for an ion, the greater the deflection.*

■ Radioactivity

■ Types of radioactivity

Alpha radiation is a stream of alpha particles. An alpha particle is a helium nucleus. Beta particles are fast-moving electrons. The electrons are emitted from unstable nuclei as the result of a neutron being converted into a proton and an electron (and an anti-neutrino not shown).

$$^{1}_{0}n \rightarrow \,^{1}_{1}p + \,^{0}_{-1}e$$

Gamma radiation is high-energy electromagnetic radiation. Gamma rays are produced due to transitions involving nuclear energy levels. Gamma emission usually follows other types of radioactive decay that leave the nucleus in an excited state.

■ Radioactive decay

In alpha decay the proton number of the nucleus decreases by two, and the nucleon number decreases by four. In beta decay a daughter nuclide is formed with the proton number increased by one, but with the same nucleon number. In gamma emission no particles are emitted and there is no change to the proton number or nucleon number of the parent nuclide.

Table 2.3 Types of radiation summary

Type of decay	Radiation	Change in atomic number	Change in nucleon number	Type of nuclide
alpha	α	-2	-4	Heavy isotope, $Z > 83$
beta	β	$+1$	0	$\dfrac{\text{neutron}}{\text{proton}}$ ratio too large
Gamma	γ	0	0	excited nucleus

■ Nuclear reactions

Nuclear equations are used to describe alpha decay, beta decay and transmutation reactions. Nuclear equations must balance: the sum of the nucleon numbers and the proton numbers must be equal on both sides of the equation. In nuclear notation the emissions are represented as: alpha particle $^{4}_{2}He$; beta particle $^{0}_{-1}e$; gamma radiation $^{0}_{0}\gamma$.

■ Uses of radioisotopes

Radioactive isotopes are used as tracers and followed through a process (e.g. ^{32}P to study nutrient uptake in plants) and nuclear medicine (e.g. ^{131}I to study thyroid function. ^{60}Co is used in radiotherapy; see Option D, Medicinal Chemistry). ^{14}C is used in radiocarbon dating and ^{235}U is used in fission reactors and ^{3}H in fusion reactors (Option C, Energy).

2.2 Electron configuration

Revised ☐

Essential idea: The electron configuration of an atom can be deduced from its atomic number.

Emission spectra

Revised ☐

- Emission spectra are produced when photons are emitted from atoms as excited electrons return to a lower energy level.
- The line emission spectrum of hydrogen provides evidence for the existence of electrons in discrete energy levels, which converge at higher energies.

■ The nature of light

Much of the understanding of the electronic structure of atoms, molecules and ions has come from spectroscopy – the study of how light and matter interact. Energy in the form of electromagnetic radiation is applied to a sample of a substance. Either the energy absorbed by the particles in the sample or the energy the particles release is studied (Figure 2.7).

Figure 2.7 The principles of absorption and emission spectroscopy

Two models are used to describe the behaviour of light: the wave model and the particle model. Some properties are best described by the wave model, for example, diffraction and interference; the particle model is better for others.

Light is one form of electromagnetic radiation and behaves like a wave with a characteristic wavelength and frequency. A wave of light will travel the distance between two points in a certain time. The speed of light is the same for visible light and all other types of electromagnetic radiation. It has a value of $3.00 \times 10^8 \, m \, s^{-1}$ when the light is travelling in a vacuum.

Like all waves, the light wave (and other electromagnetic waves) has a wavelength (symbol, λ) and frequency (symbol, ν). Different colours of light have different wavelengths and different frequencies (energies).

Wave A has twice the wavelength of wave B, but wave B has twice the frequency of wave A (Figure 2.8). The frequency is the number of cycles a wave goes through every second and is measured in units of reciprocal seconds (s^{-1}) or Hertz (Hz).

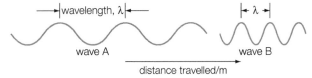

Figure 2.8 The wavelengths and frequencies of waves A and B, where wave B has twice the frequency of wave A; both waves travel at the same speed in a vacuum

The product of wavelength and frequency of an electromagnetic wave is a fundamental physical constant (the speed of light, c) and are related by the wave equation: $c = \nu \times \lambda$.

Light refers to the visible part of the electromagnetic spectrum (Figure 2.9) which includes all the electromagnetic radiation, which differ in frequency, wavelength and energy. All of these forms of radiation can be used in emission and absorption spectroscopy.

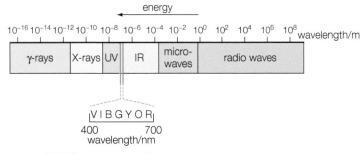

Figure 2.9 Electromagnetic spectrum

The visible region of the electromagnetic spectrum is an example of a continuous spectrum. Radiations corresponding to all wavelengths (within a certain range) are present. There are no lines present in a continuous spectrum (Figure 2.10).

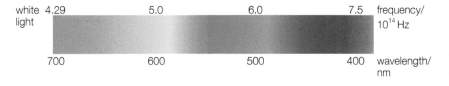

Figure 2.10 Continuous spectrum of white light

■ **QUICK CHECK QUESTIONS**

10 Compare the relative wavelength, frequency and energy of red and blue lights from a rainbow.

11 **a** Calculate the frequency of green light with a wavelength of 590 nm.

   b Light of a particular colour has a frequency of $5.75 \times 10^{14} \, s^{-1}$. Calculate its wavelength (in nm) and state its colour.

12 Based on the information in the table below, explain why UV-A rays are less dangerous than UV-B rays to the skin and eyes.

Type of radiation	Wavelength/nm
UV-A	320–380
UV-B	290–320

◼ Formation of emission spectra

According to Bohr's atomic model, electrons move in specific electron shells around the nucleus. These orbits are energy levels (Figure 2.11) and an electron is located in a certain energy level.

Hence, an electron has a fixed amount of energy (quantized). Each energy level is described by an integer, n, known as the principal quantum number ('shell number'). The energy levels become closer as n increases and finally converge.

Figure 2.11 Energy levels in a hydrogen atom (a quantum system)

In a hydrogen atom, the single electron is located in the lowest energy level ($n = 1$) or ground state. When the electron absorbs energy (thermal or electrical) in the form of a photon, it will move up to a higher energy level (an unstable excited state).

The excited state is unstable and the electron will lose energy and fall back to a lower energy level, radiating the energy as light. Since an electron can only be in a specific energy level the energy absorbed or radiated in each electron transition is in a discrete amount.

The light emitted with a discrete amount of energy is called a photon (Figure 2.12).

light as a wave

light as a stream of photons (packets of energy)

Figure 2.12 Wave-particle duality of light

Each photon of light has a fixed frequency, v, that can be calculated from Planck's equation:

$$\Delta E = hv = \frac{hc}{\lambda}$$

where ΔE = energy difference (J) between energy levels

h = Planck's constant = 6.63×10^{-34} J s

v = frequency of light (Hz or s^{-1})

c = speed of light (3.00×10^{8} m s^{-1}) and

λ = wavelength.

◼ **QUICK CHECK QUESTION**

13 Distinguish between the ground state and excited state of a gaseous atom.

Figure 2.13 shows how the absorption and emission of light by an atom is quantized and how the difference between the energy levels determines the energy absorbed or released. The amount of energy is determined by Planck's equation. The two energy levels are represented here by circles rather than lines.

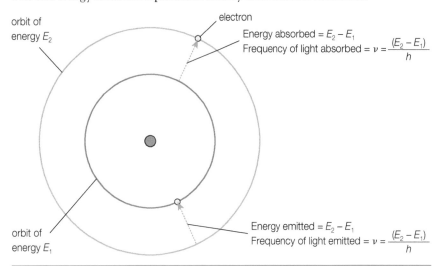

Figure 2.13 The origin of spectral lines

■ **QUICK CHECK QUESTION**

14 An electron undergoes a transition from a higher energy level to a lower energy level. Radiation with a frequency of $6.60 \times 10^{15}\,s^{-1}$ is emitted.

  a Calculate the wavelength of the radiation and the energy emitted by 1 mole of electrons for this transition.

  b Calculate the energy difference (in joules) corresponding to an electron transition that emits a photon of light with a wavelength of 600 nm.

Each electron transition from a higher energy level to a lower energy level corresponds to radiation with a specific frequency or wavelength. Therefore, an emission spectrum is obtained (Figure 2.14) consisting of a series of sharp coloured lines on a black background. The lines start to converge at higher frequency (energy) or shorter wavelength because the energy levels converge.

Figure 2.14 The Balmer series: the line spectrum in the visible region for the atomic emission spectrum of hydrogen

■ **QUICK CHECK QUESTION**

15 Distinguish between a continuous spectrum and line spectrum.

When an electron undergoes a transition from higher energy levels to $n = 1$, the frequency of the light (photons) emitted is in the ultraviolet region of the electromagnetic spectrum. This series of lines is known as the Lyman series (Figure 2.15).

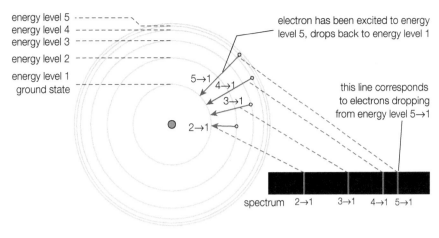

energy level 5
energy level 4
energy level 3
energy level 2
energy level 1
ground state

electron has been excited to energy
level 5, drops back to energy level 1

5→1
4→1
3→1
2→1

this line corresponds
to electrons dropping
from energy level 5→1

spectrum   2→1   3→1   4→1 5→1

Figure 2.15 How the energy levels
in the hydrogen atom give rise to
the Lyman series

When an electron undergoes a transition from higher energy levels to $n = 2$,
the frequency of the light (photons) emitted is in the visible region of the
electromagnetic spectrum. This series of lines is known as the Balmer series. Figure
2.16 shows the formation of the Lyman series, Balmer series and other spectral series.

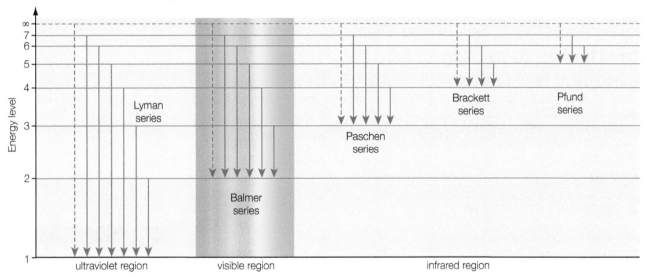

Figure 2.16 The origin of all the major spectral series in the emission spectrum of a hydrogen atom

The important feature to remember is that the lines in any series in the emission
spectrum of an atom are produced when the excited electron *falls down* from a
high-energy 'excited' level to a lower energy level (*specifically energy level 2 for the
visible spectrum*).

■ **QUICK CHECK QUESTION**

16 a Explain why an atomic emission spectrum from gaseous atoms consists of
a series of lines.

b Draw and label an energy level diagram for the hydrogen atom. Show how
the series of lines in the visible region of the emission spectrum are produced.

Every element has its own unique emission (line) spectrum. Hence, an element
can be identified from its atomic emission spectrum (Figure 2.17). The helium
emission spectrum is different from that of hydrogen because of the differences in
the energy levels (due to the increase in nuclear charge) and the larger number of
transitions possible due to the additional electron.

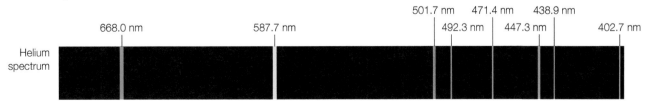

Figure 2.17 Emission spectrum of helium atoms

■ Orbitals

Bohr's concept of circular orbits for electrons has been replaced with the idea of an orbital. This is derived from a quantum mechanical model of an atom that regards the electron as behaving like a stationary wave.

An atomic orbital (Figure 2.18) is a region in an atom where there is a high chance of finding an electron. Each atomic orbital can hold up to a maximum of two electrons (as a spin pair). There are four different types of orbitals: s, p, d and f. They have different shapes and different energies.

s orbitals are spherical in shape and symmetrical around the nucleus. Every shell has an s orbital. There are three p orbitals in each shell, after the first shell. The three p orbitals of a sub-level are aligned along the *x*-, *y*- or *z*-axes. There is zero electron density at the centre of an atom. Each p orbital has two lobes.

■ **QUICK CHECK QUESTION**

17 State two deductions about the energy levels of the helium atom from the emission spectrum.

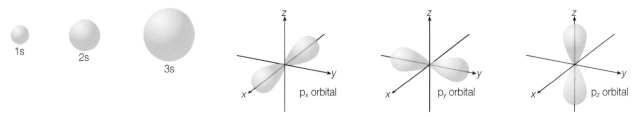

Figure 2.18 Shapes of s and p orbitals

As with s orbitals, the p orbitals also increase in size as the shell number (main energy level) increases. The three p orbitals can each hold up to two electrons and a p sub-level can hold up to six electrons.

Expert tip

Take care that you are clear about the distinction between the several technical terms used in this context – such as energy levels, sub-levels and orbitals.

■ **QUICK CHECK QUESTIONS**

18 Define the term orbital.

19 a Draw the shape of an *s* orbital and a p_x orbital. Label the *x*-, *y*- and *z*-axes on each diagram.

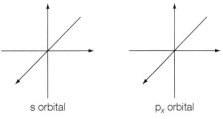

s orbital p_x orbital

Figure 2.19 An *s* orbital and a p_x orbital

  b State the similarities and the differences between a 2s orbital and a 3s orbital.

Expert tip

You will need to know, sketch and describe the shape and orientation of the s and p orbitals of an atom. Note that the nucleus of the atom is at the centre of all these axes.

● An s orbital is spherical about the nucleus.

● A p orbital has two lobes and is directed along one of the three axes – hence the name p_x, p_y or p_z.

You will not be asked to draw the shapes of the d orbitals.

There are five types of d orbitals and their shapes are shown in Topic 3 Periodicity. The five different types of 3d orbitals can each hold up to two electrons and a d sub-level can hold up to 10 electrons. An f sub-level has 7 orbitals and can hold up to 14 electrons.

Table 2.4 summarizes the maximum number of electrons in each shell. The maximum number of electrons is $2n^2$, where *n* is the shell (main energy level) number.

Table 2.4 Sub-levels in each shell and maximum number of electrons in each shell

Shell	Sub-levels in each shell	Maximum number of electrons
1st	1s	2 = **2**
2nd	2s 2p	2 + 6 = **8**
3rd	3s 3p 3d	2 + 6 + 10 = **18**
4th	4s 4p 4d 4f	2 + 6 + 10 + 14 = **32**

The energy level diagram in Figure 2.20 shows the relative energies of the sub-levels. Note that the 4s sub-level is of lower energy than the 3d sub-level (for elements up to calcium) and hence is filled first.

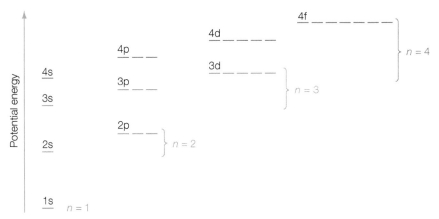

Figure 2.20 Orbital structure of atoms

Three rules control how electrons fill the atomic orbitals.

An electron enters the lowest available energy level (nearest the nucleus). This building-up process is known as the Aufbau principle. The order of filling orbitals is shown in Figure 2.21. An atom whose electrons are in orbitals of the lowest available energy is in its ground state.

The Pauli exclusion principle states that each orbital can hold a maximum of two electrons (with opposite spins). When there are two or more orbitals with the same energy they fill singly, with parallel spins, before the electrons pair up. This is known as Hund's rule.

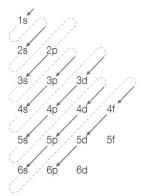

Figure 2.21 The Aufbau principle for filling atomic orbitals with electrons

The electron configuration of an element describes the number and arrangement of electrons in the orbitals of its atom. The electron arrangement describes the number and arrangement of electrons in the shells (main energy levels) of its atom.

Figure 2.22 shows three ways of representing the full electron configuration of a nitrogen atom, $1s^2\,2s^2\,2p^3$. Note the total number of electrons equals the atomic number of the element. Note that the three p orbitals are filled singly (Hund's rule).

A condensed electron configuration uses the nearest noble gas to represent the core electrons. Note how the electron configuration relates to the nitrogen atom, with an electron arrangement of 2,5.

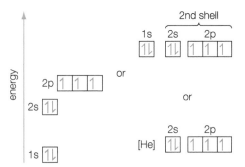

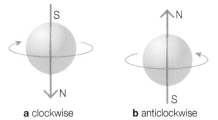

Figure 2.22 Ways of representing the electron configuration of the nitrogen atom (in the ground state) by means of orbital diagrams

The spin of an electron (Figure 2.23) makes it behave like a tiny magnet that can line up in a magnetic field, either with the field (spin up, ↑) or against the magnetic field (spin down, ↓).

a clockwise   **b** anticlockwise   **Figure 2.23** Electron spin

> **Expert tip**
>
> You should be aware of the difference between the *electron arrangement* of an atom (which you may have met in earlier courses), written in terms of just the main shells (principal energy levels) of an atom or ion, and the *electron configuration*, written in terms of s, p, d and f orbitals.
>
> Sodium, for instance, has an electron arrangement 2.8.1 *but* its electron configuration is $1s^2\ 2s^2\ 2p^6\ 3s^1$. While the arrangement is a useful shorthand when discussing bonding and the octet rule, be careful to give the configuration when asked for in an exam.

■ Chromium and copper

The electronic configurations of the transition elements copper and chromium do not follow the expected patterns. The chromium atom has the electron configuration [Ar] $3d^5\ 4s^1$ (rather than the expected [Ar] $3d^4\ 4s^2$). The copper atom has the electronic configuration [Ar] $3d^{10}\ 4s^1$ (rather than the expected [Ar] $3d^9\ 4s^2$). These two exceptions must be learnt, but are explained in more advanced courses.

> **Common mistake**
>
> Superscripts are essential in electron configurations; for example, if you write $2p_3$ or $2p_3$ you will not be awarded marks.

> **■ QUICK CHECK QUESTION**
>
> 22 Write out the electron configuration (full and condensed) and orbital diagram for a selenium atom, Se, and selenide ion, Se^{2-}.

> **Expert tip**
>
> The term electron configuration may refer to a condensed or full electron configuration. If you are in doubt, write out the full configuration.
>
> When working out the electron configurations of the elements of the first row transition metals, it is crucial to remember that *the 4s orbital is filled before* any electrons are placed in the 3d orbitals.

Table 2.5 Detailed electron configurations of gaseous isolated atoms in the ground state

Atomic number	Chemical symbol of element	Electron configuration	Atomic number	Chemical symbol of element	Electron configuration	Atomic number	Chemical symbol of element	Electron configuration
1	H	$1s^1$	13	Al	$[Ne]3s^23p^1$	25	Mn	$[Ar]4s^23d^5$
2	He	$1s^2$	14	Si	$[Ne]3s^23p^2$	26	Fe	$[Ar]4s^23d^6$
		$[He]2s^1$	15	P	$[Ne]3s^23p^3$	27	Co	$[Ar]4s^23d^7$
3	Li	$[He]2s^2$	16	S	$[Ne]3s^23p^4$	28	Ni	$[Ar]4s^23d^8$
4	Be	$[He]2s^22p^1$	17	Cl	$[Ne]3s^23p^5$	29	**Cu**	$\mathbf{[Ar]4s^13d^{10}}$
5	B	$[He]2s^22p^2$	18	Ar	$[Ne]3s^23p^6$	30	Zn	$[Ar]4s^23d^{10}$
6	C	$[He]2s^22p^2$	19	K	$[Ar]4s^1$	31	Ga	$[Ar]4s^23d^{10}4p^1$
7	N	$[He]2s^22p^3$	20	Ca	$[Ar]4s^2$	32	Ge	$[Ar]4s^23d^{10}4p^2$
8	O	$[He]2s^22p^4$	21	Sc	$[Ar]4s^23d^1$	33	As	$[Ar]4s^23d^{10}4p^3$
9	F	$[He]2s^22p^5$	22	Ti	$[Ar]4s^23d^2$	34	Se	$[Ar]4s^23d^{10}4p^4$
10	Ne	$[He]2s^22p^6$	23	V	$[Ar]4s^23d^3$	35	Br	$[Ar]4s^23d^{10}4p^5$
11	Na	$[Ne]3s^1$	24	**Cr**	$\mathbf{[Ar]4s^13d^5}$	36	Kr	$[Ar]4s^23d^{10}4p^6$
12	Mg	$[Ne]3s^2$				37	Rb	$[Kr]5s^1$

■ **QUICK CHECK QUESTIONS**

23 Copper is the ninth element in the first row of the d-block of the periodic table.

   a Write the outer electronic configuration of copper according to the Aufbau principle and the actual configuration as determined by experiment.

   b Give a simple reason to explain the anomaly.

24 Name another transition element in the first row of the d-block which shows a similar anomaly, and write the electronic configuration showing the anomaly.

■ Electronic configuration of ions

Cations are formed when electrons are removed from atoms. Anions are formed when atoms gain electrons. Many simple ions have a noble gas configuration. Figure 2.24 shows the formation of a positive and negative ion from a hydrogen atom.

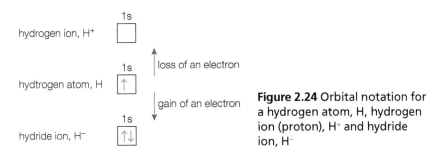

Figure 2.24 Orbital notation for a hydrogen atom, H, hydrogen ion (proton), H^+ and hydride ion, H^-

In general, electrons in the outer sub-level (valence shell) are removed when metal atoms form their cations. However, the d-block metals behave differently. The 4s sub-level fills before the 3d sub-level, but when atoms of a d-block metal lose electrons to form cations, the 4s electrons are lost first.

Although the 4s sub-level is filled before the 3d sub-level; cation formation in transition metals involves the removal of the 4s electrons before the 3d. The filling of the 4s sub-level alters the relative energy levels of the 3d and 4s sub-levels (Figure 2.25).

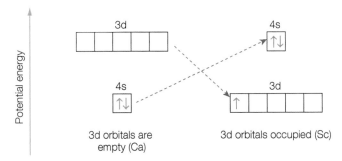

Figure 2.25 The 3d–4s sub-shell overlap

Expert tip

It is important you realize the following basic rule concerning the electron arrangements of transition metals and their ions:

- The 4s orbital is *filled first* when assigning electrons to the orbitals of the atoms.

- But it is the 4s electrons that are *removed first* when these atoms are ionized to form positive ions.

■ **QUICK CHECK QUESTION**

25 Write condensed electron configurations for the following species: V, V^{2+}, V^{3+} and V^{4+}.

Topic 3 Periodicity

3.1 Periodic table

Revised ☐

Essential idea: The arrangement of elements in the periodic table helps to predict their electron configuration.

The periodic table

Revised ☐

- The periodic table is arranged into four blocks associated with the four sub-levels – s, p, d and f.
- The periodic table consists of groups (vertical columns) and periods (horizontal rows).
- The period number (n) is the outer energy level that is occupied by electrons.
- The number of the principal energy level and the number of the valence electrons in an atom can be deduced from its position on the periodic table.
- The periodic table shows the positions of metals, non-metals and metalloids.

In the periodic table elements are placed in order of increasing atomic (proton) number. The chemical properties of elements are largely determined by the electrons in the outer (valence) shell.

Elements are arranged into four blocks associated with the four sub-levels: s, p, d and f. The s-block consists of groups 1 and 2 with hydrogen and helium. The d-block consists of groups 3 to 12. The p-block consists of groups 13 to 18. These blocks are based upon the configuration of the electrons in the outer (valence) sub-level (Figure 3.1). The f-block consists of two series of metals: the lanthanoids and actinoids.

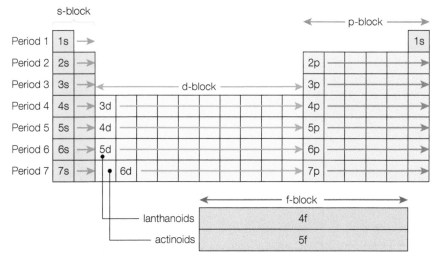

Figure 3.1 A diagram showing electron sub-shell filling in periods 1 to 7

The period number is the outer energy level (valence shell) that is occupied by electrons. Across each period (horizontal row) there is a decrease in metallic character or increase in non-metallic character (due to an increase in electronegativity). Each period starts with the filling up of a new shell and is continued until the p orbitals of the same shell are filled according to the Aufbau principle.

Elements in the same group (vertical column) contain the same number of electrons in the outer energy level. They will have similar chemical properties. Down each group, there is an increase in metallic character or decrease in the non-metallic character.

The number of main energy levels (shells) and the number of the valence electrons in an atom can be deduced from its position on the periodic table. For example, the element chlorine is in group 17 and period 3. Hence, it has seven valence electrons and three main energy levels (shells).

■ QUICK CHECK QUESTIONS

1. Explain why elements of the same group often show similar chemical properties.

2. Deduce the number of valence electrons in an atom of arsenic and the number of shells (main energy levels) in calcium. Explain your answers in terms of the positions of these two elements in the periodic table. Write out the electron configurations of both elements and classify them in terms of blocks in the periodic table.

3. Determine the period, block and group for atoms of each element with the following electronic configurations:

   $1s^2\ 2s^2\ 2p^3$

   $1s^2\ 2s^2\ 2p^6\ 3s^2\ 3p^6$

   $1s^2\ 2s^2\ 2p^6\ 3s^2$

   $1s^2\ 2s^2\ 2p^6\ 3s^2\ 3p^6\ 4s^2\ 3d^3$

The periodic table shows the position of metals (on the left), non-metals (on the right) and metalloids (near the border between metals and non-metals). The elements in the second and third periods change across the periods from metallic, to metalloid and to non-metallic (Figure 3.2).

Metalloids such as boron, silicon and germanium share some of the physical properties of metals but have chemical properties intermediate between those of a metal and a non-metal.

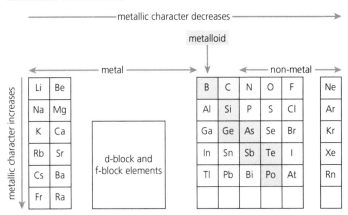

Figure 3.2 The positions of metals, non-metals and metalloids in the periodic table

The oxides of the elements in the second and third periods change across the periods from being ionic and basic (reacting with acids) to being covalent and acidic (reacting with bases). This correlates with the decrease in metallic character across the periods.

Certain groups have their own name, for example, group 1 is known as the alkali metals, group 17 as the halogens and group 18 as the noble gases. The d-block elements (groups 3 to 12) are known as the transition metals (with the exception of zinc) and the f-block elements form two distinct groups of metals – the lanthanoids (from La (57) to Lu (71)) and the actinoids (from Ac (89) to Lr (103)) (Figure 3.3). Table 3.1 indicates the orbitals being filled in each row and the number of elements in each period.

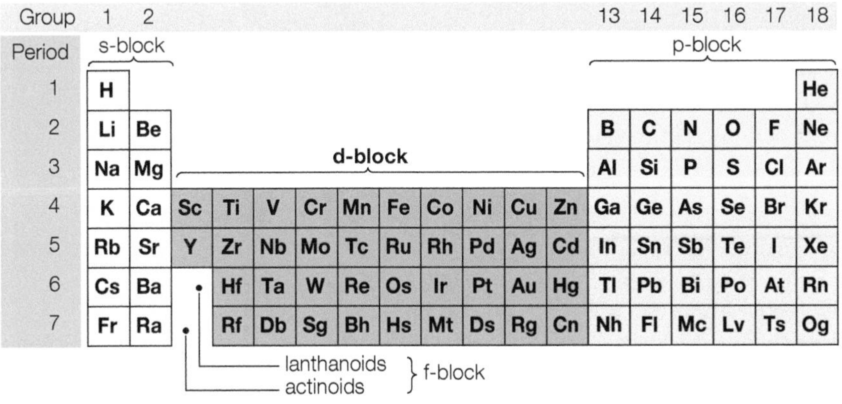

Figure 3.3 The short form of the periodic table showing the four blocks

Table 3.1 Period numbers with the orbitals being filled and the number of elements

Period number	Orbitals being filled	Number of elements
1	1s	2
2	2s, 2p	8
3	3s, 3p	8
4	4s, 3d, 4p	18
5	5s, 4d, 5p	18
6	6s, 4f, 5d, 6p	32
7	7s, 5f, 6d, 7p	32 to date

Expert tip

Your pre-IB chemistry course may have numbered the groups from 1 to 8. Be careful not to mislabel groups 13 to 18 and remember that the number of outer, valency, electrons of atoms in these groups is equal to (group number − 10).

■ **QUICK CHECK QUESTION**

4   Gadolinium (Gd) is a lanthanoid with the valence configuration [Xe] $4f^7$ $5d^1$ $6s^2$.

   a Draw an orbital diagram to show the valence electrons. Lanthanoids are metals.

   b State two expected physical properties and two expected chemical properties of gadolinium.

Common mistake

The lanthanoids and actinoids are groups of metals with some properties similar to those of transition metals; for example, variable oxidation states, but they are not classified as transition metals.

NATURE OF SCIENCE

The first serious classification of the elements was carried out by Mendeleev in 1869. His periodic table was similar to the short form used today and although there were gaps in his table, he was able to predict the existence and properties of some elements that at that time had not been discovered. One such gap was between silicon and tin in group 14 (then group 4). This was filled 15 years later by the discovery of germanium whose properties he had accurately predicted. Mendeleev actually used relative atomic masses (then atomic weights) as the basis for his classification because the concept of atomic numbers was not then known and, because of this, several elements did not fit by mass the positions indicated by their properties.

3.2 Periodic trends

Revised ☐

Essential idea: Elements show trends in their physical and chemical properties across periods and down groups.

Periodicity

Revised ☐

- Vertical and horizontal trends in the periodic table exist for atomic radius, ionic radius, ionization energy, electron affinity and electronegativity.
- Trends in metallic and non-metallic behaviour are due to the trends above.
- Oxides change from basic through amphoteric to acidic across a period

Elements in the same groups, especially groups 1, 2, 17 and 18, have similar chemical properties and clear trends in physical properties. There is a change in chemical and physical properties across a period from metallic to non-metallic behaviour. The repeating pattern of physical, chemical and atomic properties shown by elements in the different periods is known as *periodicity*.

■ Atomic radius

There are three definitions for 'atomic radius' (Figure 3.4), depending on the type of element. The van der Waals' radius is the radius of an imaginary hard sphere which can be used to model individual atoms, such as those of a noble gas. The metallic radius is half the shortest inter-nuclear distance between two atoms in a metallic crystal (lattice). The covalent bond radius is half the inter-nuclear distance between two identical atoms joined by a single covalent bond.

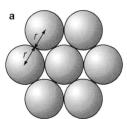

a

metallic radius

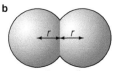

b

covalent radius

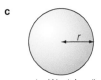

c

van der Waals' radius
(for group 18)

Figure 3.4 Atomic radius

Expert tip

Covalent radii can be obtained for most elements, but atoms of the noble gases in group 18 do not have a covalent radius, as they do not form bonds with each other. Their atomic radii can be determined from their van der Waals' radius.

Atomic radius increases down a group because of the balance between the following effects:

- The addition of extra protons causes an increase in nuclear charge and this causes all electrons to experience stronger electrostatic forces of attraction. Each member of a group has one more electron shell than the previous element. This increases the electron–electron repulsion. Each electron shell is located further away from the nucleus.

■ **QUICK CHECK QUESTION**

5   Explain what is meant by the term 'periodicity of elements'.

- The effect of the increase in nuclear charge is more than outweighed by the addition of an extra electron shell and additional electron–electron repulsion (also known as shielding), hence atoms increase in size down a group (Figure 3.5). Effective nuclear charge decreases down a group; this is the average electrostatic attraction experienced by the electrons in the atoms.

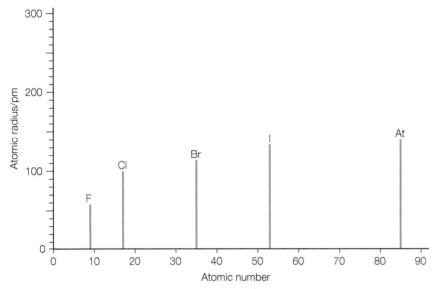

Figure 3.5 Bar chart showing the variation of atomic radii in group 17

Atomic radius decreases across a period (Figure 3.6) because the number of protons (and hence the nuclear charge), and the number of electrons, increases by one with each successive element.

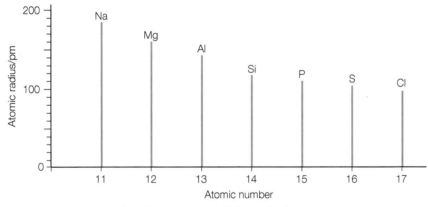

Figure 3.6 Bar chart showing atomic radii in period 3

The additional electron added to the atom of each successive element occupies the same main energy level (shell). This means that the shielding effect (electron–electron repulsion) only increases slightly.

Hence, the greater attractive electrostatic force exerted by the increasing positive nuclear charge on all the electrons pulls them in closer to the nucleus. Hence, the atomic radius decreases across a period.

Effective nuclear charge increases across a period; this is the average electrostatic attraction experienced by the electrons in the atoms.

Common mistake

Atomic radii decrease across a period (Figure 3.7) as electrons and protons are added.

Figure 3.7 Variation of atomic radii across period 3

They do not increase; remember an atom is mainly empty space and its radius and volume are determined by electrostatic forces.

■ QUICK CHECK QUESTION

6 State and explain the variation of the atomic radius of atoms of elements across a period and down a group.

Ionic radius

Cations

Cations contain fewer electrons than protons so the electrostatic attraction between the nucleus and all the electrons is greater and the ion is always smaller than the parent atom (Figure 3.8). The effective nuclear charge (experienced by all the electrons) has increased.

It is also smaller because the number of electron shells has decreased by one. The removal of electrons results in a decrease of electron–electron repulsion (shielding) which contributes to the decrease in radius.

Across period 3 the ions Na^+, Mg^{2+}, Al^{3+} contain the same number of electrons (isoelectronic), but an increasing number of protons, so the ionic radius decreases due to the increase in nuclear charge (effective nuclear charge).

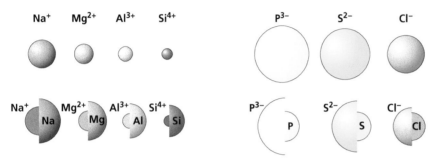

Figure 3.8 Relative sizes of atoms and ions of period 3 elements (note that the size of the silicon cation is a hypothetical estimate).

Anions

Anions contain more electrons than protons so are always larger than the parent atom (Figure 3.8). The addition of each electron increases the electron–electron repulsion. Each remaining electron experiences a smaller electrostatic attraction. The effective nuclear charge (experienced by all the electrons) has decreased.

Across period 3, the size of the anions N^{3-}, S^{2-} and Cl^- decreases because the number of electrons remains constant (isoelectronic) but the number of protons increases. This increase in nuclear charge more than outweighs the increase in electron–electron repulsion (shielding).

■ QUICK CHECK QUESTIONS

7 a On axes of the type shown below, sketch separately the general trends (using straight lines) of the two atomic properties, atomic radius and ionic radius, for the elements in the third period of the periodic table.

b Explain the periodic trends in the atomic and ionic radii shown in the graph you have drawn.

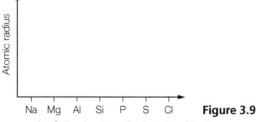

Figure 3.9

8 Arrange the following ions (isolated and in the gaseous state) in ascending order of their ionic radii. Explain your answer.

Na^+, Mg^{2+} and Al^{3+}

Mg^{2+}, S^{2-} and Cl^-

▩ Melting points

Melting points depend both on the structure of the element and on the type of attractive forces holding the atoms or molecules together (Figure 3.10).

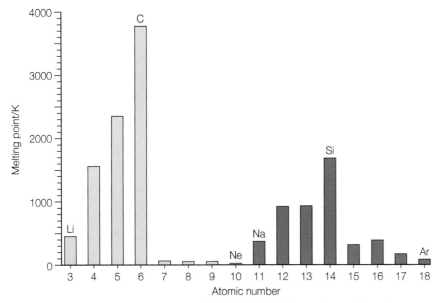

Figure 3.10 A graph showing periodicity in the melting points of the elements of Periods 2 and 3

At the left of period 3 the elements sodium, magnesium and aluminium are metals whose solids contain metallic bonding. This increases in strength as the number of delocalized valence electrons increases (which also increases the charge on the metal cation) and as the cation radius decreases.

Silicon, the metalloid, in the middle of period 3 has a three-dimensional macromolecular covalent structure with very strong covalent bonds, resulting in the highest melting point.

Elements in groups 15, 16 and 17 are simple molecular structures, P_4, S_8 and Cl_2, with weak intermolecular forces (London or dispersion forces) of attraction between the non-polar molecules. The strengths of these weak intermolecular forces increase with number of electrons (molar mass).

Argon and the other noble gases (group 18) exist as single atoms, with extremely weak forces of attraction (London or dispersion forces) operating between the atoms in the solid state.

In group 1, the alkali metals (Figure 3.11), the melting point decreases down the group as the cation radius increases (due to additional electron shells) and the strength of the metallic bonding decreases due to decreased charge density (charge per volume) in the cations.

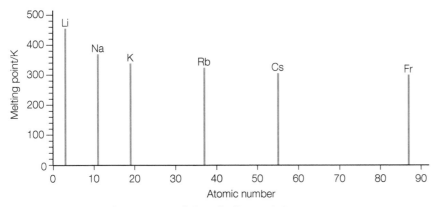

Figure 3.11 The melting point of the alkali metals in group 1

In group 17, the halogens, the intermolecular attractive forces between the diatomic molecules increase down the group (due to greater numbers of electrons and hence stronger London forces) so the melting and boiling points increase (Figure 3.12).

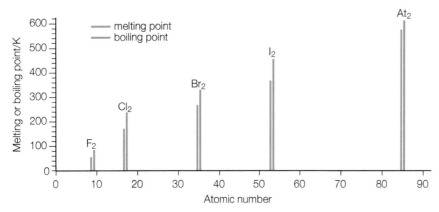

Figure 3.12 The melting and boiling points of the halogens in group 17

■ QUICK CHECK QUESTION

9   Explain why magnesium has a higher melting point than sodium, and why silicon has a much higher melting point than sulfur.

▨ First ionization energy

The first ionization energy is the minimum amount of energy required to remove one mole of electrons from one mole of gaseous atoms (in their ground state) to form one mole of unipositive ions (under standard thermodynamic conditions):

$$M(g) \rightarrow M^+(g) + e^-$$

In general, the first ionization energy decreases within a group (from top to bottom) because the electron being removed is in a shell which is located progressively further from the nucleus.

The increase in nuclear charge (proton number) down a group is more than outweighed by the increase in electron–electron repulsion (shielding) due to the presence of extra electrons in a new shell.

■ QUICK CHECK QUESTION

10 a Define the term first ionization energy.
   b Write an equation describing the first ionization energy of potassium.
   c Write the full electron configurations for potassium and its ion.

Within a period, there is a general tendency for the ionization energy to increase across the period from left to right. This is due to the increase in effective nuclear charge.

However, there is a decrease between groups 2 and 3 because elements of group 3 contain one extra electron in a p orbital, compared with those in group 2. The p electron is well shielded by the inner electrons and the ns^2 electrons. Hence, less energy is required to remove a single p electron than to remove a pair of s electrons from the same shell.

Expert tip

The p orbital is more diffuse than the s orbital because of the nodal plane between the lobes of the p orbital. This means there is zero probability of finding the electron(s) near the nucleus in a p orbital. That is not the case in a spherical s orbital.

Elements of group 16 contain one extra electron in a p orbital compared with the stable half-filled p orbitals of group 15. Hence, less energy is required to remove this additional electron compared to the removal of an electron from a half-filled p orbital.

■ **QUICK CHECK QUESTION**

11 The graph of the first ionization energy plotted against atomic number for the first 20 elements (from hydrogen to calcium) shows periodicity.

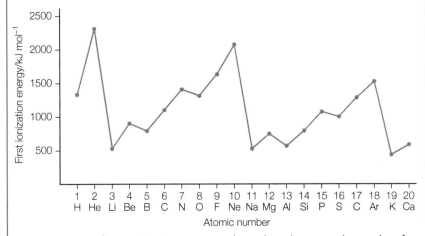

Figure 3.13 The first ionization energy plotted against atomic number for the first 20 elements

a Explain how information from this graph provides evidence supporting the idea of main energy levels and sub-levels within atoms.

b Account for the decreases observed between elements 12 and 13 and between elements 15 and 16 in period 3.

■ Electronegativity

Electronegativity is a measure of the relative tendency of an atom to attract an electron pair in a covalent bond. The Pauling scale is used which assigns a value of 4.0 to fluorine, the most electronegative element.

Electronegativity increases across a period and decreases down a group, due to changes in size of the atoms, which decrease across a period, but increase down a group (Figure 3.14).

The attraction between the outer (valence) electrons and the nucleus increases as the atomic radius decreases in a period. The electronegativity also increases. However, electronegativity values decrease with an increase in atomic radii down a group. The trend is similar to that of first ionization energy.

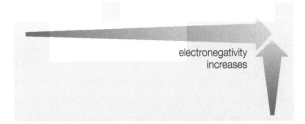

electronegativity increases

Figure 3.14 Trends in electronegativity for s- and p-block elements

The increase in electronegativity is accompanied by an increase in non-metallic properties (or decrease in metallic properties) of elements. The decrease in electronegativity down a group is accompanied by a decrease in non-metallic properties (or increase in metallic properties) of elements.

■ Electron affinity

The first electron affinity is the energy change when one mole of electrons is added to one mole of gaseous atoms to form one mole of singly charged negative ions. Electron affinity is a measure of the ease with which an atom accepts an electron to form an anion.

$$M(g) + e^- \rightarrow M^-(g)$$

Expert tip

Unlike ionization energy and electron affinity, electronegativity is not a measurable quantity with units. It cannot be directly measured, and is instead calculated from other atomic and molecular properties. It is a dimensionless quantity without units. It provides a means of predicting the polarity of a covalent bond.

Depending on the element, the process of adding an electron to the atom can either be exothermic or endothermic. For many atoms, such as the halogens (Figure 3.15), the first electron affinity is exothermic and given a negative value. This means that the anion is more energetically stable than the atom.

Common mistake

Electron affinity values are generally exothermic, so care must be taken with the terms 'larger' and 'smaller', because, for example, −450 kJ mol⁻¹ is a smaller value than −400 kJ mol⁻¹. It is preferable to use the terms 'more exothermic' and 'less exothermic'.

In general, the first electron affinity increases (becomes more negative) from left to right across a period. This is due to the decrease in atomic radius caused by the increase in effective nuclear charge. The nuclear charge increases more rapidly and outweighs the increase in shielding. The additional electron will enter a shell that is closer to the nucleus and experience a stronger force of electrostatic attraction.

In general, the first electron affinity decreases (becomes less negative) moving down a group. This is due to the increase in atomic radius caused by the addition of electron shells. The shielding (electron repulsion) increases more rapidly and outweighs the increase in nuclear charge. The additional electron will enter a shell that is further from the nucleus and experience a weaker force of electrostatic attraction.

First electron affinities may be endothermic or exothermic, but second electron affinities are always endothermic (energy absorbed) due to repulsion: between the like charges on the anion and electron.

$$M^-(g) + e^- \rightarrow M^{2-}(g)$$

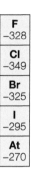

Figure 3.15 Values of first electron affinity for the halogens (group 17)

Common mistake

Electronegativity and electron affinity are both measures of an atom's ability to attract electrons. However, they have different definitions and must not be used interchangeably.

■ QUICK CHECK QUESTION

12 a Distinguish between electron affinity and electronegativity.

  b Write an equation showing the first and second electron affinities of fluorine.

  c Explain why the first electron affinity is exothermic but the second electron affinity is endothermic.

  d Outline why values of first electron affinity and electronegativity both decrease down group 17.

Figure 3.16 and Table 3.2 are summaries of the vertical and horizontal trends in properties in the periodic table which can be used to predict and explain the metallic and non-metallic behaviour of an element based on its position in the periodic table.

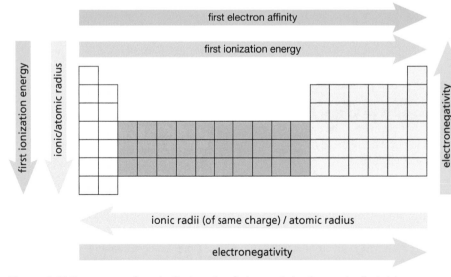

Figure 3.16 Summary of periodic trends of elements in the periodic table

Table 3.2 Summary of trends in key physical properties of the elements

Physical property	Horizontal trend (left to right)	Vertical trend (downwards)
Atomic radius	Decreases	Increases
Cation radius	Decreases	Increases
Anion radius	Decreases	Increases
First ionization energy	Increases	Decreases
First electron affinity	Increases	Decreases
Electronegativity	Increases	Decreases

Trends across period 3

Metals are located on the left of the periodic table (see Figure 3.16). Metals are shiny solids that are excellent thermal and electrical conductors. They are ductile and malleable. Metals are reducing agents and form cations; their oxides and hydroxides behave like bases and neutralize acids. In period 3, sodium, magnesium and aluminium are metals.

$$Na_2O(s) + H_2O(l) \rightarrow 2NaOH(aq)$$

$$NaOH(aq) + HCl(aq) \rightarrow NaCl(aq) + H_2O(l)$$

$$MgO(s) + H_2O(l) \rightarrow Mg(OH)_2(aq)$$

$$Mg(OH)_2(aq) + 2HCl(aq) \rightarrow MgCl_2(aq) + 2H_2O(l)$$

$$Al_2O_3(s) + 6HCl(aq) \rightarrow 2AlCl_3(aq) + 3H_2O(l)$$

Aluminium oxide, however, is insoluble in water and, importantly, is amphoteric, meaning it not only reacts with acids as above but also with strong bases:

$$Al_2O_3(s) + 2NaOH(aq) + 3H_2O(l) \rightarrow 2NaAl(OH)_4(aq)$$

Non-metals are located on the right of the periodic table. Non-metals are solids, liquids or gases. If solids, they are insulators and brittle. Non-metals tend to be oxidizing agents and form anions; their oxides tend to be acidic and are neutralized by bases. Phosphorus, sulfur, chlorine and argon are non-metals.

$$SO_2(g) + H_2O(l) \rightarrow H_2SO_3(aq)$$

$$H_2SO_3(aq) + 2NaOH(aq) \rightarrow Na_2SO_3(aq) + 2H_2O(l)$$

$$SO_3(g) + H_2O(l) \rightarrow H_2SO_4(aq)$$

$$H_2SO_4(aq) + 2NaOH(aq) \rightarrow Na_2SO_4(aq) + 2H_2O(l)$$

$$P_4O_{10}(s) + 6H_2O(l) \rightarrow 4H_3PO_4(aq)$$

Silicon is a metalloid and its oxide is acidic: it reacts with hot concentrated alkalis.

$$SiO_2(s) + 2NaOH(aq) \rightarrow Na_2SiO_3(aq) + H_2O(l)$$

Nitrogen forms a number of oxides, but most are acidic. For example, nitrogen dioxide reacts with water (in a redox reaction) to form nitric acid.

$$3NO_2(g) + H_2O(l) \rightarrow 2HNO_3(aq) + NO(g)$$

Table 3.3 summarizes the trend in behaviour of the oxides of the elements of period 3 and the link to the transition from metallic to non-metallic character of the elements when moving across the period.

Expert tip

Hydrolysis is a specific chemical reaction in which water breaks down a molecule into two parts.

■ QUICK CHECK QUESTIONS

13 a Gallium oxide, like aluminium hydroxide, is amphoteric. Explain the meaning of this term.

 b Write equations for the reaction of solid gallium oxide with aqueous hydrochloric acid and concentrated aqueous potassium hydroxide.

14 Nitrogen(V) oxide reacts with water to form nitric(V) acid. Write a balanced equation and state the type of reaction shown.

Common mistake

A basic oxide will react with an acid to form salt and water only. Oxides of reactive metals may react with water to form a basic solution. An acidic oxide will react with a base to form a salt and water only. Some oxides of non-metals will react with water to form an acidic solution.

Table 3.3 Trends in structure and nature of the oxides of the elements of Period 3

Group	1	2	3	4	5	6	7	0
Element	Na	Mg	Al	Si	P	S	Cl	Ar
Structure of element	← giant metallic →			giant covalent	← simple molecular →			
Formula of oxide	Na_2O	MgO	Al_2O_3	SiO_2	P_4O_{10} P_4O_6	SO_3 SO_2	Cl_2O_7 Cl_2O	
Structure of oxide	← giant ionic →			giant covalent	← simple molecular →			
Acid–base character of oxide	← basic →		amphoteric	← acidic →				no oxide

Group 1 – the alkali metals

Revised ☐

The group 1 elements are all reactive metals and so are stored under oil. They are soft and silvery and have excellent thermal and electrical conductors. Their compounds are ionic and all contain the group 1 metal as a cation in the +1 oxidation state (number). All group 1 compounds are soluble in water.

All the group 1 elements react rapidly with water releasing hydrogen and forming alkaline solutions:

$$2M(s) + 2H_2O(l) \rightarrow 2MOH(aq) + H_2(g)$$

The oxides are basic and react with water to form solutions of the hydroxides:

$$M_2O(s) + H_2O(l) \rightarrow 2MOH(aq)$$

The group 1 metals are strong reducing agents and react with most non-metals forming ionic compounds, for example:

$$2Na(s) + X_2 \rightarrow 2NaX(s), \text{ where X is a halogen}$$

$$4Na(s) + O_2(g) \rightarrow 2Na_2O(s)$$

Reducing agents are substances that release electrons and cause another substance to be reduced (Topic 9 Redox processes). The reaction between a group 1 metal and water is also a redox reaction.

Group 17 – the halogens

Revised ☐

The halogens consist of diatomic molecules, X_2, linked by a single covalent bond. The halogens are all volatile and intermolecular forces increase down the group as the number of electrons in the atoms increases. Hence melting points and boiling points rise down the group. Fluorine and chlorine are gases at room temperature. Bromine is a liquid and iodine is a solid which sublimes on heating.

The halogens are highly electronegative and form ionic compounds or compounds with polar bonds. Electronegativity decreases down the group. The halogens are powerful oxidizing agents and oxidizing power (reactivity) decreases down the group.

In group 17 a more reactive halogen displaces a less reactive halogen in a redox reaction. The order of reactivity (oxidizing strength) for the halogens is Cl > Br > I and the more reactive halogen oxidizes the halide ions of a less reactive halogen. For example:

$$Br_2(aq) + 2I^-(aq) \rightarrow 2Br^-(aq) + I_2(aq)$$

In group 17, iodine is the weakest oxidizing agent so it has the least tendency to form negative ions. Hence iodide ions are the ones that most readily give up electrons and form iodine molecules.

Expert tip

Remember that the halogens become more deeply coloured as you progress down the group.

Common mistake

Replacement (or displacement) reactions are not an additional type of chemical reaction. They are all examples of redox reactions (Figure 3.17), which involve electron transfer.

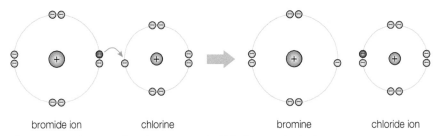

bromide ion   chlorine   bromine   chloride ion

Figure 3.17 The reaction between a halide ion and a halogen atom

NATURE OF SCIENCE

Chemists look for patterns or trends in physical and chemical data and then formulate theories that can be used to make testable predictions. Examination questions may expect a prediction of chemical and physical properties from data about unfamiliar elements, such as those at the bottom of groups 1 and 17, or compounds.

Insoluble silver salts precipitate on mixing silver nitrate solution with solutions of a soluble chloride, bromide or iodide. Silver fluoride is soluble so there is no precipitate on adding silver nitrate to a solution of fluoride ions.

$Ag^+(aq) + X^-(aq) \rightarrow AgX(s)$, where X = Cl, Br or I

This precipitation reaction is used as a test for halide ions. The three silver compounds can be distinguished by their colour. Silver chloride is white; silver bromide is cream and silver iodide is yellow.

Common mistake

Remember that iodine molecules will be coloured in water (brown) or non-polar solvents (purple), but iodide ions will be colourless.

■ QUICK CHECK QUESTION

18 a Write an ionic equation for the reaction that occurs when bromine solution is added to potassium iodide solution.

  b Identify the oxidizing and reducing agents in the reaction.

  c State the colour change that would be observed if silver nitrate solution is added to potassium iodide solution.

Group 18 – the noble gases

Revised ☐

The elements in group 18 were once called the inert gases ('inert' meaning unreactive). However, since a few compounds of these elements have been prepared, they are now termed the noble gases.

The melting and boiling points of the noble gases are very low because the London (dispersion) forces are extremely small. They increase as the group is descended because there are more electrons that are easily polarized on the outside of the atoms.

The elements are monoatomic. Their ionization energies are too high for them to form cations. However, xenon can use its d orbitals to form bonds with fluorine and oxygen.

Expert tip

The simplest compounds of xenon are fluorides, XeF_2, XeF_4 and XeF_6, and their oxides XeO_3 and XeO_4. Only when xenon is bonded to the small and electronegative atoms of fluorine are the covalent bonds strong enough to supply the energy to use 5d orbitals. The shapes of these molecules are consistent with those predicted by VSEPR theory (Topic 4 Chemical bonding and structure).

Topic **4** Chemical bonding and structure

A chemical bond results from an attractive electrical force which holds two or more atoms or oppositely charged ions together. All chemical reactions involve bond breaking followed by bond making. The breaking of bonds requires (absorbs) energy – an endothermic process. The formation of bonds releases energy – an exothermic process (Topic 5 Energetics/thermochemistry). The feasibility of a reaction depends, partly, on the energy changes associated with the breaking and making of bonds. All types of chemical bonding and intermolecular forces (Figure 4.1) involve interactions between opposite charges and they are electrostatic in nature.

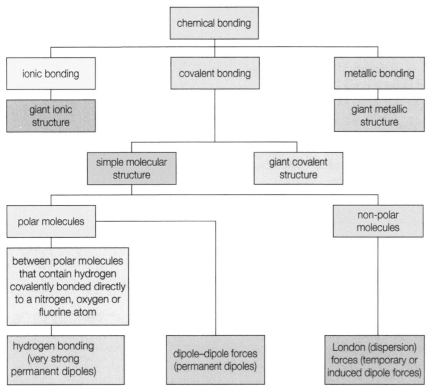

Figure 4.1 A summary of bonding and intermolecular forces

4.1 Ionic bonding and structure

Revised ☐

Essential idea: Ionic compounds consist of ions held together in lattice structures by ionic bonds.

Ionic bonding and structure

Revised ☐

- Positive ions (cations) form by metals losing valence electrons.
- Negative ions (anions) form by non-metals gaining electrons.
- The number of electrons lost or gained is determined by the electron configuration of the atom.
- The ionic bond is due to electrostatic attraction between oppositely charged ions.
- Under normal conditions, ionic compounds are usually solids with lattice structures.

Formation of ions

Ionic bonding is formed by the transfer of one or more electrons from the outer orbital of one atom to the outer orbital of another atom. It generally occurs between atoms of metals and non-metals, with metals forming positive and non-metals negative ions (Table 4.1). The atom losing an electron or electrons becomes a positively charged ion (cation) and the atom gaining an electron or electrons becomes a negatively charged ion (anion).

Table 4.1 A broad summary of the formation of ions by the elements in different groups of the Periodic Table; the final row gives an example of the ions formed

Group 1	Group 2	Group 3	Transition metals	Group 15	Group 16	Group 17
Lose 1 electron	Lose 2 electrons	Lose 3 electrons	Can form more than one ion	Gain 3 electrons	Gain 2 electrons	Gain 1 electron
+1	+2	+3	+1 to +3	−3	−2	−1
Li^+	Ca^{2+}	Al^{3+}	Sc^+/ Fe^{3+}	N^{3-}	O^{2-}	F^-

The formation of ionic bonding is favoured by a large difference in electronegativity between the two atoms concerned. However, it is possible for compounds to exhibit varying degrees of ionic/covalent character depending on circumstances. Aluminium is a case in point here, with aluminium chloride being a compound with a high degree of covalent character.

In many cases the resulting ions have filled outer shells (main energy levels) with a stable noble gas electronic structure (Table 4.2), often an octet (four electron pairs). This is known as the octet rule but atoms in period 3, such as phosphorus and sulfur can 'expand' their octet. Table 4.2 gives a more detailed summary of the simple ions formed by the elements of period 3.

Table 4.2 Electron arrangements of the atoms and simple ions of the elements in period 3

Group	1	2	13	14	15	16	17	18
Element	Sodium	Magnesium	Aluminium	Silicon	Phosphorus	Sulfur	Chlorine	Argon
Electron arrangement and configuration	2,8,1 $1s^2\,2s^2\,2p^6\,3s^1$	2,8,2 $1s^2\,2s^2\,2p^6\,3s^2$	2,8,3 $1s^2\,2s^2\,2p^6\,3s^2\,3p^1$	2,8,4 $1s^2\,2s^2\,2p^6\,3s^2\,3p^2$	2,8,5 $1s^2\,2s^2\,2p^6\,3s^2\,3p^3$	2,8,6 $1s^2\,2s^2\,2p^6\,3s^2\,3p^4$	2,8,7 $1s^2\,2s^2\,2p^6\,3s^2\,3p^5$	2,8,8 $1s^2\,2s^2\,2p^6\,3s^2\,3p^6$
Number of electrons in outer shell	1	2	3	4	5	6	7	8
Common simple ion	Na^+	Mg^{2+}	Al^{3+}	–	P^{3-} (phosphide)	S^{2-} (sulfide)	Cl^- (chloride)	–
Electron arrangement and configuration of ion	2,8 $1s^2\,2s^2\,2p^6$	2,8 $1s^2\,2s^2\,2p^6$	2,8 $1s^2\,2s^2\,2p^6$	–	2,8,8 $1s^2\,2s^2\,2p^6\,3s^2\,3p^6$	2,8,8 $1s^2\,2s^2\,2p^6\,3s^2\,3p^6$	2,8,8 $1s^2\,2s^2\,2p^6\,3s^2\,3p^6$	–

Ionic bonding is the result of electrostatic attraction between two oppositely charged ions. The formation of an ionic compound involves a metal with a low ionization energy and a non-metal with a high electron affinity. The atomic ratio of two elements that react to form an ionic compound is determined by the number of valence electrons that has to be transferred or received by the reacting element in order to achieve a stable noble gas configuration.

Expert tip

In general, ionic compounds are formed between a metal and a non-metal. However, be careful, ammonium chloride does not contain a metal but it is an ionic compound containing the ammonium ion (NH_4^+) and the chloride ion (Cl^-).

Expert tip

A Lewis diagram can be used to represent the transfer of valence (outer) electrons that occurs during the formation of ionic bonds. Figure 4.2 shows the formation of the constituent ions in sodium chloride.

Figure 4.2 Ionic boding in sodium chloride (showing just the outer valence electrons)

Expert tip

Aluminium fluoride is an ionic compound whereas anhydrous aluminium chloride, $AlCl_3$, is a compound that sublimes and is best regarded as being covalent (in the solid and gaseous states).

■ **QUICK CHECK QUESTION**

1 Lithium fluoride is an ionic compound. Write the electron configurations of the atoms and ions and describe the formation of the compound using a Lewis diagram. Deduce the formula and outline the nature of the bonding present in the compound.

Formulas of ionic compounds

The charge carried by an ion depends on the number of electrons the atom needs to gain or lose to get a full outer shell. Transition metals can form more than one stable cation (Table 4.3). For example, iron can form iron(II), Fe^{2+}, and iron(III), Fe^{3+}, and copper can form copper(I), Cu^+, and copper(II), Cu^{2+}.

Table 4.3 Charges on selected transition element cations

Name of transition metal	Simple positive ions
Silver	Ag^+
Iron	Fe^{2+}, Fe^{3+}
Copper	Cu^+, Cu^{2+}
Manganese	Mn^{2+}, Mn^{3+} and Mn^{4+}
Chromium	Cr^{3+} and Cr^{2+} (not stable in air)

Ionic compounds are electrically neutral because the total positive charge is balanced by the total negative charge. For example, aluminium oxide, Al_2O_3, $[2Al^{3+}\ 3O^{2-}]$ and copper(II) chloride, $CuCl_2$, $[Cu^{2+}\ 2Cl^-]$.

> **Expert tip**
>
> The formulas of ionic compounds are always empirical formulas and show the simplest ratio of cations and anions. Note that a shortcut to working out the formula of an ionic compound is to switch over the number of the charge on the ions to generate the formula.

■ **QUICK CHECK QUESTIONS**

2   Deduce the formulas of the following binary ionic compounds:

sodium oxide, magnesium oxide, aluminium oxide, magnesium phosphide, sodium chloride, magnesium chloride, aluminium chloride, chromium(III) chloride, copper(I) oxide, iron(III) sulfide, aluminium sulfide

3   Classify the following substances as whether they are predicted to have ionic or covalent bonding:

CO, Li_2S, PCl_5, F_2O, S_2Cl_2, H_2O_2, CuO, MgO

Polyatomic ions (Table 4.4) contain two or more elements; many involve oxygen. Compounds with polyatomic ions are electrically neutral and brackets are used to show that the subscript coefficient refers to all the elements inside the brackets; for example, copper nitrate, $Cu(NO_3)_2$.

Table 4.4 Common polyatomic ions

Name of polyatomic ion	Formula	Structure	Comment
Hydroxide	OH^-		
Nitrate	NO_3^-	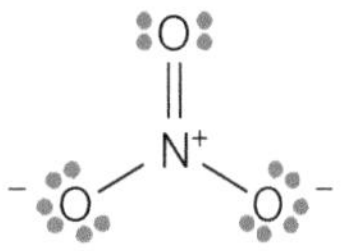	Derived from nitric(V) acid, HNO_3
Sulfate	SO_4^{2-}	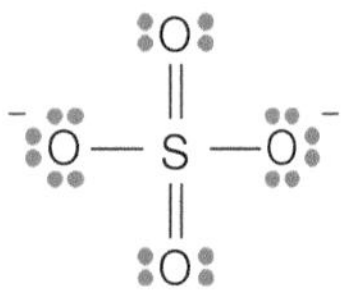	Derived from sulfuric(VI) acid (second ionization), H_2SO_4
Hydrogen sulfate	HSO_4^-	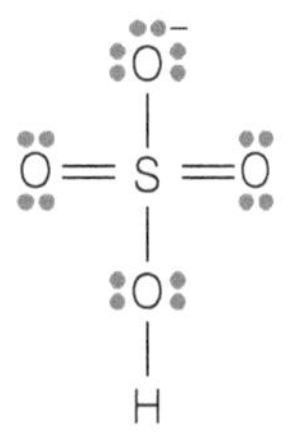	Derived from sulfuric(VI) acid (first ionization), H_2SO_4
Carbonate	CO_3^{2-}	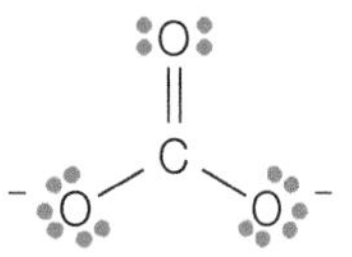	Derived from carbonic acid (second ionization), H_2CO_3

Name of polyatomic ion	Formula	Structure	Comment
Hydrogen carbonate	HCO_3^-	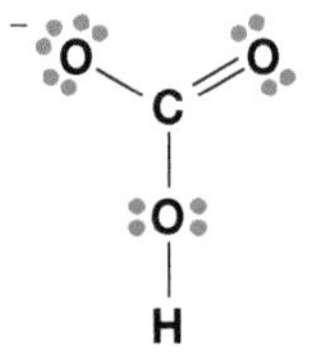	Derived from carbonic acid (first ionization), H_2CO_3
Phosphate	PO_4^{3-}	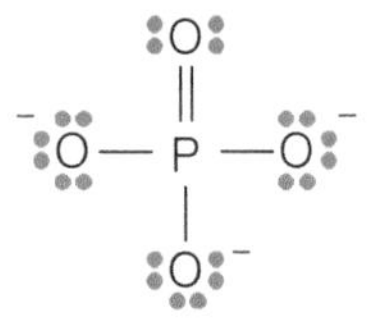	Derived from phosphoric(V) acid (third ionization), H_3PO_4
Ammonium	NH_4^+	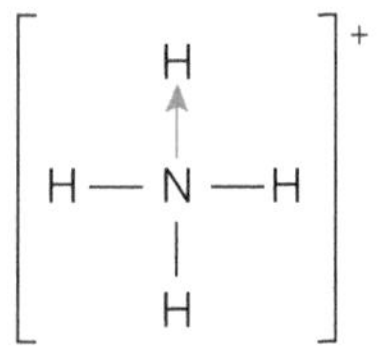	Derived from the protonation of an ammonia molecule

Common mistake

Compounds containing polyatomic ions have ionic bonding and covalent bonding. For example, ammonium sulfate, $(NH_4)_2SO_4$, does not contain a metal, but there are ionic attractions between the NH_4^+ and SO_4^{2-} ions. Both ions contain covalent bonds.

■ QUICK CHECK QUESTION

4 Deduce the names and formulas of the compounds formed from the following combinations of ions:

[$Na^+ SO_4^{2-}$], [$Na^+ OH^-$], [$Na^+ HSO_4^-$], [$Na^+ CO_3^{2-}$], [$Na^+ HCO_3^-$], [$Na^+ PO_4^{3-}$], [$Mg^{2+} NO_3^-$], [$Mg^{2+} PO_4^{3-}$], [$Al^{3+} OH^-$], [$Al^{3+} PO_4^{3-}$], [$NH_4^+ SO_4^{2-}$]

Ionic giant structures

Revised ☐

Ionic compounds are generally solids under standard conditions. They contain cations and anions packed together into a three-dimensional lattice. The ionic bonding is the sum of all the electrostatic attractions (between ions of the opposite charge) and electrostatic repulsion (between ions of the same charge). The attractive forces outweigh the repulsive forces.

The structure of the lattice depends on the relative sizes of the anion and cation and the stoichiometry of the compound. The unit cell is the smallest repeating unit of the lattice and lithium fluoride adopts the simple cubic unit cell (Figure 4.3).

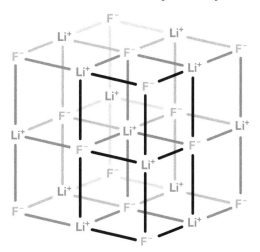

Figure 4.3 Ionic lattice for lithium fluoride

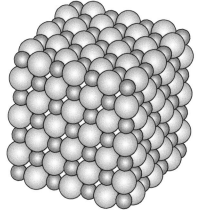

Figure 4.4 Arrangement of ions in lithium fluoride (small red spheres represent lithium ions; larger green spheres represent fluoride ions)

In ionic crystals the ions behave like charged spheres in contact held together by their opposite charges (Figure 4.4). In lithium fluoride each lithium ion is in contact with six fluoride ions, each lithium ion is in contact with six fluoride ions. Each fluoride ion touches six lithium ions.

The typical physical properties of ionic compounds and their explanation in terms of ionic bonding are shown in Table 4.5.

Table 4.5 Physical properties of typical ionic compounds

Physical property	Explanation
High melting and boiling points (low volatility)	Strong non-directional ionic bonding throughout the lattice of the crystal.
Brittle and hard crystals shatter if crushed	If layers of ions slip against each other then ions with the same charge come up against each other – like charges repel and break up the crystal (Figure 4.5).
Non-conductors of electricity when solid but conduct when liquid	Charged ions can move when the compound is molten (electrolysis) but not when it is solid.
Soluble in water but insoluble in organic solvents	Polar water molecules hydrate ions via ion-dipole forces. The attraction between water molecules and ions is strong enough to break up the lattice and release hydrated ions (Figure 4.6).

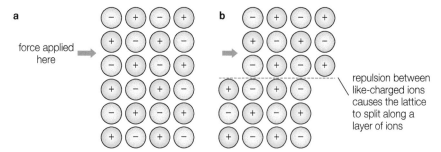

a force applied here

b repulsion between like-charged ions causes the lattice to split along a layer of ions

Figure 4.5 Cleavage in ionic solids: a layer of ions **a** before and **b** after force is applied (and cleavage occurs)

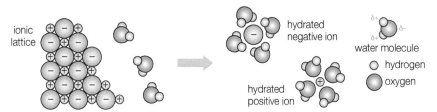

ionic lattice

hydrated negative ion

water molecule

○ hydrogen
○ oxygen

hydrated positive ion

Figure 4.6 An ionic solid dissolving in water

The strength of ionic bonding depends on the radii (size) of the ions (as well as the type of lattice). Small ions attract more strongly than large ions, due to their closer approach. Highly charged ions attract more strongly than ions with less charge.

Ionic bonding strength is determined by the lattice enthalpy (Topic 15 Energetics/thermochemistry). The more endothermic the lattice energy is, the stronger is the ionic bonding. The lattice enthalpy of an ionic compound is defined as the amount of energy required to convert one mole of the ionic solid into its constituent gaseous ions under standard conditions (Figure 4.7).

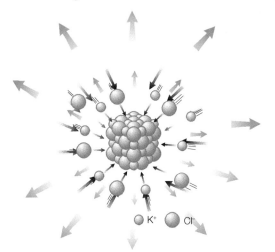

K$^+$ Cl$^-$

Figure 4.7 The lattice enthalpy for potassium chloride is the heat energy which would be absorbed from the surroundings (black arrows) if one mole of an ionic compound could directly form infinitely free gaseous ions moving apart (red arrows)

NATURE OF SCIENCE

The ionic bonding model is based on a range of experimental observations and data. X-ray diffraction data of ionic compounds shows little electron density between the two nuclei. Lattice energy can be calculated by assuming that ions are present. Strong evidence that a compound may be ionic include: it conducts and undergoes electrolysis when molten, but not as a solid. This behaviour supports the idea that ions are present in the compound.

Common mistake

The 'driving force' behind the formation of an ionic lattice is the energy released when oppositely charged ions attract each other. All ionization energies are always endothermic and some first electron affinities are also endothermic. It is therefore wrong to think that isolated atoms will spontaneously form ions.

QUICK CHECK QUESTION

5   Lithium fluoride is an ionic compound. Explain the following physical properties in terms of its structure:

   a volatility

   b electrical conductivity

   c solubility.

4.2 Covalent bonding

Revised ☐

Essential idea: Covalent compounds form by sharing electrons.

Covalent bonding

Revised ☐

- A covalent bond is formed by the electrostatic attraction between a shared pair of electrons and the positively charged nuclei.
- Single, double and triple covalent bonds involve one, two and three shared pairs of electrons respectively.
- Bond length decreases and bond strength increases as the number of shared electrons increases.
- Bond polarity results from the difference in electronegativities of the bonded atoms.

Lewis structures

Revised ☐

A Lewis structure (electron dot structure) shows all the valence electrons of the atoms or ions involved in a compound (Table 4.6). They can be used to describe covalent and ionic compounds.

Table 4.6 The Lewis structures of some elements in period 2 and 3

Group	Electronic configuration	Lewis diagram	Electronic configuration	Lewis diagram
1	Li $1s^2 2s^1$	Li•	Na [Ne] $3s^1$	Na•
2	Be $1s^2 2s^2$	•Be•	Mg [Ne] $3s^2$	•Mg•
13	B $1s^2 2s^2 2p^1$	•B̊•	Al [Ne] $3s^2 3p^1$	•Ål•
14	C $1s^2 2s^2 2p^2$	•C̈•	Si [Ne] $3s^2 3p^2$	•S̈i•
15	N $1s^2 2s^2 2p^3$	•N̈•	P [Ne] $3s^2 3p^3$	•P̈•
16	O $1s^2 2s^2 2p^4$	•Ö•	S [Ne] $3s^2 3p^4$	•S̈•
17	F $1s^2 2s^2 2p^5$	:F̈•	Cl [Ne] $3s^2 3p^5$	:C̈l•

A variety of notations are used to represent a pair of electrons: a line, dots, crosses or dots and crosses, or a combination. However, the use of dots and crosses (and stars) allows the tracking of electrons from each atom. The four methods shown in Figure 4.8 are all correct and equivalent ways of representing the Lewis structure of the iodine molecule.

Figure 4.8 Lewis structures (electron dot diagrams) for the iodine molecule, I_2

> **Expert tip**
>
> Electron pairs may be shown using dots, crosses or lines (especially for lone pairs) and there is no need to show the origin of the electrons. However, it may be helpful to use dots and crosses to keep track of which atom provided the electrons. An asterisk may be used to represent any additional electrons added to a molecular structure.

■ Covalent bonding

A single covalent bond is a shared pair of electrons between two non-metal atoms. The attraction between the two positively charged nuclei and the electron pair(s) in between holds the atoms together (Figure 4.9).

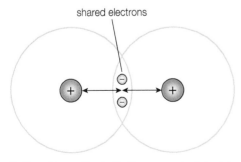

shared electrons

Both nuclei are attracted to the same pair of shared electrons. This holds the nuclei together.

Figure 4.9 A simple electrostatic model of the covalent bond in the hydrogen molecule, H_2

The shared electrons are normally localized. These electrons are held between the two atoms and are not free to move from one atom to the next (they are localized).

In many common molecules the number of bonds formed is such that each atom in the molecules has the same number of electrons in its outer shell as the nearest noble gas (Figure 4.10), usually eight. This is the octet rule.

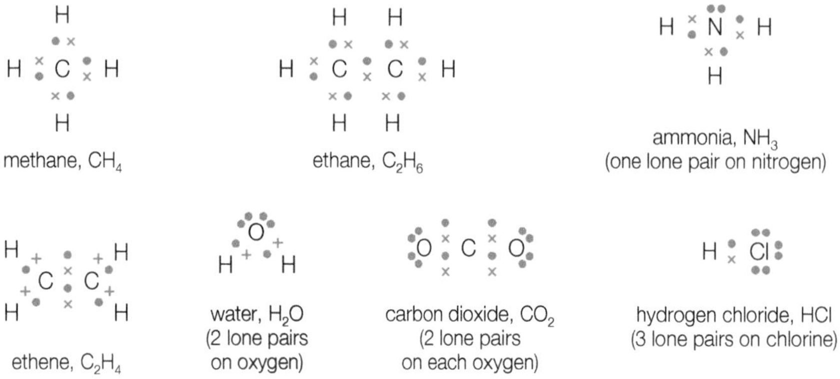

Expert tip

An octet of eight electrons refers to four pairs of valence electrons around an atom.

Figure 4.10 Lewis structures (electron dot diagrams) for a selection of simple covalent compounds (with single bonds)

Expert tip

When discussing covalent bonding, the term the **valency** of an atom refers to the number of bonds that atom can form. Table 4.7 shows the valency of some common atoms.

Table 4.7 The valencies of some common atoms

Element	Number of covalent bonds formed by atom
Hydrogen	1
Oxygen	2
Nitrogen	3
Carbon	4

Multiple covalent bonding

One shared pair of electrons forms a single covalent bond. Double or triple bonds involve two or three shared pairs of electrons (Figure 4.11).

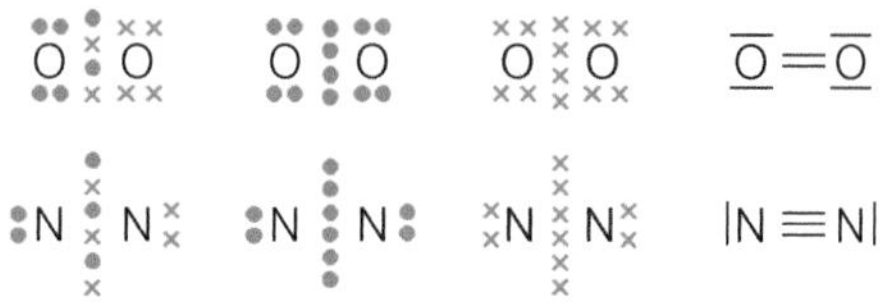

Figure 4.11 Lewis structures (electron dot diagrams) for oxygen and nitrogen molecules

Coordinate (dative) covalent bonding

In normal covalent bonding, each atom contributes one electron to the electron pair to form a single bond. However, in some molecules or ions, both electrons in the covalent bond are supplied by one of the bonded atoms. A bond formed in this way is called a *coordinate bond*.

There is no difference between a coordinate (dative) bond and a single covalent bond except that in the coordinate (dative) bond, only one of the bonded atoms provides the two electrons which are shared. The atom which donates the lone pair is called the electron-pair donor. The atom accepting the pair of electrons is called the electron-pair acceptor.

For an atom to act as a donor, it must have at least one pair of unshared electrons (a lone pair) and the acceptor must have at least one vacant orbital in its outer shell to receive the electron pair. The process of forming a dative bond (Figure 4.12) is called coordination.

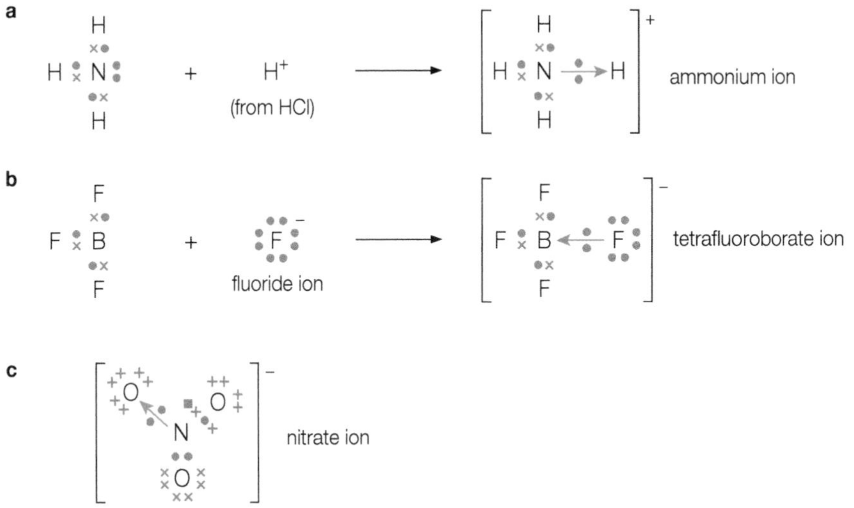

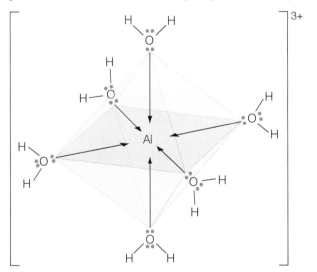

Figure 4.12 Formation of the **a** ammonium ion **b** tetrafluoroborate ion, **c** structure of the nitrate ion

Sulfur dioxide and sulfur trioxide may be represented by structures with a coordinate bond between sulfur and oxygen (to preserve the octet rule). The molecules may also be represented with double bonds between the sulfur and oxygen atoms (expanded octets). Both representations (Figure 4.14) are acceptable.

Complex ions

A *complex ion* is an ion that consists of a central cation (usually a transition metal) which forms coordinate bonds with one or more molecules or anions. These molecules or anions contain lone pairs of electrons.

The molecule or anion that uses the lone pair of electrons to form a coordinate bond with the central metal cation is called a ligand. Figure 4.15 shows the hydrated aluminium ion, which consists of a central aluminium cation, Al^{3+}, accepting lone pairs from six water molecules, acting as ligands (electron pair donors).

Figure 4.15 Structure of the hexaaquaaluminium ion, $[Al(H_2O)_6]^{3+}(aq)$

Common mistake

Do not forget that the following common species contain a coordinate bond: CO (Figure 4.13), NH_4^+ and H_3O^+.

carbon monoxide

Figure 4.13 Lewis structure for the carbon monoxide molecule, CO

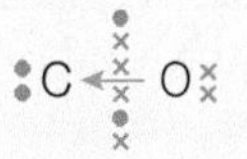

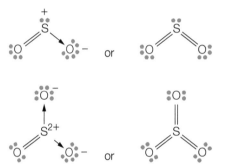

Figure 4.14 Alternative Lewis structures representing sulfur dioxide and sulfur trioxide molecules

Expert tip

If the species is an ion then square brackets should be placed around the structure and the charge indicated by a number then + or –.

QUICK CHECK QUESTIONS

6 Draw the Lewis structures for $[SiF_6]^{2-}$ and NO_2^+.

7 Show the formation of coordinate bonds in the following reactions involving aluminium compounds: $2AlCl_3 \rightarrow Al_2Cl_6$ and $AlCl_3 + Cl^- \rightarrow AlCl_4^-$. Identify the donor and acceptor atoms.

One simple model (ignoring hybridization) of this complex ion has the aluminium ion using its empty 3s, 3p and 3d orbitals accepting electron pairs from water molecules (Figure 4.16).

1s	2s	2p	3s	3p	3d

Figure 4.16 Orbital description of the aluminium ion in the hexaqua ion

Factors affecting covalent bond strength

Bond enthalpy is a measure of covalent bond strength. It is a measure of the energy required to break the bond in the gaseous state and release gaseous atoms (Topic 5 Energetics/thermochemistry). Bond strength (bond enthalpy) is affected by the following factors.

Bond length

The longer the bond length, the further from the shared pair of electrons the nuclei are. Therefore, a longer bond results in a lower bond energy and a lower strength.

Bond multiplicity

This refers to the presence of multiple covalent bonds between two atoms. The presence of multiple bonds increases the overall bond strength between the two atoms. The greater the number of shared pairs of electrons the greater the attraction for them by the nuclei, resulting in a shorter and hence stronger bond (Table 4.8).

Table 4.8 Bond lengths and bond enthalpies in carbon–carbon bonds

	C—C	C=C	C≡C
Bond length/pm	154	134	120
Bond enthalpy/kJ mol^{-1}	346	614	839

Bond polarity

An increase in polarity also means an increase in the ionic character of the bond. This usually results in an increase in bond strength and energy.

Polarity

Bond polarity

In diatomic molecules containing the same element, for example, hydrogen, H_2, the electron pair will be shared equally (Figure 4.17): the nuclei of both atoms exert an identical attraction because they have the same value of electronegativity. The electron density will be symmetrically distributed between the two bonded hydrogen atoms.

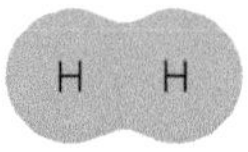

However, when the atoms of the elements are different the more electronegative atom has a stronger attraction for the shared electron pair(s). One end of the molecule (with the more electronegative element) will be more electron rich than the other end (with the less electronegative element), which will be electron deficient.

A *polar covalent bond* is formed and the small difference in charge (the dipole) is represented by the fractional charges, δ^+ and δ^-. The separation of the two electrical charges creates a dipole: two equal and opposite charges separated by a small fixed distance.

The size of a dipole is measured by its dipole moment. This is a product of charge and distance. It is a measure of the polarity of the bond and is a vector: it has direction and size. The greater the difference in electronegativities the greater the polarity and ionic character of that bond.

QUICK CHECK QUESTION

8 a Draw Lewis (electron dot structures) for molecules of carbon monoxide, CO, carbon dioxide, CO_2 and propan-1-ol, $CH_3CH_2CH_2OH$.

b List with explanation, the three molecules in order of increasing carbon to oxygen bond length (shortest first).

Figure 4.17 The electron density of the hydrogen molecule

Figure 4.18 shows how the polarity and dipole moment of the hydrogen chloride molecule, HCl, can be represented.

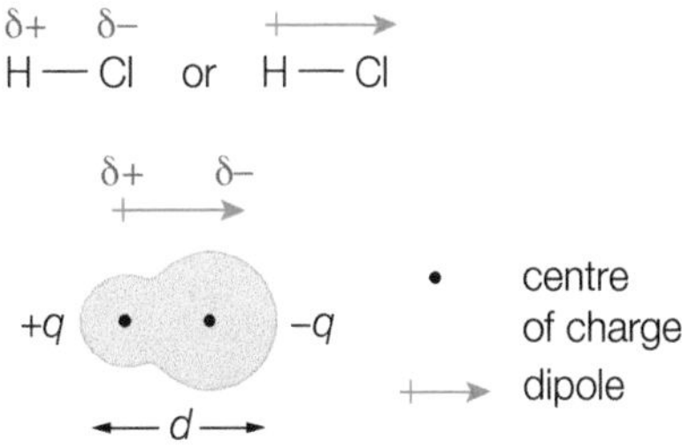

$$\overset{\delta+}{H}\underset{}{\longrightarrow}\overset{\delta-}{Cl} \quad or \quad H\overset{\longrightarrow}{\longrightarrow}Cl$$

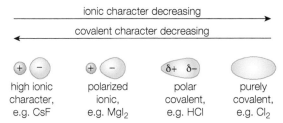

- centre of charge
- dipole

Figure 4.18 Illustration of the polarity of the hydrogen chloride molecule

■ Molecular polarity

Many molecules with polar bonds are themselves polar and have a permanent molecular dipole. Molecular polarity is a vector quantity. Therefore, vector addition can be used to sum up all the dipole moments present in the bonds to determine whether there is a net dipole moment for the molecule. If there is a net dipole, the molecule is polar.

Figure 4.19 shows the polar molecules: hydrogen fluoride, chloromethane and water. The carbon–hydrogen bond polarity is ignored since the electronegativities of carbon and hydrogen are relatively similar.

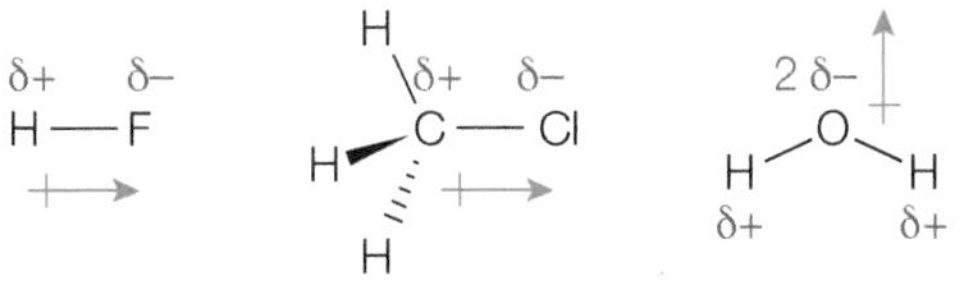

Figure 4.19 Molecular polarities of HF, CH_3Cl and H_2O molecules

However, some molecules with polar bonds are non-polar (Figure 4.20) as their bond polarities (dipole moments) cancel each other out. So the net molecular dipole is equal to zero. Hence, a molecule can possess polar bonds and still be non-polar. If the polar bonds are evenly (or symmetrically) distributed, the bond dipoles cancel and do not create a molecular dipole, meaning that the molecule will not orientate itself in a magnetic field (it has no overall dipole moment).

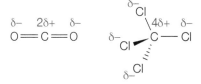

Figure 4.20 Molecular polarities of CO_2 and CCl_4 molecules

■ Polar covalent bonding

In a compound formed between a reactive metal, such as potassium, and the most reactive non-metal, fluorine, the bonding is essentially ionic (high percentage of ionic character). As a rough 'rule of thumb' compounds tend to be ionic when the electronegativity difference is greater than 1.8.

In a molecule, such as fluorine (F_2) where both atoms (and values of electronegativity) are the same, the electron pair is shared equally and the bonding is purely covalent (100% covalent character).

In many compounds, the bonding is neither purely ionic nor purely covalent, but intermediate in nature (Figure 4.21).

ionic character decreasing →

← covalent character decreasing

⊕ ⊖	⊕ ⊖	δ+ δ–	
high ionic character, e.g. CsF	polarized ionic, e.g. MgI_2	polar covalent, e.g. HCl	purely covalent, e.g. Cl_2

Figure 4.21 A spectrum of bonding

The ionic model of bonding regards ions as spherical, but the cation will distort electron density of the neighbouring anions, giving rise to some degree of electron sharing, that is, some degree of covalent bonding. If there is substantial covalent bonding then it is known as polar covalent bonding.

Chemists have developed theories of chemical bonding based on various models to explain the physical properties of different substances and to classify substances into groups of substances with related physical and chemical properties. Many substances can be classified as ionic (electron transfer) or covalent (electron pair sharing), but this is a simplistic model and many substances have both covalent and ionic character.

A spherical anion is non-polar and when a cation pulls electron density towards itself the anion becomes polarized and a dipole is formed. The smaller a cation and the larger its charge, the greater the extent to which it tends to polarize an anion (Figure 4.22).

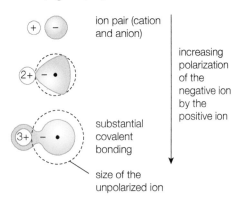

Figure 4.22 Ionic bonding with increasing covalent character (sharing of electron density) because the cation has distorted the neighbouring anion (dotted circles show the unpolarized anions)

Polarizing power increases along the isoelectronic series Na^+, Mg^{2+} and Al^{3+} as the cations get smaller and more highly charged. Their charge density increases, which increases their ability to polarize anions.

The larger the anion and the larger its charge, the more polarizable it becomes. Hence iodide ions are more polarizable than smaller fluoride ions. Sulfide ions, S^{2-}, are more polarizable than chloride ions, Cl^-, largely due to the higher charge.

> **Expert tip**
>
> The polarizing power of a cation is proportional to its charge, but inversely proportional to its size.

> ■ **QUICK CHECK QUESTIONS**
>
> 12 State whether the following compounds are ionic, covalent or strongly polar covalent:
>
> potassium chloride, KCl, silicon(IV) hydride, SiH_4, hydrogen iodide, HI, calcium oxide, CaO, beryllium chloride, $BeCl_2$
>
> 13 State and explain which ionic compound is expected to have a larger degree of covalent character: sodium oxide or sodium fluoride.

4.3 Covalent structures

Revised ▢

Essential idea: Lewis (electron dot) structures show the electron domains in the valence shell and are used to predict molecular shape.

Covalent structures

Revised ▢

- Lewis (electron dot) structures show all the valence electrons in a covalently bonded species.
- The 'octet rule' refers to the tendency of atoms to gain a valence shell with a total of eight electrons.
- Some atoms, like Be and B, might form stable compounds with incomplete octets of electrons.
- Resonance structures occur when there is more than one possible position for a double bond in a molecule.
- Shapes of species are determined by the repulsion of electron pairs according to VSEPR theory.
- Carbon and silicon form giant covalent/network covalent structures.

■ Giant covalent structures

In giant covalent structures, the strong covalent bonds extend between all atoms throughout the solid. Their physical properties are determined by the nature of the strong covalent bonds. Diamond and silicon(IV) oxide (Figure 4.23) are

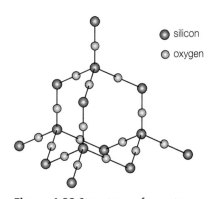

● silicon
○ oxygen

Figure 4.23 Structure of quartz (silicon dioxide, SiO_2)

examples of three-dimensional giant covalent structures. Graphite and graphene are examples of two-dimensional giant covalent structures.

In three-dimensional giant covalent structures, the strong covalent bonds extend between all atoms throughout the solid. Their physical properties are determined by the nature of the strong covalent bonds:

- hard, because strong covalent bonds hold all the atoms in place (except for between the layers in graphite)
- high melting and boiling points, because a large amount of thermal energy is needed to break all the strong covalent bonds holding the atoms in place. This is necessary in order to form a liquid
- totally insoluble in water and other polar solvents – solvent molecules cannot interact with the non-metal atoms strongly enough to break up the giant structure
- non-conductors of electricity, owing to the absence of delocalized electrons or mobile ions, except in graphite and graphene (Figure 4.24) (where electrons move between, or above and below, the monolayer(s)).

Allotropes of carbon

Diamond, graphite and carbon-60 are different forms (allotropes) of crystalline carbon. Their skeletal structures and important physical properties are summarized below in Table 4.9.

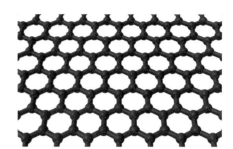

Figure 4.24 Graphene showing its atomic-scale honeycomb lattice formed of carbon atoms

> ■ **QUICK CHECK QUESTION**
>
> **14** Compare the structure and bonding in carbon dioxide and silicon dioxide.

Common mistake

Giant covalent structures and covalent (simple) molecular substances both contain covalent bonds, but have very different physical properties.

Table 4.9 Structure, bonding and properties of the allotropes of pure carbon

Diamond	Graphite	Carbon-60
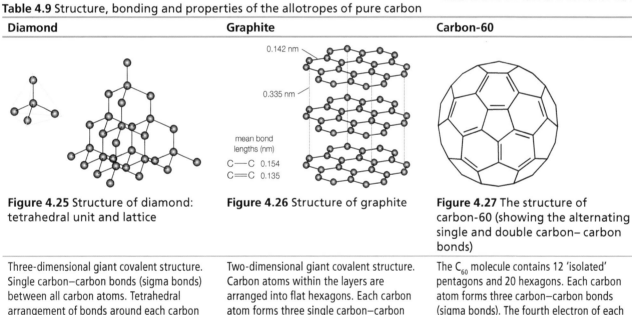 **Figure 4.25** Structure of diamond: tetrahedral unit and lattice	**Figure 4.26** Structure of graphite 0.142 nm 0.335 nm mean bond lengths (nm) C—C 0.154 C=C 0.135	**Figure 4.27** The structure of carbon-60 (showing the alternating single and double carbon–carbon bonds)
Three-dimensional giant covalent structure. Single carbon–carbon bonds (sigma bonds) between all carbon atoms. Tetrahedral arrangement of bonds around each carbon atom.	Two-dimensional giant covalent structure. Carbon atoms within the layers are arranged into flat hexagons. Each carbon atom forms three single carbon–carbon bonds (sigma bonds). The fourth electron of each carbon atom is delocalized.	The C_{60} molecule contains 12 'isolated' pentagons and 20 hexagons. Each carbon atom forms three carbon–carbon bonds (sigma bonds). The fourth electron of each carbon atom is delocalized (to some extent) over the surface of the C_{60} molecule.
Very high melting point, very high boiling point and very hard. Insulator – no delocalized electrons.	High sublimation point (under standard conditions). Conducts electricity along the planes of carbon atoms when a voltage is applied. Good lubricant because adjacent layers of carbon atoms can readily slide over each other.	Significantly lower sublimation point than graphite because only weak London (dispersion) forces need to be broken between the discrete C_{60} molecules. Insulator because the electron delocalization is only within C_{60} molecules.

Silicon has the same structure as diamond but the covalent bonds are weaker (because of the poor overlap between silicon atoms with larger and more diffuse orbitals than carbon atom).

Graphene consists of a single or monolayer of graphite. It has a high tensile strength and is an electrical conductor (Figure 4.28) (across, that is, parallel to its layers). It has physical properties similar to that of graphite but they are modified because of quantum effects.

Expert tip

Graphite is the most energetically and thermodynamically stable form of carbon (under standard conditions). It is a two-dimensional giant covalent structure (Figure 4.28).

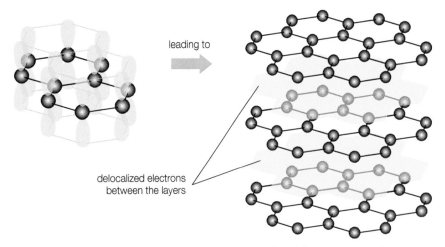

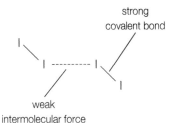

Figure 4.28 The structure of graphite showing the delocalization of electrons between the layers

■ Simple covalent molecules

These compounds, such as iodine (Figure 4.29) and carbon dioxide, exist as simple discrete covalent molecules. The strong covalent bonds are within the molecules. Between the molecules, there are only intermolecular forces which are weaker than covalent bonds.

These compounds have the following properties:
- usually soft or brittle
- low melting and boiling points because the intermolecular forces of attraction are easy to overcome (Figure 4.30)
- their solubilities in water and organic solvent depend on the nature of the molecules (whether they are polar or non-polar)
- low electrical conductivity, owing to absence of mobile charged particles (ions or electrons).

Figure 4.30 Strong covalent and weak intermolecular forces (London (dispersion) forces)

■ Resonance

The theory of resonance involves using two or more Lewis structures to describe a molecule or polyatomic ion. Resonance is invoked when a single Lewis structure is incompatible with the experimental data of the molecule or ion. Resonance leads to charge being spread more evenly over an ion or molecule, and leads to increased energetic stability (the energy is lowered).

For example, the gas trioxygen (ozone), O_3, consists of molecules that can be represented by two Lewis (electron dot) structures as shown in Figure 4.31. The double-headed arrow represents resonance and the curly arrows show (imaginary) movement of electron pairs. Both structures obey the octet rule and are equivalent.

Figure 4.31 Lewis structures for the ozone molecule, O_3

Expert tip

Remember that fullerene-60 (C_{60}) is a simple molecular substance not giant covalent despite the number of atoms involved.

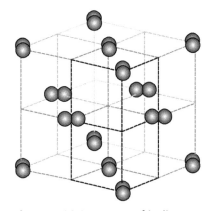

Figure 4.29 Structure of iodine: a simple molecular substance

In both structures there is an O–O single bond and an O=O double bond. The O–O and O=O bond lengths are 148 pm and 121 pm, respectively. However, the experimentally determined oxygen–oxygen bond lengths in O_3 are the same (128 pm). This cannot be represented by either of the two Lewis structures.

Hence the oxygen–oxygen bonds in the O_3 molecule are intermediate between a double and a single bond. The O_3 molecule can be described as a resonance hybrid (Figure 4.32) where the two resonance structures make equal contributions to the hybrid.

According to the theory of resonance, whenever a single Lewis structure cannot describe a molecule, a number of structures with similar energy, positions of nuclei, bonding and lone pairs are taken as resonance structures of the hybrid. A formal charge can be used to choose the more important resonance structures, those that make major contributions to the hybrid.

Delocalization of π (pi) electrons

Resonance structures are used to describe certain molecules and polyatomic ions where the structure can only be adequately described by two or more resonance structures. These species can also be described by an equivalent bonding model involving the delocalization of π (pi) electrons over three or more atoms.

For example, in the ozone molecule each of the three oxygen atoms have an unhybridized p orbital which can merge to form a delocalized pi bond extending over all three oxygens (Figure 4.33). The bond order of 1.5 can also be shown with a dotted line.

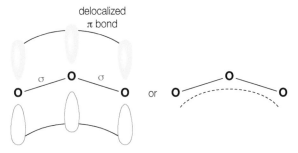

Figure 4.33 Delocalized π (pi) bond in the ozone molecule, O_3

QUICK CHECK QUESTIONS

17 a Draw two Lewis (electron dot) structures describing the ozone molecule, O_3 and state the approximate value of the bond angle.

   b Explain why the oxygen–oxygen bond lengths in the ozone molecule are equal.

18 Baking soda is sodium hydrogencarbonate ($NaHCO_3$). It contains sodium ions, Na^+, and resonance stabilized hydrogen carbonate ions, HCO_3^-.

   a Explain the meanings of the term resonance and resonance hybrid.

   b Draw two equivalent Lewis structures describing the two resonance structures of HCO_3^-.

Extensive delocalization of pi electrons occurs in conjugated systems, such as buta-1,3-diene, which contain a series of alternating carbon–carbon double and single bonds (Figure 4.34).

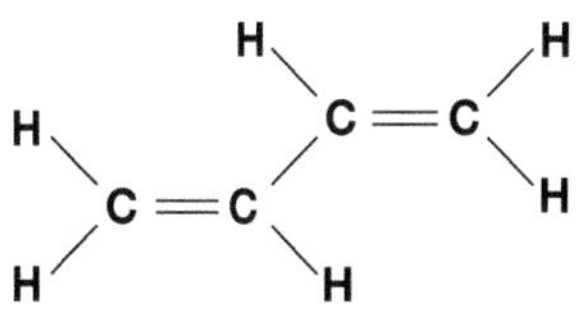

All the unhybridized p orbitals on the carbon atoms overlap sideways and a delocalized pi bond is formed, extending across all four carbon atoms (Figure 4.35). The delocalization of pi electrons in conjugated systems lowers the overall energy of the molecule and increases its stability.

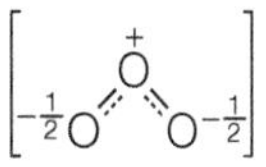

Figure 4.32 Resonance hybrid for ozone, O_3

Common mistake

Resonance structures (Lewis structures) must show lone pairs and any charges. The resonance hybrid may have fractional charges but does not include lone pairs (unless they are identical in all resonance structures).

Figure 4.34 Structure of the buta-1,3-diene molecule

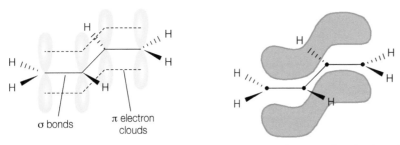

Figure 4.35 π (pi) bond formation in the buta-1,3-diene molecule showing the delocalized pi clouds above and below the carbon skeleton

■ Molecular shape

■ VESPR theory

The shapes of simple molecules and ions can be predicted by the valence-shell electron-pair repulsion (VSEPR) theory. If the Lewis (electron dot) diagram can be drawn for a molecule or a polyatomic ion, the shape of this molecule or ion can be predicted using this theory.

According to VSEPR theory:
■ The electron pairs around the central atom repel each other.
■ Bonding pairs and lone pairs of electrons (known as electron domains) arrange themselves to be as far apart as possible.
■ Bond angles are maximized.
■ Molecules and ions take up the shape that minimizes the repulsion between bonding and lone pairs of electron.

■ Basic molecular shapes

There are five basic molecular geometries: linear, trigonal planar, tetrahedral, trigonal bipyramid and octahedral: but at this point in the syllabus you need to be aware of the first three (those involving 2, 3 and 4 electron pairs, or domains, around the central atom). Table 4.10 shows the arrangement of the electron pairs that results in minimum repulsion and the shapes of the molecules.

Table 4.10 Basic molecular shapes of molecules and ions involving 2, 3, and 4 electron domains around a central atom

Molecule shape	Number of electron pairs	Description	Example
	2	Linear	$BeCl_2(g)$
	3	Triangular planar (trigonal planar)	$BF_3(g)$
	4	Tetrahedral	$CH_4(g)$

> **Expert tip**
>
> Remember it is only the electron pairs around the *central* atom that determine the shape of a molecule or polyatomic ion. The key words when answering questions on the VSEPR approach are 'the central atom' of the molecule or ion – it is the arrangement around this atom that is important. When two or more resonance structures can represent a molecule, the VSEPR model is applicable to any resonance structure.

■ Shapes of molecules and bond angles

The shapes and bond angles of molecules and ions are determined by the number of electron pairs (electron domains). In the VSEPR theory, a lone pair of electrons repels other electron pairs more strongly than a bonding pair. This is because the region in space occupied by a lone pair of electrons is smaller and closer to the nucleus of an atom than a bonding pair.

Bonding pairs of electrons are spread out between the nuclei of the two atoms which they bind together. Hence lone pairs can exert a greater repelling effect than a bonding pair. The order of the repulsion strength of lone pairs and bonding pairs of electrons is shown in Table 4.11.

Table 4.11 Order of repulsion strength for electron pairs

Bond pair--bond pair repulsion	Lone pair–bond pair repulsion	Lone pair–lone pair repulsion
Weakest	Medium	Strongest

These repulsion effects result in deviations from idealized shapes and changes in bond angles in molecules (Figure 4.36). The bond angle is the angle between two adjacent covalent bonds around a central atom.

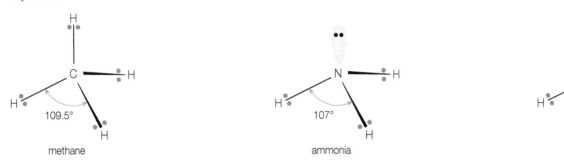

Figure 4.36 The shapes and bond angles of the methane, ammonia and water molecules

A multiple bond (double or triple bond) is treated as if it were a single electron pair (electron domain). Hence the carbon dioxide molecule, CO_2, has a linear structure like the beryllium chloride molecule (Table 4.11) and the ethene molecule, C_2H_4, is trigonal planar around each of the two carbon atoms (Figure 4.37).

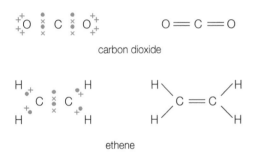

carbon dioxide

ethene

Figure 4.37 Lewis structure and molecular shapes of the carbon dioxide and ethene molecules

Table 4.12 summarizes the geometry of some simple molecules with central atoms having one or more lone (non-bonding) pairs of electrons.

Table 4.12 Geometry of common molecules (where E represents a lone pair of electrons)

Molecule type	Number of bonding pairs	Number of lone pairs	Arrangement of electron pairs	Shape	Examples
AB_2E	2	1	Trigonal planar	Bent	SO_2, O_3
AB_3E	3	1	Tetrahedral	Trigonal pyramidal	NH_3
AB_2E_2	2	2	Tetrahedral	Bent	H_2O

Expert tip

Multiple bonds have the same effect on bond angles because the electron pairs occupy the same space and direction as those in a single bond.

■ **QUICK CHECK QUESTION**

19 PH_3, PH_2^- and PH_4^+ are three examples of phosphorus-containing species. They have different H–P–H bond angles. Arrange the bond angles in ascending order. Explain your answer.

Expert tip

When dealing with structures containing lone pairs of electrons be careful of the difference between the geometry of the electron domains present and the actual *shape of the molecule*.

Expert tip

Covalent bonds are directional since the molecules have specific bond angles and shapes determined by VESPR theory. There is a sharing of electron density between atoms within the molecule.

■ **QUICK CHECK QUESTIONS**

20 a Draw a Lewis (electron dot) structure for a nitric(III) acid, HONO, molecule. Identify which nitrogen–oxygen bond is the shorter.

b Deduce the approximate value of the hydrogen–oxygen–nitrogen bond angle in nitric(III) acid and explain your answer.

21 The Lewis (electron dot) structure of the drug paracetamol (acetaminophen) is shown below.

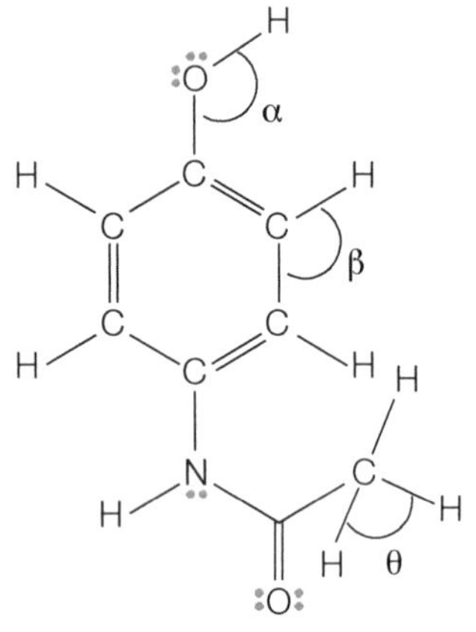

Figure 4.38 The Lewis (electron dot) structure of the drug paracetamol (acetaminophen)

Deduce the approximate values of the bond angles.

22 Draw the Lewis structure for BCl_4^-. Using the VSEPR theory, deduce the shape and the bond angle(s) of BCl_4^-.

23 In the gaseous state beryllium chloride exists as discrete molecules. In the solid state, $BeCl_2$ molecules polymerize to form molecular chains via coordinate bonding.

a Draw the Lewis structure of a $BeCl_2(g)$ molecule.

b State how the central atom breaks the octet rule.

c Using four $BeCl_2$ molecules, show part of the polymeric structure in $BeCl_2(s)$ using Lewis structures. State the Cl–Be–Cl angle in this polymer.

4.4 Intermolecular forces

Revised ☐

Essential Idea: The physical properties of molecular substances result from different types of forces between their molecules.

Intermolecular forces

Revised ☐

■ Intermolecular forces include London (dispersion) forces, dipole–dipole forces and hydrogen bonding.

■ The relative strengths of these interactions are London (dispersion) forces < dipole–dipole forces < hydrogen bonds.

Intermolecular forces

Revised ☐

As well as ionic and covalent bonding there are a number of weaker attractive forces that exist between molecules. These are London (dispersion) forces, dipole-dipole forces and hydrogen bonds.

■ London (dispersion) forces

These are the generally the weakest of the intermolecular forces. They act between all molecules whether they are polar or non-polar. They exist due to the movement of electrons that in turn causes the formation of an instantaneous or temporary dipole. These induce dipoles in neighbouring molecules (Figure 4.39). London (dispersion) forces increase in strength with the number of electrons available for polarization in a molecule (or atom). This explains why the boiling

Expert tip

Van der Waals' forces is a general term used to collectively describe intermolecular forces. An electric dipole is a separation of positive and negative charges. The simplest example of this is a pair of electric charges of equal magnitude but opposite sign, separated by a distance.

points rise down group 17 (halogens), group 18 (the noble gases) and with increasing molar mass in the alkanes.

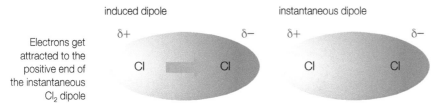

Figure 4.39 An instantaneous dipole–induced dipole attraction

For example, although H–Cl is a more polar molecule than H–Br or H–I, the boiling point of H–Cl is lower than that of H–Br or H–I. This shows that the effect of temporary induced dipoles is more significant than permanent dipoles when comparing molecules with significantly different molar masses.

The strength of London (dispersion) forces is also proportional to the total surface area of the molecule. The larger the total surface area the greater is the contact between the molecules, the greater the London (dispersion) forces and the higher the boiling point of a sample of the molecules (Figure 4.40).

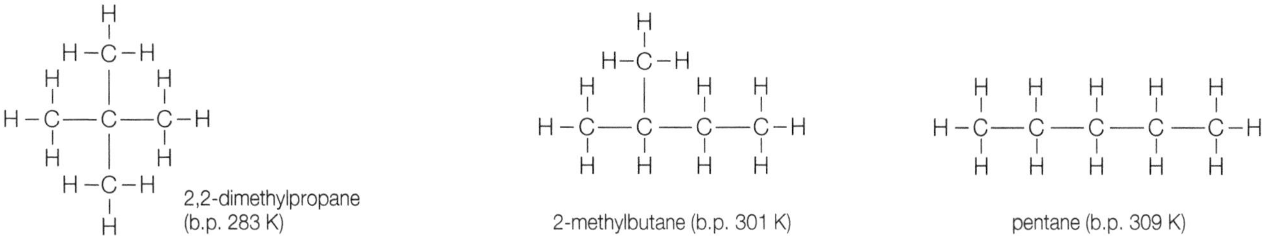

Figure 4.40 Branched chain alkanes have lower boiling points than straight-chain isomers

NATURE OF SCIENCE

Intermolecular forces are weak attractive forces between molecules. Without intermolecular forces there could be no molecular liquids or solids. Trends in boiling points of molecular liquids of wide range of inorganic and organic liquids provided experimental data supporting the concept that the various intermolecular forces arise from electrostatic attraction between permanent or temporary dipoles.

Dipole–dipole interactions

These occur between polar covalent molecules, that is, those containing different elements (atoms). A permanent dipole–dipole interaction exists between polar molecules because the positive end of one dipole will attract the negative end of another molecule (Figure 4.41).

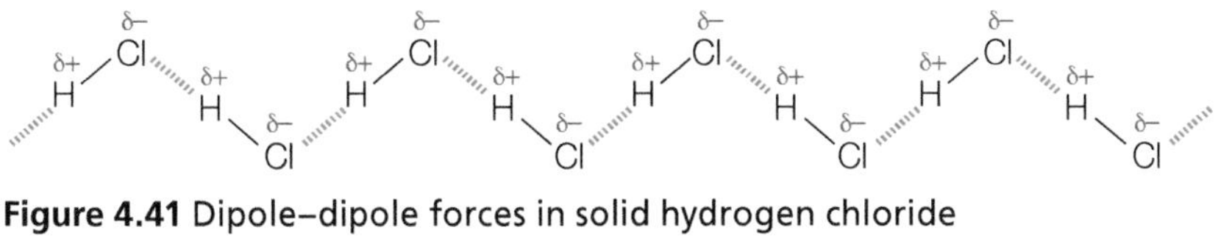

Figure 4.41 Dipole-dipole forces in solid hydrogen chloride

Hydrogen bonds

These are very strong dipole–dipole interactions between molecules containing hydrogen bonded to nitrogen, oxygen or fluorine (small and electronegative atoms). These intermolecular forces results from the interaction between the partially positive hydrogen atoms with lone pairs of electrons on the nitrogen (Figure 4.42), oxygen or fluorine atoms, so the hydrogen atom can be considered as acting as a 'bridge' between two electronegative atoms.

This form of intermolecular force can have significant effects on the physical properties of the compounds concerned, such as increasing boiling point (Figure 4.43), melting point and viscosity.

Expert tip

When molecules differ significantly in their molar mass, London (dispersion) forces are more significant than dipole–dipole forces. Therefore, the molecule with the largest molar mass has the strongest attractive forces.

Expert tip

When molecules have similar molar masses and shapes, dipole–dipole forces are significant. Therefore, the most polar molecule has the strongest attractive forces.

Expert tip

Gases consisting of atoms or non-polar molecules deviate from ideal behaviour (as described by the ideal gas equation and gas laws) because of the presence of London (dispersion) forces between gaseous molecules.

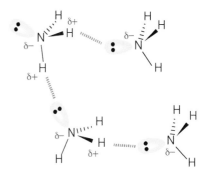

Figure 4.42 Hydrogen bonding in ammonia

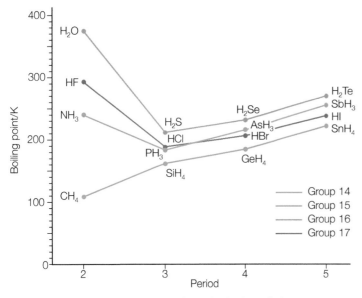

Figure 4.43 The boiling points of the hydrides of elements in groups 14 to 17

■ **QUICK CHECK QUESTIONS**

24 The boiling points of the hydrides of the group 16 elements increase in the order: $H_2S < H_2Se < H_2Te < H_2O$. Explain the trend in the boiling points in terms of intermolecular forces.

25 Explain the following observations: tetrachloromethane, CCl_4, is immiscible in water but methanol, CH_3OH, is miscible with water in all proportions. Methanoic acid, HCOOH, is soluble in water but octanoic acid ($C_7H_{15}COOH$) is much less soluble.

4.5 Metallic bonding

Revised ☐

Essential idea: Metallic bonds involve a lattice of cations with delocalized electrons.

Metallic bonding

Revised ☐

■ A metallic bond is the electrostatic attraction between a lattice of positive ions and delocalized electrons.
■ The strength of a metallic bond depends on the charge of the ions and the radius of the metal ion.
■ Alloys usually contain more than one metal and have enhanced properties.

In a metal crystal each metal atom forms a positive ion (cation) by releasing its valence electrons into a 'cloud' or 'sea' of delocalized electrons. Metallic bonding results from the attraction between the positive metal ions and the delocalized electrons (Figure 4.44). The electrons move in an overall random direction through the crystal.

A sample of a metal consists of a large number of very small crystal grains (Option A Materials). Each grain is a giant structure of metal atoms held together by metallic bonding. Metallic bonding is non-directional: all of the valence electrons are attracted to the nuclei of all the metal ions. The attraction decreases with distance between the nuclei and delocalized electrons (electron cloud).

In most metals the atoms are packed together as closely as possible so that every metal atom is touching 12 other metal atoms. A small number of less dense metals have a more open cubic structure in which each metal atom has a coordination number of eight. Metallic structures (Figure 4.45) are studied using X-ray diffraction (Option A Materials).

nuclei and inner-shell electrons, i.e. cations   delocalized outer-shell electrons

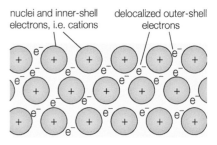

Figure 4.44 Structure of a generalized metallic lattice (in two dimensions)

 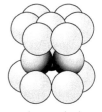

body-centred cubic (not close-packed)   close-packed

Figure 4.45 Two ways of packing the atoms in metal crystals

Typical properties of metals

Table 4.13 relates the physical properties of a typical metal to the structure and bonding within a metallic lattice.

Table 4.13 Physical properties of metals and their relation to the 'electron sea' model of metallic bonding

Property	Explanation
High melting and boiling points	Strong metallic bonding through the crystal; strong electrostatic forces of attraction between the close packed cations and delocalized electrons.
Malleability and ductility (Figure 4.46)	Metal atoms are always held within the cloud of the delocalized electrons; the bonding is three-dimensional (non-directional).
Good conductors of electricity when in the solid and liquid states (Figure 4.47)	Delocalized valence electrons are free to move (as an electric current) in metal crystal, when a voltage (potential difference) is applied.
	The mobile electrons also conduct heat by carrying kinetic energy (vibrations) from a hot part of the metal lattice to a colder one. The ions in the lattice also vibrate and spread thermal energy when there is a temperature difference.
Insoluble in water and organic solvents	Solvent molecules cannot bond with metal atoms strongly enough to break up the giant metallic structure; reactive metals in groups 1 and 2 react with water to form a soluble product.
Become harder on alloying with another metal	The regular lattice is disrupted and the layers of atoms will not slide over each other.

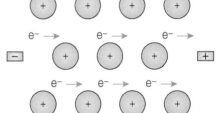

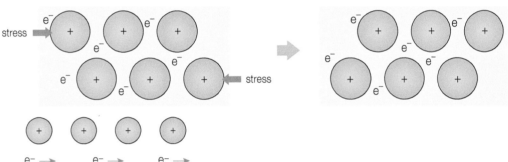

Figure 4.46 The application of a shear force to a metallic lattice: adjacent layers can slide over each other

Figure 4.47 The flow of an electric current in a metal when a voltage is applied

Melting points of metals

Generally the strength of metallic bonding depends on the charge of the metal cation and the radius of the metal ion. (The type of lattice also has an effect of the strength of metallic bonding.) The melting and boiling points can be used to assess the strength of metallic bonding. The stronger the metallic bonding the higher the melting and boiling points.

The strength of metallic bonding decreases down groups 1 and 2 because the ionic radius of the cation increases. Hence the distance between the delocalized electrons and cation nucleus increases and hence the electrostatic forces of attraction decreases. Melting and boiling points decrease down groups 1 and 2.

The strength of metallic bonding increases across period 3 from sodium to magnesium to aluminium (Figure 4.48). This is due to two factors: an increase in cation charge from +1, to +2 and +3, and a decrease in ionic radius. Both factors increase the electrostatic forces of attraction between the cations and delocalized electrons.

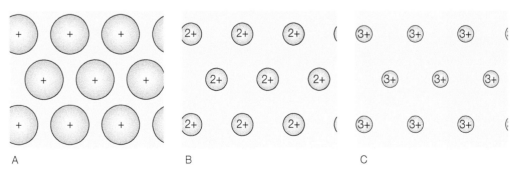

Figure 4.48 Metals with small highly charged ions form stronger metallic bonds (metallic bond strength increases A < B < C)

A   B   C

■ Alloys

Alloys are homogeneous mixtures (solid solutions) of two or more metals, or a metal with a non-metal, usually carbon. For example, brass is an alloy of copper and zinc and bronze is an alloy of copper and tin. Steels are alloys containing iron, carbon and other metals, for example, stainless steel is an alloy of iron, carbon and chromium.

The addition of small amounts of another metal to a pure metal will alter its physical properties. The additional metal atoms will enter the lattice and form a cation with a different ionic radius and perhaps a different ionic charge.

Generally, alloys are less ductile and less malleable than pure metals since the added metal atoms prevent the sliding of adjacent layers of metal atoms across each other (Figure 4.49). The addition of carbon to iron results in steel with a high tensile strength. The addition of chromium produces stainless steel which resists corrosion.

Expert tip

The transition elements are hard, strong and have high melting and boiling points. This is due to the strong metallic bonding which involves the 3d and 4s electrons.

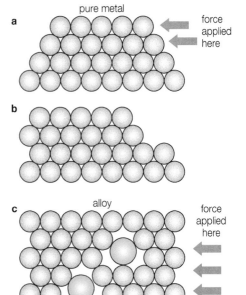

Figure 4.49 a The position of the atoms in a pure metal before a force is applied; **b** after the force is applied, slippage has taken place; **c** in an alloy, slippage is prevented because the atoms of different size cannot easily slide over each other

---

■ QUICK CHECK QUESTIONS

26 Describe the type of bonding present in caesium metal and explain why it is an excellent conductor of electricity.

27 Duralumin is an alloy containing aluminium and small amounts of copper. Explain why aluminium metal is malleable, but the alloy is much less malleable. Explain why the electrical conductivity of aluminium is almost unaffected by the formation of the alloy.

28 a The variations of melting point and the densities of three metals, potassium (K), calcium (Ca) and titanium (Ti) are shown. Outline reasons for these variations.

Element	Melting point/°C	Density/g cm^{-3}
K	64	0.86
Ca	838	1.55
Ti	1500	3.00

b State and explain the expected trend in the melting and boiling points of the group 2 elements: calcium, strontium and barium.

Summary

Table 4.14 presents a summary of the different possible types of structure and the bonding or inter-molecular forces associated with them.

Table 4.14 Bonding, structure and properties: a summary

	Ionic	Giant covalent	Metallic	Macromolecular	Simple molecular
What substances have this type of structure?	Compounds of metals with non-metals, especially reactive elements	Some elements in group 14 and some of their compounds	Metals	Polymers	Some non-metal elements and some non-metal/non-metal compounds
Examples	Sodium chloride, NaCl, calcium oxide, CaO	Silicon(IV) oxide, SiO_2, diamond, C, silicon, Si, graphite, C, and graphene	Copper, Cu, iron, Fe, sodium, Na, and magnesium, Mg	Polythene, terylene, proteins, cellulose and DNA	Carbon dioxide, CO_2, chlorine, Cl_2, water, H_2O, and ammonia, NH_3
What type of particle does it contain?	Ions: cations and anions	Atoms	Positive ions (cations) surrounded by delocalized valence electrons	Long-chain molecules	Small molecules
How are the particles bonded together?	Strong ionic bonding; electrostatic attraction between oppositely charged ions	Strong covalent bonds; attraction of atomic nuclei for shared pairs of electrons	Strong metallic bonding; attraction of, nuclei for delocalized electrons	Weak intermolecular forces between molecules; strong covalent bonds between the atoms within each molecule	Weak intermolecular forces between molecules; strong covalent bonds between the atoms within each molecule
Melting and boiling points	High	Very high	Generally high	Moderate (often decompose on heating)	Low
Hardness	Hard but brittle	Very hard	Hard but malleable	Many are soft but often flexible	Soft
Electrical conductivity	Conduct when molten or dissolved in water; electrolytes	Do not normally conduct	Conduct when solid or liquid	Do not normally conduct	Do not conduct
Solubility in water	Often soluble	Insoluble	Insoluble (but some react to form the metal hydroxide and hydrogen gas)	Usually insoluble	Usually insoluble, unless molecules contain groups which can hydrogen bond with water or they react with water
Solubility in non-polar solvents	Insoluble	Insoluble	Insoluble	Some soluble	Usually soluble

---

■ QUICK CHECK QUESTIONS

29 The types of bonding and structures in different compounds can be used to explain a variety of physical properties.

   a State whether CH_3CN and CHI_3 are polar molecules and describe the electrostatic attraction between molecules in CH_3CN and CHI_3, respectively.

   b Explain why only CH_3CN has significant solubility in water.

30 The boiling points of two nitrides from period 3 of the Periodic Table are given in the table below.

Formula of period 3 nitrides	Boiling point/°C
S_4N_4	187
Ca_3N_2	1572

Using bonding and structure, explain the large difference in the boiling points of these two nitrides.

Topic 5 Energetics/thermochemistry

5.1 Measuring energy changes

Essential idea: The enthalpy changes from chemical reactions can be calculated from their effect on the temperature of their surroundings.

Thermodynamics

- Heat is a form of energy.
- Temperature is a measure of the average kinetic energy of the particles.
- Total energy is conserved in chemical reactions.
- Chemical reactions that involve transfer of heat between the system and the surroundings are described as endothermic or exothermic.
- The enthalpy change (ΔH) for chemical reactions is indicated in $kJ\,mol^{-1}$.
- ΔH values are usually expressed under standard conditions, given by $\Delta H^\ominus$, including standard states.

The enthalpy changes from chemical reactions can be calculated from their effect on the temperature of their surroundings. Thermodynamics is the study of energy changes during chemical reactions and physical processes, for example, dissolving. The universe is divided into the system and the surroundings (Figure 5.1). The system refers to that part of the universe where a process is being studied: universe = system + surroundings. In a chemical reaction the system is generally the reactants plus products.

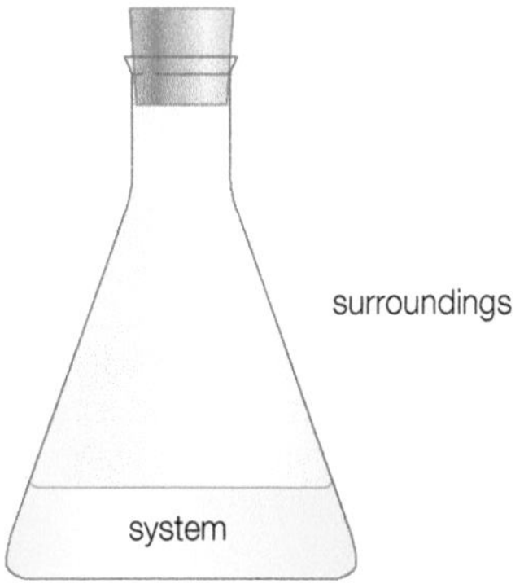

Energy is the capacity to do work. The energy of a chemical system is the sum of the kinetic and potential energies of the particles in the system. A good example of a thermodynamic system is gas confined by a piston in a cylinder. If the gas is heated or gas is produced by a reaction (Figure 5.2), it will expand, doing work on the piston.

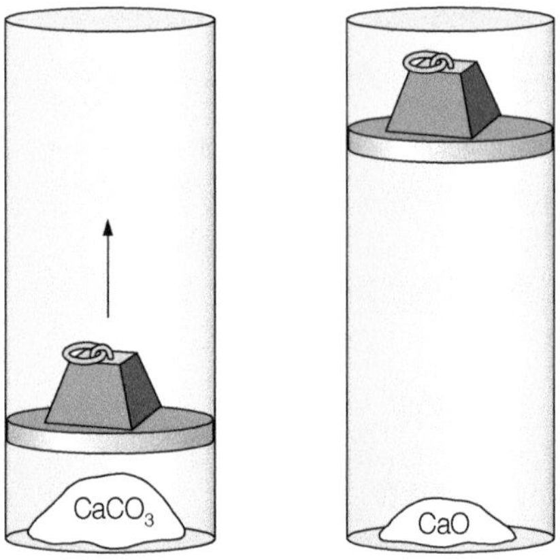

surroundings

system

Figure 5.1 A diagram illustrating the thermodynamic concepts of system and surroundings

NATURE OF SCIENCE

The first law of thermodynamics states that the amount of energy in an isolated system (where neither matter nor energy can enter or leave) is constant: when one form of energy is transferred, an equal amount of energy in another form is produced. A law is a well-tested theory that, so far, has not been falsified by experimental data.

CaCO₃

CaO

Figure 5.2 The decomposition of calcium carbonate (within a closed system) illustrating expansion work

■ Potential energy

Particles (atoms, ions or molecules) have potential energy (Figure 5.3) due to their relative positions. The closer two attracting particles are to each other, the lower is their potential energy (unless they get so close that the repulsion outweighs the attraction).

These potential energies may be associated with the (relatively strong) chemical bonds which hold atoms together in molecules, or with the (weaker) intermolecular forces which hold molecules together in the liquid and solid states. In both cases energy is needed to pull the particles apart and is released as they come together.

Figure 5.3 Potential energy in particles arises from their relative positions. Potential energy is released as particles move together; energy, or work, must be put in to move particles apart from each other, so they gain potential energy

Kinetic energy

Particles have kinetic energy because of the rotational and especially vibrational movements in all three states of matter. Particles in liquids and gases also have translational energy. Kinetic energy = $\frac{1}{2}mv^2$, where m is mass and v is velocity.

Because energy is dispersed through so many forms – translations, rotations, vibration of particles, intermolecular forces, chemical bonds, etc. – it is essentially impossible to measure the energy of a system. However, when the system *changes*, energy may be absorbed or released. This change in energy usually appears as heat or work and is easy to measure.

Temperature

The temperature of a system is a measure of the average kinetic energy of the particles in the system on an arbitrary scale. Temperature differences determine which way heat will flow. Heat always flows from higher to lower temperature. At absolute zero (−273.15 °C) all motion of the particles stops (theoretically). The absolute temperature expressed in kelvin is directly proportional to the average kinetic of the particles in a sample of matter.

Heat and enthalpy

When a system changes, by some physical or chemical process, the new system is likely to have a different energy because new substances may have formed and/ or the intermolecular forces/bonding may have changed. This energy change typically appears as absorption or release of heat. The heat absorbed or released in a process at constant pressure (which is how most laboratory experiments are done) measures the change in the *enthalpy*, or 'heat content', of the system. The heat content has the symbol H.

The heat content of system is the total amount of energy (kinetic and potential) in the system. The higher the enthalpy of a substance, the less energetically stable it is.

There is no way of measuring all the energies of all the particles (the enthalpy) of system, so it has been agreed that elements in their normal states at 298 K and 100 kPa (approximately 1 atmosphere) have zero enthalpy. Changes in enthalpy, ΔH, are measured relative to this standard.

Exothermic and endothermic reactions

In an exothermic reaction a system loses heat (enthalpy) to its surroundings, which may increase in temperature unless some means are provided to take away the energy released. To avoid confusion, scientists have agreed to adopt the convention that enthalpy changes for exothermic changes are shown as negative. In an endothermic change a system gains heat from its surroundings, which may cool down unless energy is provided, so the enthalpy changes for endothermic reaction are shown as positive (Figure 5.4).

Expert tip

The size of a kelvin is the same as a degree Celsius. A temperature change of 5 K is the same as 5 °C.

Common mistake

Heat is best regarded as a form of energy transfer, rather than a form of energy.

Expert tip

Enthalpy is a state function, that is, the change in its value depends on the initial and final states only, not on what happens between those states.

Common mistake

It is important to remember that the 'surroundings' include all the apparatus being used. The thermometer is part of this and so if the temperature goes down then the reaction is endothermic. Conversely, if the temperature measured by a thermometer rises then the reaction is exothermic.

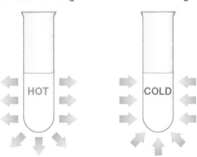

Exothermic reactions give out heat. This warms the mixture and then heat is lost to the surroundings

Endothermic reactions take in heat. This cools the mixture at first and then heat is gained from the surroundings

Figure 5.4 The directions of heat flow during exothermic and endothermic reactions

Endothermic and exothermic reactions can be represented by energy or enthalpy level diagrams or potential energy profiles (Figure 5.5).

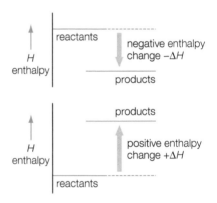

H enthalpy

reactants

negative enthalpy change $-\Delta H$

products

H enthalpy

products

positive enthalpy change $+\Delta H$

reactants

Figure 5.5 Generalized energy level diagrams for an exothermic and endothermic reaction

■ Standard enthalpy changes

When a system changes, by a physical or chemical process, the new system is likely to have a different energy and this energy change will depend upon the conditions of the process. For example the same reaction carried out at different concentrations of reactants in solution may have different values of ΔH. To overcome this problem, other variables are kept constant and chemists define a *standard state*.

All standard enthalpy changes, $\Delta H^\ominus$, are measured for a mole of substance reacting under the standard conditions of 289 K and 100 kPa (approximately 1 atm). If solutions are involved then concentrations are 1 mol dm^{-3}.

■ QUICK CHECK QUESTIONS

3   Calculate the energy, in kJ, released when 1.00 mol of carbon monoxide is burned according to the following equation:

   $2CO(g) + O_2(g) \rightarrow 2CO_2(g)$; $\Delta H^\ominus = -283$ kJ

4   Hydrogen gas burns in air to form water:

   $2H_2(g) + O_2(g) \rightarrow 2H_2O(l)$; $\Delta H^\ominus = -564$ kJ

   Calculate how much heat energy (in kJ) is released if a rocket carrying 10.10 kg of hydrogen gas is burnt in excess oxygen.

Large numbers of values for enthalpy changes have now been measured for chemical reactions. In principle they could all be listed in a giant database and searched but this would be quite inconvenient and would not help with reactions whose enthalpy changes have not been measured. It is much easier to use the idea of an *enthalpy of formation*.

This works because chemists choose to define the enthalpy of formation of any element as zero for the most stable state of that element under standard conditions. For example the standard state of zero enthalpy for oxygen is O_2 because that is the most stable form at 298 K and 100 kPa.

■ Enthalpy of formation, $\Delta H_f^\ominus$

The enthalpy of formation is the enthalpy change when one mole of a pure compound is made from its elements in their standard state. For example, the enthalpy of formation of silicon(IV) hydride is described by the following thermochemical equation:

   $Si(s) + 2H_2(g) \rightarrow SiH_4(g)$ $\Delta H_f^\ominus = +33$ kJ mol^{-1}

Enthalpies of formation can be either positive or negative. A negative value means that under standard conditions the compound is energetically more stable than its elements, whereas a positive value means that it is less energetically stable.

The enormous usefulness of enthalpies of formation is the fact that the enthalpy change of any reaction can be determined by calculation, from the enthalpy changes of formation of all the substances in the chemical equation, using Hess's law. In symbols:

   $\Delta H = \sum \Delta H_f^\ominus [\text{products}] - \sum \Delta H_f^\ominus [\text{reactants}]$

The calculation can be carried out using an energy level diagram or a Hess's law cycle or with algebra. Standard enthalpies of formation have been measured or deduced for very many compounds and are available as tables. They are very useful because, as long as we know the values of $\Delta H_f^\ominus$ for products and reactants, we can calculate ΔH for any reaction, even one which would be difficult or impossible to measure.

Expert tip

The standard state refers to the pure substance under standard conditions. Graphite is the standard state of carbon.

Expert tip

Values and signs of enthalpy of formations indicate the energetic stability of compounds. However, they cannot be used to predict the rates of reactions. Many endothermic compounds are kinetically stable: they have a high activation energy barrier to decomposition.

Common mistake

The enthalpy change of formation of an element in its most stable form in the standard state is zero by definition.

■ QUICK CHECK QUESTIONS

5   Write thermochemical equations showing the enthalpy change of formation for the following compounds: nitrogen dioxide, sodium chloride and hydrogen sulfide.

6   Hydrogen sulfide can be oxidized by using a controlled amount of air to give steam and solid sulfur. Using the data below, find the standard enthalpy change of this reaction per mole of gaseous hydrogen sulfide.

	$\Delta H_f^\ominus$ kJ mol^{-1}
$H_2S(g)$	−20.5
$H_2O(g)$	−243.0

■ Enthalpy of combustion, $\Delta H_c^\ominus$

The enthalpy of combustion is the enthalpy change in standard conditions when one mole of substance is completely combusted in excess oxygen. For example, the enthalpy of combustion of aluminium is described by the following thermochemical equation:

$$Al(s) + \frac{3}{4}O_2(g) \rightarrow \frac{1}{2}Al_2O_3(s) \qquad \Delta H_c^\ominus = -835\,kJ\,mol^{-1}$$

Enthalpy changes of combustion are always negative as heat is released during combustion processes.

The standard enthalpy of combustion of carbon is the heat released when one mole of solid carbon (graphite) is completely combusted to form carbon dioxide gas under standard conditions:

$$C(s) + O_2(g) \rightarrow CO_2(g) \qquad \Delta H_c^\ominus = -393\,kJ\,mol^{-1}.$$

The standard enthalpy of combustion of ethanol is defined as the heat released when one mole of liquid ethanol is completely combusted to form carbon dioxide and water (under standard conditions):

$$CH_3CH_2OH(l) + 3O_2(g) \rightarrow 2CO_2(g) + 3H_2O(l) \qquad \Delta H_c^\ominus = -1371\,kJ\,mol^{-1}$$

The complete combustion of carbon and hydrogen present in any compound yields carbon dioxide and water.

> ### ■ QUICK CHECK QUESTIONS
>
> 7 Write equations describing the standard enthalpy of combustion of hydrogen, potassium, ethene (C_2H_4), benzene (C_6H_6) and methanol (CH_3OH).
>
> 8 When an unknown mass of methanol, CH_3OH, was burnt, the temperature of 500.00 cm³ water contained in a copper beaker increased by 5.40 °C. If the enthalpy of combustion of methanol is −726 kJ mol⁻¹, determine the mass of methanol burnt.
>
> 9 When 0.02 mol octane, $C_8H_{18}(l)$, was completely combusted in a bomb calorimeter, the temperature of 1000.00 cm³ water increased by 24.20 °C. Calculate the enthalpy of combustion of octane.

■ Enthalpy of neutralization, $\Delta H_{neut}^\ominus$

Neutralization refers to the reaction between an acid and an alkali (soluble base). The enthalpy of neutralization is defined as the heat released when one mole of water is formed, when an acid reacts with an alkali (under standard conditions). For standard conditions, the concentration of the acid and alkali are 1 mol dm⁻³.

For example, hydrochloric acid reacting with potassium hydroxide:

$$HCl(aq) + NaOH(aq) \rightarrow NaCl(aq) + H_2O(l)$$

and sulfuric acid reacting with sodium hydroxide:

$$\frac{1}{2}H_2SO_4(aq) + NaOH(aq) \rightarrow \frac{1}{2}Na_2SO_4(aq) + 2H_2O(l)$$

For neutralization between an alkali (strong base) and a strong acid, the value of the enthalpy of neutralization is approximately −57 kJ mol⁻¹. This is because strong acids and alkalis are completely ionized in aqueous solution to hydrogen ions and hydroxide ions. Hence, neutralization reactions can be described by the following ionic equation:

$$H^+(aq) + OH^-(aq) \rightarrow H_2O(l)$$

Expert tip

Note that the enthalpy of combustion of carbon is the same as the enthalpy of formation of carbon dioxide, because they are described by the same thermochemical equation.

Expert tip

There is a linear relationship between the enthalpy of combustion of the alkanes and the number of carbon atoms because the compounds in the homologous series have an extra $-CH_2-$ group between successive members.

However, the nature of the acids and bases involved will affect the value of $\Delta H^{\ominus}_{neut}$. Weak acids and weak bases are only partially ionized in solution. Part of the heat released from the neutralization process is used to dissociate the weak acid and/or weak base for further neutralization.

A simple method for determining the enthalpy change of neutralization involves mixing equal volumes of dilute solutions of a strong acid and a strong base of known concentration and measuring the temperature rise. The maximum temperature can be deduced from the graph (see Figure 5.6) by extrapolating both lines back to find out where they intersect.

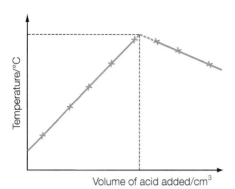

Figure 5.6 A graph of temperature versus volume of acid for an investigation to determine the enthalpy change of neutralization for an acid

■ QUICK CHECK QUESTIONS

10 a Write an equation showing the enthalpy of neutralization between propanoic acid, C_3H_7COOH, and sodium hydroxide.

b Explain why the enthalpy change is $-55.2\,kJ\,mol^{-1}$.

11 When $50.00\,cm^3$ of $1.0\ mol\,dm^{-3}$ nitric acid solution, $HNO_3(aq)$, is added to $50.00\,cm^3$ of $1.0\ mol\,dm^{-3}$ potassium hydroxide solution, $KOH(aq)$, the temperature of the mixture increases by $6.4\,°C$. Deduce the temperature change when $25.00\,cm^3$ of each of these solutions are mixed together.

12 $150.00\,cm^3$ of potassium hydroxide solution, $KOH(aq)$, of concentration $2.00\ mol\,dm^{-3}$ and $250.00\,cm^3$ of $1.50\,mol\,dm^{-3}$ hydroiodic acid, $HI(aq)$, were mixed in a calorimeter. If the temperature rise is $10.20\,°C$, calculate the enthalpy of neutralization for this reaction.

■ Enthalpy of solution

The enthalpy of solution is the enthalpy change when one mole of an ionic compound is dissolved in a large excess of water (infinite dilution). Its value may be positive or negative. For example, the following thermochemical equation describes the enthalpy of solution of sodium chloride:

$$NaCl(s) + (aq) \rightarrow NaCl(aq) \qquad \Delta H^{\ominus}_{solution} = +3.9\,kJ\,mol^{-1}$$

■ QUICK CHECK QUESTION

13 When $18.70\,g$ sodium chloride was dissolved in $400.00\,cm^3$ of distilled water, the temperature of the solution decreased by $1.00\,°C$. Calculate the enthalpy of solution of sodium chloride.

5.2 Hess's law

Revised

Essential idea: In chemical transformations energy can neither be created nor destroyed (the first law of thermodynamics).

Hess's law

Revised

■ The enthalpy change for a reaction that is carried out in a series of steps is equal to the sum of the enthalpy changes for the individual steps.

The first law of thermodynamics (the law of conservation of energy) states that energy can neither be created nor destroyed, but only transferred from one form to another. This concept is applied to calorimetry, Hess's law and energy cycles.

■ Calorimetry

When a reaction is carried out in a calorimeter the heat lost/gained by the reacting system is equal to the heat gained/lost by the calorimeter and its contents. Changes in heat content are calculated using:

quantity of heat (J) = mass(g) × specific heat capacity (J/g/°C) × the change in temperature (°C or K)

$$q\ (J) = m\ (g) \times c\ (J/g/°C) \times \Delta T\ (°C\ or\ K)$$

Enthalpies of neutralization and enthalpies for redox reactions involving metals and metal ions can be measured by performing the reaction in an insulated plastic cup (Figure 5.7). The calculation should be performed with the limiting reagent. Enthalpies of combustion can be measured by burning a liquid in a spirit burner which transfers the heat to a metal can (Figure 5.8) or a flame combustion calorimeter (Figure 5.9). The most accurate measurements are from a bomb calorimeter where known masses are burnt in excess oxygen.

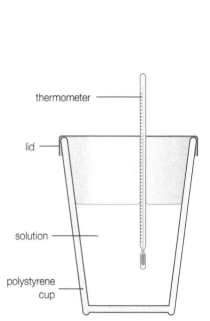

Figure 5.7 A simple calorimeter: polystyrene cup, lid and thermometer

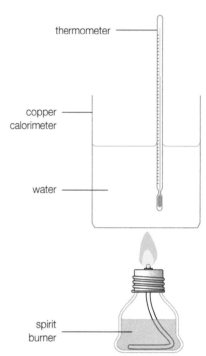

Figure 5.8 Simple apparatus used to measure enthalpy changes of combustion of

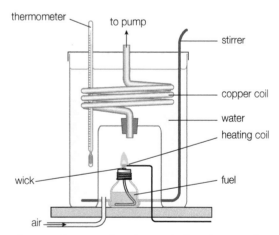

Figure 5.9 Measuring the enthalpy change of combustion of a liquid fuel using a flame combustion calorimeter

Expert tip

The specific heat capacity of a substance is the amount of thermal energy required to raise the temperature of a unit mass of a substance by one kelvin or one degree Celsius. The specific heat capacity of water = $4.18 \, kJ \, kg^{-1} \, K^{-1}$ = $4.18 \, J \, g^{-1} \, K^{-1}$.

■ QUICK CHECK QUESTIONS

14 The specific heat capacity of iron is 0.450 J g^{-1}K^{-1}. Calculate the energy, in J, needed to increase the temperature of 50.0 g of iron by 20.0 K.

15 1.0 g of solid sodium hydroxide, NaOH, was added to 99.0 g of water. The temperature of the solution increased from 19.0 °C to 21.5 °C. The specific heat capacity of the solution is 4.18 J g^{-1} °C^{-1}. Calculate the enthalpy change (in kJ mol^{-1}) for the dissolving process.

16 To determine the enthalpy change of combustion of methanol, $CH_3OH(l)$, 0.230 g of methanol was combusted in a spirit burner. The heat released increased the temperature of 50.00 cm³ of water from 25.5 °C to 46.8 °C.

  a Calculate the standard enthalpy change of combustion of methanol.

  b State one reason it was less than the literature value of −726 kJ mol^{-1} and state one improvement that could improve the accuracy of the experiment.

Common mistake

During calorimetry experiments to find the enthalpy of combustion water is heated and its temperature increases. The mass of the water must be used in the equation, $q = mc\Delta T$, not the mass of the substance being combusted.

Accurate results can be obtained by using simple calorimeters for fast reactions, such as neutralizations or precipitations. However, for slower reactions the results will be less accurate with the same apparatus.

This is because there is heat loss to the calorimeter water surroundings, which will increase if the reaction is slow because the heat will be lost over a longer period of time. Hence the temperature rise observed in the calorimeter is less than it should be. However, an allowance can be made for this by plotting a temperature–time graph (or cooling curve) (Figure 5.10).

Expert tip

Experimentally determined enthalpies of combustion may be lower than expected because of incomplete combustion, volatility of the substance being combusted, and transfer of heat to the surroundings, calorimeter and thermometer.

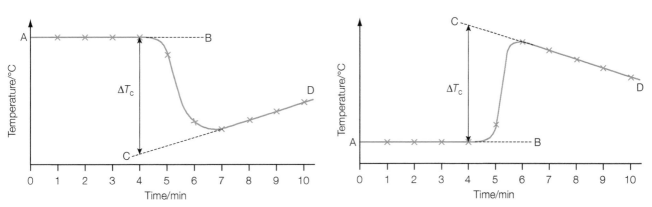

Figure 5.10 Temperature correction curves for exothermic (left) and endothermic (right) reactions

■ Hess's law

Hess's law (Figure 5.11) is a version of the first law of thermodynamics in a convenient form for chemists. It states that the enthalpy change in a chemical reaction depends only on the initial and final states and is independent of the reaction pathway.

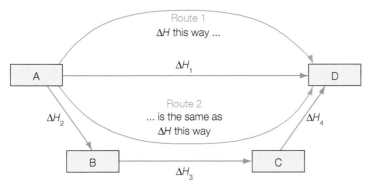

Figure 5.11 An illustration of the principle of Hess's law

Worked example

Given the following enthalpies of reaction:

$$P_4(s) + 5O_2(g) \rightarrow P_4O_{10}(s) \qquad \Delta H = -2900\,kJ\,mol^{-1}$$

$$P_4O_6(s) + 2O_2(g) \rightarrow P_4O_{10}(s) \qquad \Delta H = -1300\,kJ\,mol^{-1}$$

calculate the enthalpy change of formation of phosphorus(III) oxide, $P_4O_6(s)$.

$$P_4(s) + 5O_2(g) \rightarrow P_4O_{10}(s) \qquad \Delta H = -2900\,kJ\,mol^{-1}$$

$$P_4O_{10}(s) \rightarrow P_4O_6(s) + 2O_2(g) \qquad \Delta H = +1300\,kJ\,mol^{-1}$$

(reverse of second equation, so sign of enthalpy change reversed)

Adding and cancelling the $P_4O_{10}(s)$ and simplifying the $O_2(g)$:

$$P_4(s) + 3O_2(g) \rightarrow P_4O_6(s)$$

$$\Delta H^\ominus = [-2900 + 1300]\,kJ = -1600\,kJ\,mol^{-1}$$

Energy changes are often expressed in the form of a triangle, the sides of which represent the different reaction pathways. These triangles or energy cycles (Figures 5.12 and 5.13) are used to find enthalpy changes that cannot be measured directly in the laboratory. For example, they can be based on enthalpies of formation or enthalpies of combustion.

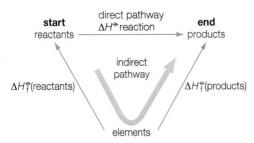

Figure 5.12 Hess's law cycles based on enthalpies of formation

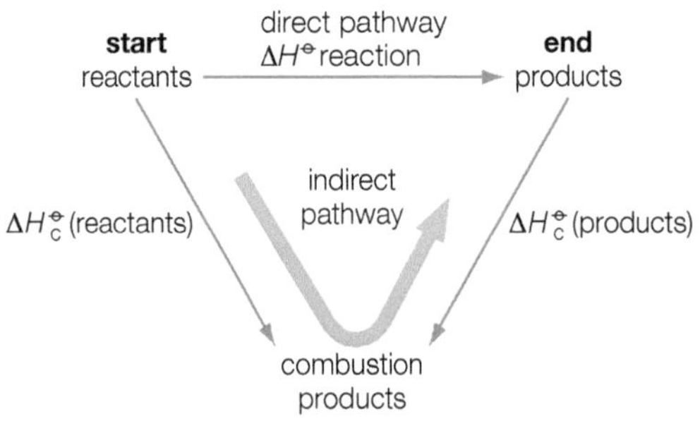

Figure 5.13 Hess's law cycles based on enthalpies of combustion

To calculate the enthalpy change using a Hess's law cycle involving *enthalpies of formation* use the following procedure:

1 Write the balanced equation at the top of the cycle.

2 Draw the cycle with the elements at the bottom.

3 Draw in all arrows, making sure they point in the correct directions.

4 Apply Hess's law (direct route = indirect route), taking into account the amounts of each reactant and product.

Figure 5.14 shows an example of an enthalpy cycle for calculating an enthalpy of reactions from enthalpies of formation.

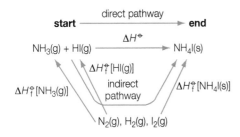

Figure 5.14 Hess's law cycle to calculate the enthalpy change between ammonia and hydrogen iodide to form ammonium iodide

Expert tip

All energetics cycles or algebraic working should include balanced equations, state symbols and, where relevant, electrons.

To calculate the enthalpy change using a Hess's law cycle involving *enthalpies of combustion* use the following procedure (see Figure 5.13):

1 Write the equation for the enthalpy change of formation at the top, add oxygen on both sides of the equation to balance the combustion reaction.

2 Draw the cycle with the combustion products at the bottom.

3 Draw in all arrows, making sure they point in the correct directions.

4 Apply Hess's law (direct route = indirect route), taking into account the amounts of each reactant and product.

Figure 5.15 shows an example of an enthalpy cycle for calculating an enthalpy of reaction from enthalpies of combustion.

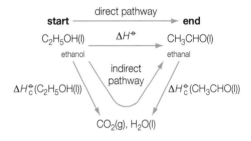

Figure 5.15 Hess's law cycle to calculate the enthalpy change for the conversion of ethanol to ethanal

Hess's law can also be used to calculate the enthalpy change when an anhydrous salt becomes hydrated. The standard enthalpy changes of solution can be used to complete the enthalpy cycle (Figure 5.16).

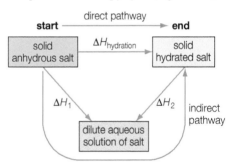

Figure 5.16 A generalized enthalpy cycle to calculate the enthalpy change of hydration of an anhydrous salt

---

■ **QUICK CHECK QUESTIONS**

17 Consider the following two thermochemical equations.

$$Ca(s) + \frac{3}{2}O_2(g) \rightarrow CaO + O_2(s) \qquad \Delta H^\ominus = +x\,kJ\,mol^{-1}$$

$$Ca(s) + \frac{1}{2}O_2(g) + CO_2(g) \rightarrow CaCO_3(s) \qquad \Delta H^\ominus = +y\,kJ\,mol^{-1}$$

Deduce the standard enthalpy change in $kJ\,mol^{-1}$, for the formation of calcium carbonate: $CaO(s) + CO_2(g) \rightarrow CaCO_3(s)$.

18 It is very difficult to measure the enthalpy change when an anhydrous salt such as anhydrous sodium thiosulfate becomes hydrated:

$$Na_2S_2O_3(s) + 5H_2O(l) \rightarrow Na_2S_2O_3.5H_2O(s)$$

The enthalpies of solution of $Na_2S_2O_3(s)$ and $Na_2S_2O_3.5H_2O(s)$ are $-7.7\,kJ\,mol^{-1}$ and $+47.4\,kJ\,mol^{-1}$.

Draw a labelled enthalpy level diagram and determine the enthalpy of hydration of anhydrous sodium thiosulfate.

5.3 Bond enthalpies

Revised ☐

Essential idea: Energy is absorbed when bonds are broken and is released when bonds are formed.

Bond enthalpies

Revised ☐

- Bond-forming releases energy and bond-breaking requires energy.
- Average bond enthalpy is the energy needed to break one mole of a bond in a gaseous molecule averaged over similar compounds.

■ Bond enthalpy

The bond enthalpy is the enthalpy change when one mole of covalent bonds are broken in the gas phase to form atoms. This is an endothermic process.

$$X–Y(g) \rightarrow X(g) + Y(g)$$

In this process the bond is breaking homolytically with one electron from the bonding pair going to each atom.

$$X \overset{\curvearrowleft}{\underset{\curvearrowright}{\;\overset{x}{\bullet}\;}} Y \longrightarrow X^x + Y^\bullet$$

■ Average bond enthalpy

In a polyatomic molecule, such as methane (Figure 5.17), bonds will break one after another, each one breaking off a different fragment of the original molecule:

$$CH_3 – H(g) \longrightarrow CH_3(g) + H(g); \Delta H_1^\ominus$$
$$CH_2 – H(g) \longrightarrow CH_2(g) + H(g); \Delta H_2^\ominus$$
$$CH – H(g) \longrightarrow CH(g) + H(g); \Delta H_3^\ominus$$
$$C – H(g) \longrightarrow C(g); \Delta H_4^\ominus$$

Figure 5.17 Successive bond enthalpies of methane

The average C–H bond enthalpy is given by the expression:
$$\frac{(\Delta H_1^\ominus + \Delta H_2^\ominus + \Delta H_3^\ominus + \Delta H_4^\ominus)}{4}$$

Each of these reactions will have different value of ΔH and these are not easy to measure. For many purposes it is useful to consider the average value, so for CH_4 the average C—H bond enthalpy is $[\frac{(\Delta H_1^\ominus + \Delta H_2^\ominus + \Delta H_3^\ominus + \Delta H_4^\ominus)}{4}]$ and for all polyatomic molecules, the bond energies quoted are the mean or average values of all the bonds between the same two atoms unless otherwise stated.

NATURE OF SCIENCE

The validity of calculating enthalpy changes based on average bond energies can be assessed by comparison with accurate experimental values. The model will have less validity if the molecule is strained or exists as a resonance hybrid. The accuracy of the calculation can be improved by using bond energies specific to that reaction.

■ QUICK CHECK QUESTIONS

19 Enthalpies of combustion can be calculated using average bond enthalpies or enthalpies of formation. These two methods give closer results for benzene than cyclohexane. Explain this difference.

20 Define the term *average bond enthalpy*. Explain why the I–I bond cannot be used as an example to illustrate the concept of average bond enthalpy.

When chemical changes take place, bonds are broken and new bonds are formed, hence when isomers of organic compound undergo combustion, the same type and the same number of bonds are broken and formed. The enthalpy changes of combustion are approximately the same. For example, the standard

Expert tip

Understanding energetics starts with learning the definition of the various enthalpy changes. For example, the thermochemical equation $H–F(g) \rightarrow \frac{1}{2}H_2(g) + \frac{1}{2}F_2(g)$ does not represent the bond enthalpy of hydrogen fluoride. $\frac{1}{2}H_2(g)$ represents 0.5 mol of H_2 molecules; $H(g)$ represents 1 mol of H atoms.

enthalpy change of combustion of butane, $CH_3CH_2CH_2CH_3$, is $-2878\,kJ\,mol^{-1}$ while that of its isomer, methylpropane, $(CH_3)_3CH$, is $-2870\,kJ\,mol^{-1}$.

The oxygen bond (O=O) in the oxygen molecules is stronger than the oxygen–oxygen bonds in the ozone molecule, which have a strength intermediate between a double and single bond. This is important as ozone (O_3) protects the Earth's surface from harmful ultraviolet radiation by absorbing energy of the energy required to dissociate ozone molecules into an oxygen molecule and an oxygen atom.

Bond energies and enthalpy changes

Enthalpy changes are due to the breaking and forming of bonds. Breaking covalent bonds requires energy. The energy is needed to overcome the attractive forces operating between the atoms. Energy is released when new bonds are formed. Bond breaking is endothermic and bond formation is exothermic.

In a chemical reaction, if the energy needed to break new covalent bonds is less than the energy released when new covalent bonds are formed, the reaction will release energy and is exothermic. If the energy needed to break the covalent bonds is more than the energy released when new covalent bonds are formed, the reaction will absorb energy and is endothermic.

Enthalpy or energy level diagrams can be drawn to show these changes (Figure 5.18). For simplicity it can be assumed that all the covalent bonds in the reactants are broken to form atoms and then new bonds are formed in the products.

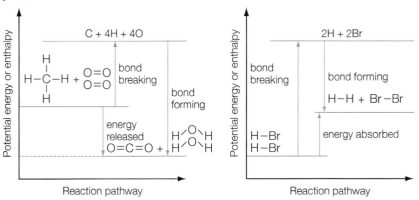

Figure 5.18 Energy level diagrams showing bond breaking and bond formation for the combustion of gaseous methane (an exothermic reaction) and bond breaking and bond formation for the decomposition of gaseous hydrogen bromide (an endothermic reaction)

Calculating enthalpy changes from bond energies

Average bond enthalpies can be used to calculate the enthalpy change for any reaction involving molecules in the gaseous state. This is done by assuming that an alternative route for all reactions can be achieved theoretically via the gaseous atoms involved in the compounds (Figure 5.19).

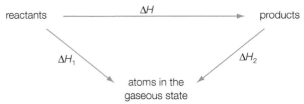

Figure 5.19 Generalized energy cycle to determine an enthalpy change from bond energies

$\Delta H = \Sigma$(average bond enthalpies of the reactants) $- \Sigma$(average bond enthalpies of the products)

i.e. $\Delta H = \Sigma$(bonds broken) $- \Sigma$(bonds made)

Bond enthalpy data can also be used to determine an unknown bond enthalpy provided that the enthalpy change and all the other bond enthalpies are known.

Expert tip

It helps to draw structural formulas for all reactants and producing when calculating an enthalpy change from bond energies.

Common mistake

Bond energy calculations can only be carried out if all reactants and products are covalent substances in the gaseous state under standard thermodynamic conditions. If there are liquids present then adjustments need to be made with enthalpies of vaporization.

■ **QUICK CHECK QUESTIONS**

21 Draw an energy or enthalpy cycle showing how bond enthalpies can be used to calculate the enthalpy change for the Haber process. Use the bond enthalpies from the IB Chemistry *data booklet* to calculate the enthalpy change.

22 Use the bond enthalpies below to calculate the enthalpy change, in kJ, for the following reaction:

$H_2(g) + I_2(g) \rightarrow 2HI(g)$

Bond	Bond enthalpy/kJ mol^{-1}
H–H	440
I–I	150
H–I	300

23 Trinitramide is a compound of nitrogen and oxygen with the molecular formula $N(NO_2)_3$. It is stable and its structure is shown below. It is a possible new oxidizing agent for rocket fuel.

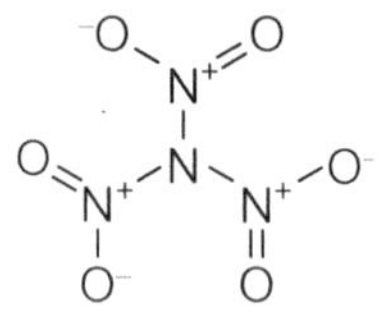

a Calculate the enthalpy change, in kJ mol^{-1}, when one mole of gaseous trinitramide decomposes to its elements, using bond enthalpy data from Table 11 of the IB Chemistry *data booklet*. Assume that all the N–O bonds in this molecule have a bond enthalpy of 305 kJ mol^{-1}.

b Outline and explain how the length of the N–N bond in the trinitramide molecule compares with the N–N bond in the nitrogen molecule, N_2.

Topic 6 Chemical kinetics

6.1 Collision theory and rates of reaction

Essential idea: The greater the probability that molecules will collide with sufficient energy and proper orientation, the higher the rate of reaction.

Collision theory and rates of reaction

■ Species react as a result of collisions of sufficient energy and proper orientation.

■ The rate of reaction is expressed as the change in concentration of a particular reactant/product per unit time.

■ Concentration changes in a reaction can be followed indirectly by monitoring changes in mass, volume and colour.

■ Activation energy (E_a) is the minimum energy that colliding molecules need in order to have successful collisions leading to a reaction.

■ By decreasing E_a, a catalyst increases the rate of a chemical reaction, without itself being permanently chemically changed.

■ Reaction rates

Chemical kinetics is the study and measurement of the **rates** of chemical reactions. It is also concerned with proposing reaction mechanisms for reactions.

During the course of a reaction, the concentrations of the reactants decrease while the concentrations of products increases (Figure 6.1).

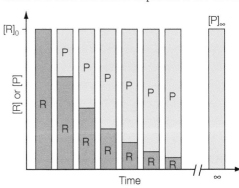

Figure 6.1 The relationship between reactant and product concentrations for a reaction that goes to completion

For example, consider a simple reaction

$$R \rightarrow P$$

where R is a single reactant and P a single product. Figure 6.2 shows the graphs of concentration versus time. In this reaction, one mole of reactant R produces one mole of product P.

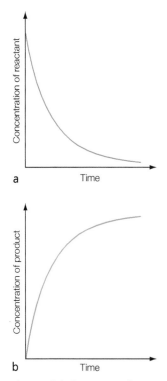

Figure 6.2 Concentration versus time graph for the reaction R → P

> **Key definition**
>
> **Rate** – the change in concentration of a reactant or product with time. The rate of a reaction can be expressed as:
>
> $$\text{rate of reaction} = \frac{\text{decrease (or increase) in the concentration of a reactant (or product)}}{\text{unit of time}}$$

The average rate of disappearance of the reactant, R, is described by the differential $-\dfrac{d[R]}{dt}$ while the average rate of formation of the product, P, is described by $+\dfrac{d[P]}{dt}$.

As the reactant R and the product P are in a 1:1 stoichiometry $-\dfrac{d[R]}{dt} = +\dfrac{d[P]}{dt}$.

There are different types of rates of reactions; the *average rate* of reaction described above, and the *instantaneous rate* at a particular point in time. A reaction is fastest at the start (the initial rate) and then gets progressively slower as the reactant(s) are used up (Figure 6.3).

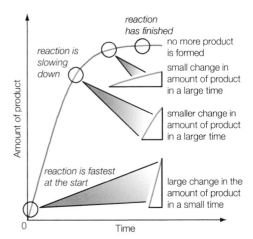

Figure 6.3 The variation in the rate of a reaction as the reaction progresses

> **Expert tip**
>
> The negative sign indicates that the concentration of the reactants decreases with time, whereas the positive sign shows that the concentration of the products increases with time.

Instantaneous rate is equal to the slope of a tangent drawn to a concentration versus time graph at a particular time for a reactant (Figure 6.4) or product (Figure 6.5).

The steeper the gradient of the graph, the faster the reaction, and the higher its rate. When the graph is horizontal (i.e. the gradient is zero) the rate of reaction is zero, indicating the reaction has finished.

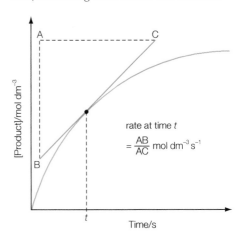

Figure 6.4 Concentration–time graph for the disappearance of a reactant. The rate of loss of reactant at time *t* is the gradient (or slope) of the curve at this point

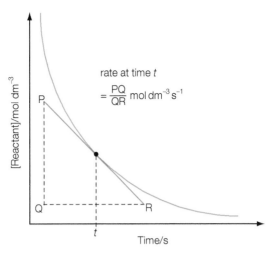

Figure 6.5 Concentration–time graph for the formation of a product. The rate of formation of product at time *t* is the gradient (or slope) of the curve at this point

In general, for any reaction the rate of consumption of reactants is related to the rate of formation of products:

aA + bB → cC + dD

The relative rates of reaction are given by the following expressions:

$$\text{rate} = \frac{-1}{a}\frac{d[A]}{dt} = \frac{-1}{b}\frac{d[B]}{dt} = \frac{+1}{c}\frac{d[C]}{dt} = \frac{+1}{d}\frac{d[D]}{dt}$$

To determine how the rate of reaction varies with concentration of each of the reactants a number of tangents are drawn on a concentration–time or product–time graph. A graph is then plotted of rate versus concentration, which will often show a linear relationship (Figure 6.6).

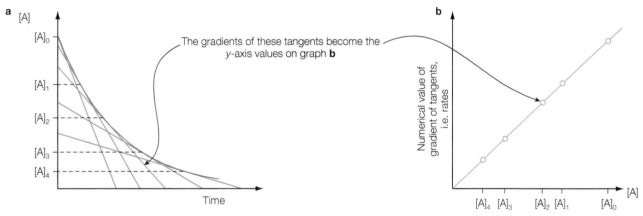

Figure 6.6 a Concentration versus time graph; **b** rate versus concentration graph

---

■ QUICK CHECK QUESTIONS

1 Which of the following are appropriate units for the rate of a reaction?

    A $mol\,dm^{-3}\,s$ **C** s

    B $mol\,dm^{-3}\,s^{-1}$ **D** $mol\,dm^{-3}$

2 What is the rate of a gaseous reaction, carried out in a $2.00\,dm^3$ vessel, if $0.08\,mol$ of a product is generated in 40 seconds?

3 Consider the reaction between bromide and bromate(V) ions under acidic conditions:

    $5Br^-(aq) + BrO_3^-(aq) + 6H^+(aq) \rightarrow 3Br_2(aq) + 3H_2O(l)$

    a How much faster is the rate of disappearance of bromide ions compared with that of bromate(V) ions?

    b What is the comparison between the rate of appearance of bromine to the rate of disappearance of hydrogen ions?

■ Measuring rates of reaction

The rate of a reaction can be measured using either a physical method or chemical method of analysis. All the methods depend on there being a change in some physical or chemical property during the course of the reaction. All of them measure either directly or indirectly a change in the concentration of either a reactant or product.

Techniques used to follow rate of reaction during one reaction include:

- regular removal of a sample from the system, stop (quench) the reaction in the sample and measure the concentration of one of the reactants or products by redox or acid–base titration (Figure 6.7)
- as the reaction proceeds, measure the change in intensity of colour of one of the reactants using a colorimeter
- continuous measurement of the change in gas volume (using a gas syringe (Figure 6.8), total gas pressure or the decrease in mass (Figure 6.9) from a flask placed on a balance (Figure 6.10)
- measure change in electrical conductance during reactions in which ions are either used up or created; or pH where the concentration of H^+ ions varies.

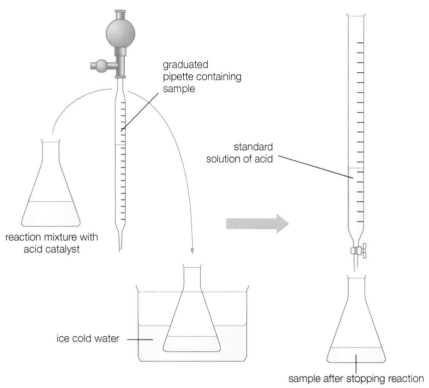

Figure 6.7 Following the course of an acid-catalysed reaction via sampling and titration with alkali

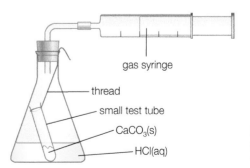

Figure 6.8 Apparatus used to study the rate of a reaction that releases a gas

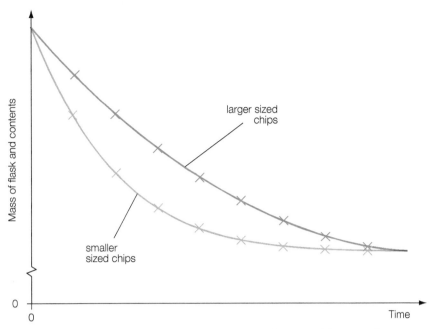

Figure 6.9 The effect of chip size on a graph of mass of flask and contents versus time for the reaction between calcium carbonate and hydrochloric acid

Figure 6.10 Following the rate of reaction between marble chips (calcium carbonate) and hydrochloric acid

All of the above methods are *continuous procedures* involving measurement of a property during the reaction at given time intervals (e.g. every minute), usually used when the investigator wants to follow the change in rate during a reaction. The alternative approach is a *discontinuous method* where the same reaction is carried out many times, each time with different starting conditions, e.g. different concentrations or different temperatures. Usually one reading or measurement is taken per experiment and the rate is the reciprocal of the time (1/t) the reaction took, or alternatively, the initial rate of each reaction is compared (an approach often used to establish the rate law of a reaction – see Topic 16 Chemical kinetics).

Examples of discontinuous methods are:
- the iodine 'clock' reaction method (Figure 6.11)
- the 'disappearing cross reaction' (Figure 6.12).

The rate of the reaction between peroxodisulfate(VI) and iodide ions can be followed using a 'clock' method. A separate experiment is set up for each determination of rate. The two reactants are mixed in the presence of a known amount of sodium thiosulfate and a little starch. Sodium thiosulfate reacts with the iodine produced converting back to iodide ions.

$$I_2(aq) + 2S_2O_3^{2-}(aq) \rightarrow 2I^-(aq) + S_4O_6^{2-}(aq)$$

The amount of sodium thiosulfate present is much smaller than the amounts of all the other reagents. When all the sodium thiosulfate has been used, the iodine (which is still being produced by the reaction) reacts with the starch molecules to give an intense blue colour. The appearance of this blue colour is timed for each experiment. For a fixed amount of sodium thiosulfate, the faster the reaction the shorter the time taken. Figure 6.11 shows two reactions, A and B, which vary either in concentration or temperature. At y, all the thiosulfate ions have been used up and the iodine reacts with the starch.

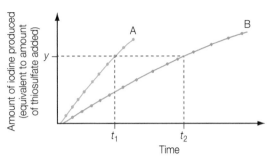

Figure 6.11 Data from two iodine 'clock' experiments

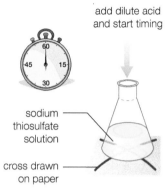

add dilute acid and start timing

sodium thiosulfate solution

cross drawn on paper

Figure 6.12 Diagram of the approach used to monitor the time of production of sufficient sulfur to hide the cross

The rate of reaction A is approximately equal to y/t_1, and the rate of reaction B is approximately equal to y/t_2. This means that the ratio of the times taken for the 'clock' to stop gives the ratio of the rates: rate of A/rate of B = t_2/t_1.

The reaction between a sodium thiosulfate solution and dilute acid produces a very fine precipitate of sulfur:

$$S_2O_3^{2-}(aq) + 2H^+(aq) \rightarrow SO_2(g) + H_2O(l) + S(s)$$

Studies of the effects of varying concentration or temperature can be carried by timing the disappearance of a cross beneath the reaction flask (Figure 6.12).

■ QUICK CHECK QUESTIONS

4 Suggest a method for measuring the rate of the following reactions:
 a $2H_2O_2(aq) \rightarrow 2H_2O(l) + O_2(g)$ with $MnO_2(s)$ as catalyst
 b $(CH_3)_2CO(aq) + I_2(aq) \rightarrow CH_3COCH_2I(aq) + HI(aq)$ with H^+ ions as catalyst
 c $C_2H_5COOC_2H_5(l) + NaOH(aq) \rightarrow C_2H_5COO^-Na^+(aq) + C_2H_5OH(l)$
 d $C_2H_5Br(l) + OH^-(aq) \rightarrow C_2H_5OH(aq) + Br^-(aq)$

5 The data presented in the table below were collected during an experiment to measure the rate of a reaction. The decreasing concentration of reactant A was followed with time, and the data used to calculate the average rate for each 50 second period.

Time/(s)	Concentration of reactant A/(mol dm⁻³)	Average rate/(mol dm⁻³ s⁻¹)
0.0	0.01200	
50.0	0.01010	3.80×10^{-5}
100.0	0.00846	3.28×10^{-5}
150.0	0.00710	2.72×10^{-5}
200.0	0.00596	2.28×10^{-5}
250.0	0.00500	1.92×10^{-5}
300.0	0.00500	

 a What happened after 300 s?
 b Calculate the average rate of the entire reaction.
 c Draw a graph of concentration of reactant A over time(s).
 d Draw a tangent at 100 s, calculate the instantaneous rate and see if it corresponds with the value in the table.

■ Simple collision theory

The important ideas of this theory, which is developed from the kinetic model of matter, are:
■ all particles move, and the average amount of kinetic energy of all the particles in a substance is proportional to the temperature in kelvin
■ reactions occur when particles (ions, atoms or molecules) collide
■ not every collision will lead to a reaction

- for a collision to be successful and result in a reaction it must:
  - ☐ occur with enough energy – this minimum amount of energy is called the activation energy, E_a. So reacting species must have $E \geqslant E_a$. Activation energy is the minimum energy that colliding molecules need in order to have successful collisions leading to a reaction. The molecules must collide with enough energy so that bonds can be broken and the reactant can be changed into product
  - ☐ during the collision, the particles must approach each other in the correct way (correct collision geometry) (Figure 6.13).

no reaction   reaction

Figure 6.13 Two nitrogen dioxide molecules, NO_2, approaching with sufficient kinetic energy to overcome the activation energy barrier must collide in the correct orientation (steric factor) in order to form dinitrogen tetroxide, N_2O_4

Factors affecting the rate of a reaction

The rate of a chemical reaction (change in concentration/time) is controlled by the following factors:

- the concentration of reactants in solution
- the pressure (or concentration) of gaseous reactants
- the surface areas of a solid reactant
- temperature
- light (if the reaction is photosensitive) and
- the presence of catalyst (this may be an acid or base).

Any factor that either increases the frequency of the collisions or increases the combined kinetic energy with which they collide will make the reaction rate increase.

An increase in concentration (for a solution), or pressure (for gases), increases the rate of reaction by increasing the *collision frequency* (Figure 6.14), the number of collisions per second.

low concentration = few collisions

high concentration = more collisions

Figure 6.14 The effect of concentration on reacting particles in solution

Only the particles at the surface area of a solid reactant can come into contact with the particles of the surrounding reactant (usually a gas or particles in aqueous solution). If a solid is powdered, the surface area is greatly increased and the rate of reaction increases with surface area (Figure 6.15).

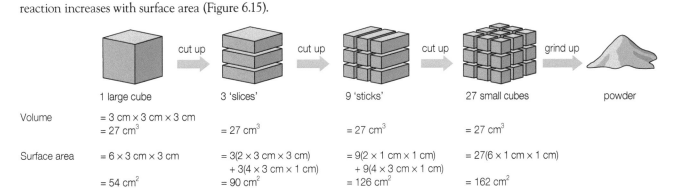

	1 large cube	3 'slices'	9 'sticks'	27 small cubes	powder
Volume	$= 3\,cm \times 3\,cm \times 3\,cm$ $= 27\,cm^3$	$= 27\,cm^3$	$= 27\,cm^3$	$= 27\,cm^3$	
Surface area	$= 6 \times 3\,cm \times 3\,cm$ $= 54\,cm^2$	$= 3(2 \times 3\,cm \times 3\,cm)$ $+ 3(4 \times 3\,cm \times 1\,cm)$ $= 90\,cm^2$	$= 9(2 \times 1\,cm \times 1\,cm)$ $+ 9(4 \times 3\,cm \times 1\,cm)$ $= 126\,cm^2$	$= 27(6 \times 1\,cm \times 1\,cm)$ $= 162\,cm^2$	

Figure 6.15 The effect of particle size on the surface area of a solid reactant

An increase in temperature increases the rate of a reaction by increasing the collision frequency (Figure 6.16), but more importantly, increasing the proportion of particles that have sufficient kinetic energy to react on collision. The combined effect of these two factors leads to an exponential increase in the rate of reaction with temperature.

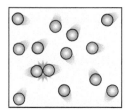

cold – slow movement,
few collisions,
little kinetic energy

hot – fast movement,
more collisions,
more kinetic energy

Figure 6.16 The effect of temperature on gaseous particles

A summary of these effects on rate of reaction (other than the role of a catalyst) is given in Table 6.1.

Table 6.1 Summary of the factors affecting the rate of reaction

Factor	Reactions affected	Changes made in conditions	Usual effect on the initial rate of reaction
Temperature	All	Increase	Increase
		Increase by 10 K	Approximately doubles
Concentration	All	Increase	Usually increases (unless zero order)
		Doubling of concentration on one of the reactants	Usually exactly doubles (if first order)
Light	Generally those involving reactions of mixtures of gases, including the halogens	Reaction in sunlight or ultraviolet light	Very large increase
Particle size	Reactions involving solids and liquids, solids and gases or mixtures of gases	Powdering the solid, resulting in a large increase in surface area	Very large increase

Expert tip

The key idea in all these examples is that the frequency of collision is increased in all cases – the particles collide more often, and therefore there is a greater chance of reaction.

Common mistake

Do not just explain the effect of temperature on reaction rate in terms of increased frequency of collision. Remember that there is a much more significant effect in this case in that the particles have greater kinetic energy, and therefore the activation energy is exceeded in a larger proportion of collisions.

■ QUICK CHECK QUESTION

6   Large granules of calcium carbonate (marble chips) are reacted with dilute hydrochloric acid and the initial rate of reaction determined. A second experiment is then carried out between the same mass of calcium carbonate powder and the same acid.

How does this change affect the collision frequency, activation energy and initial rate of reaction?

	Collision frequency	Activation energy	Initial rate of reaction
A	Increases	Increases	Increases
B	Stays constant	Stays constant	Stays constant
C	Increases	Decreases	Increases
D	Increases	Stays constant	Increases

■ Maxwell–Boltzmann distribution

The distribution of kinetic energies (velocities) of particles (atoms or molecules) in an ideal gas is described by the Maxwell–Boltzmann distribution (Figure 6.17). It is also a good description of kinetic energies of particles in systems containing real gases and species in aqueous solution. Kinetic energies are constantly changing due to collisions between particles and the average kinetic energy is directly proportional to the absolute temperature in kelvin.

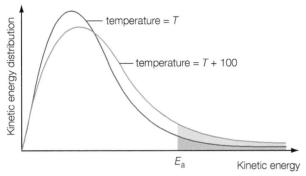

Figure 6.17 Maxwell–Boltzmann distribution of kinetic energies in a solution or gas at two different temperatures

The key features of the Maxwell–Boltzmann energy distribution curve are:
- It shows the distribution (or spread) of kinetic energy of particles in a sample of a gas at a given temperature and time (it is a purely statistical analysis).
- The highest point represents the most probable energy of a particle, the greatest fraction of molecules has this energy at a given instant.
- The curve is asymmetrical around the maximum value.
- The curve starts at the origin (no particles have zero energy), but it does not cross or touch the *x*-axis at higher energy.
- The area underneath the curve is proportional to the total number of particles in the sample of gas.
- The shape should be the same for any gas at a given temperature as, at the same temperature, the average kinetic energy should be the same.

The Maxwell–Boltzmann energy distribution curve shows why temperature has such a great effect on the rate of a reaction. As the temperature increases, the area under the curve does not change as the total number of reacting particles remains constant. The curve broadens and the most likely value of kinetic energy (the peak) has shifted to the right. The change in the Boltzmann distribution (Figure 6.17) as the temperature is increased shows how more molecules have a kinetic energy greater than the experimental activation energy (E_a). This leads to a large increase in the rate of reaction.

> **Expert tip**
>
> The vertical axis may be labelled in a variety of equally valid ways. 'Number of particles' or 'proportion of particles in sample' are possible labels. Be aware and not confused by this.

> **Expert tip**
>
> When the Boltzmann distribution curves at two different temperatures are drawn, the peak of the curve at the higher temperature must be lower than that of the lower temperature and the ending of the two curves do not intercept or meet the *x*-axis.

The effect of a catalyst

Revised ☐

Consideration of the Maxwell–Boltzmann distribution curve also helps explain why **catalysts** increase the rate of reactions. The lowering of the activation energy means that a greater proportion of the collisions between reacting molecules are successful and lead to a reaction (Figure 6.18).

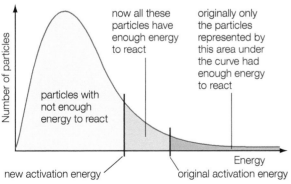

Figure 6.18 Maxwell–Boltzmann distribution of kinetic energies showing the effect of the lower activation energy of the new, catalysed pathway on the proportion of particles with sufficient energy to react

> **Key definition**
>
> **Catalyst** – a substance that increases the rate of a chemical reaction without, itself, being used up in the reaction. A catalyst increases the rate of a reaction by providing an alternative reaction pathway of lower activation energy (E_a), resulting in more successful collisions per unit time.

Catalysts function in a variety of ways including:

- making more collisions have favourable orientations
- locally increasing concentrations on solid surfaces
- providing a series of simple elementary steps rather than one direct high energy step
- increasing the reactivity of the reactive site of an organic molecule or
- providing a better attacking group which is regenerated at the end of the reaction.

■ Enthalpy level diagrams

Reactions can be described by enthalpy level diagrams (reaction profiles) which show the enthalpies (potential energies) of substances on a vertical scale and extent or progress of reaction along the horizontal scale. Figure 6.19 shows an enthalpy level diagram for a reaction where both the uncatalysed and catalysed reactions involve transition states. The profile highlights the fact that the activation energy of the new, catalysed pathway is lower than that of the uncatalysed reaction.

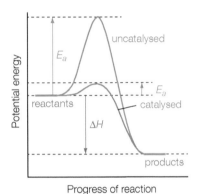

Figure 6.19 An enthalpy level diagram showing the effect of a catalyst. The new, catalysed pathway has a lower activation energy. Note that both pathways proceed via transition states

Figure 6.20 shows the energy profile for a different reaction, emphasizing that the presence of a catalyst provides an alternative reaction pathway. Here the addition of a catalyst clearly alters the mechanism as the catalysed reaction now involves two transition states and an intermediate.

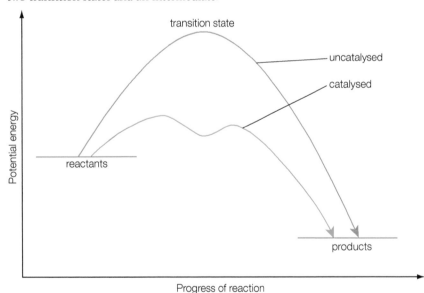

Figure 6.20 An enthalpy level diagram showing the effect of the use of a catalyst on the reaction pathway; the catalysed pathway proceeds by a transition state

■ **QUICK CHECK QUESTIONS**

7   Which of the changes listed increases the rate of the following reaction?

$C_4H_{10}(g) + Cl_2(g) \rightarrow C_4H_9Cl(l) + HCl(g)$

I   an increase in temperature

II an increase in pressure

IIIthe removal of HCl(g)

A I and III only

B I and II only

C II and III only

D I, II and III

8   Which of these statements is true about the use of sulfuric acid as a catalyst for the reaction?

$(CH_3)_2CO(aq) + I_2(aq) \rightarrow CH_3COCH_2I(aq) + HI(aq)$

I   the presence of the catalyst increases the rate of reaction

II the catalyst is used up in the reaction

IIIthe catalyst provides an alternative reaction pathway of lower activation energy

A I and III only

B I and II only

C II and III only

D I, II and III

9   Explain, using the Maxwell–Boltzmann distribution, why increasing the temperature causes the rate of a chemical reaction to increase.

Topic **7** Equilibrium

7.1 Equilibrium

Essential idea: Many reactions are reversible. These reactions will reach a state of equilibrium when the rates of the forward and reverse reactions are equal. The position of the equilibrium can be controlled by changing the conditions.

Reversible reactions and equilibrium

- A state of equilibrium is reached in a closed system when the rates of the forward and reverse reactions are equal.
- The equilibrium law describes how the equilibrium constant (K_c) can be determined for a particular chemical reaction.
- The magnitude of the equilibrium constant indicates the extent of a reaction at equilibrium and is temperature dependent.
- The reaction quotient (Q) measures the relative amounts of products and reactants present during a reaction at a particular point in time. Q is the equilibrium expression with non-equilibrium concentrations. The position of the equilibrium changes with changes in concentration, pressure and temperature.
- A catalyst has no effect on the position of equilibrium or the equilibrium constant.

▧ Dynamic equilibria

A *reversible reaction* is a chemical reaction or physical change in which the products can be changed back into reactants. If such a reaction is set up in a closed system where no reactants or products can escape, then a chemical equilibrium is reached where the rates of the forward and reverse reactions become equal (Figure 7.1). Consequently, in a chemical equilibrium, both reactants and products are present in unchanging concentrations.

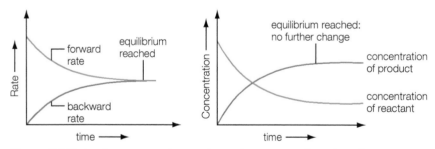

Figure 7.1 Plots of concentration and reaction rate against time for a reversible reaction involving a single reactant and product reacting in a 1:1 molar ratio (at constant temperature) to establish equilibrium

A *dynamic equilibrium* is established when the rate of the forward reaction equals that of the reverse (or backward) reaction. As a result, the concentrations of reactants and products become, and remain, constant. Such chemical equilibrium can only be achieved in a closed system in which none of the reactants or products can leave the reaction mixture.

Figure 7.2 shows graphs of data for the reversible reaction between hydrogen and iodine in the gas phase; demonstrating the achievement of the same equilibrium from either side of the equation.

$$H_2(g) + I_2(g) \rightleftharpoons 2HI(g)$$

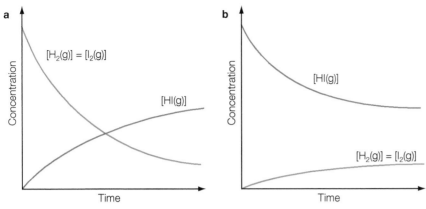

Figure 7.2 a Graph of the concentration of reactants and products with time when reacting equal amounts of hydrogen gas and iodine vapour; **b** Graph showing the achievement of the same equilibrium state by decomposition of hydrogen iodide vapour at the same temperature

Characteristics of equilibrium

A reaction at equilibrium has the following features (at constant temperature):

- It is dynamic: the particles of the reactants and products are continuously reacting.
- The equilibrium position can be reached starting from either reactants or products.
- The forward and reverse reactions occur at the same rate. Reactants are forming products and products forming reactants. The concentrations of reactants and products remain constant at equilibrium.
- The measurable (macroscopic or bulk) properties become, and remain, constant.
- It requires a closed system (Figure 7.3).

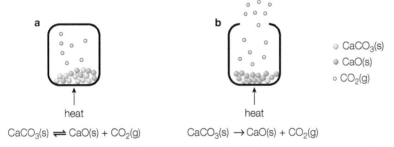

$CaCO_3(s) \rightleftharpoons CaO(s) + CO_2(g)$ $CaCO_3(s) \rightarrow CaO(s) + CO_2(g)$

Figure 7.3 The thermal decomposition of calcium carbonate. **a** A closed system where no carbon dioxide escapes and an equilibrium is established. **b** An open system. In the latter the calcium carbonate is continually decomposing as the carbon dioxide escapes. The reaction goes to completion

Expert tip

Reactions involving gases must be carried out in a closed container for equilibrium to be achieved. For reactions in a solution, an open beaker is a closed system if the solvent does not evaporate and the reactants and products are not volatile.

Establishing the dynamic nature of an equilibrium system

An equilibrium system is set up and allowed to reach equilibrium. Sampling and measuring are used to find the chemical composition of the system. An identical system is set up but where some reactant is of a heavier or radioactive isotope. The system is left to reach equilibrium and then sampling and measuring are done again to show there are products containing the detectable isotope. This supports the idea that at equilibrium both forward and reverse reactions are occurring (Figure 7.4).

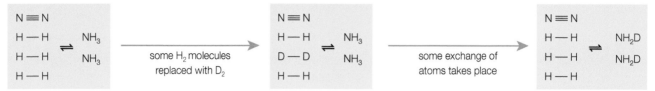

Figure 7.4 Incorporation of deuterium into ammonia within the ammonia, hydrogen and deuterium equilibrium mixture

■ QUICK CHECK QUESTION

1 Which of the following statements is true regarding a reversible chemical reaction when it has reached equilibrium?

A The concentration of the reactants and products are constantly changing.

B The concentrations of the products and reactants are equal.

C The reaction has completely stopped.

D The rate of the forward reaction is equal to the rate of the backwards reaction.

Expert tip

The main reason why some reactions are reversible is that the activation energies of the forward and reverse reactions are similar in size, and these activation energies are relatively small.

The equilibrium law

The equilibrium law is a quantitative law for predicting the amounts of reactants and products when a reversible reaction reaches a state of dynamic equilibrium.

In general, for a reversible reaction at equilibrium,

$$aA + bB \rightleftharpoons cC + dD$$

it has been found experimentally that, if the reaction is homogeneous (all species in the gas phase or in homogeneous solution) then the concentrations of the species involved are related by the equation:

$$K_c = \frac{[C]_{eqm}^c [D]_{eqm}^d}{[A]_{eqm}^a [B]_{eqm}^b}$$

The concentrations of the products appear on the top line of the equilibrium expression. The concentrations of the reactants appear on the bottom line of the equilibrium expression.

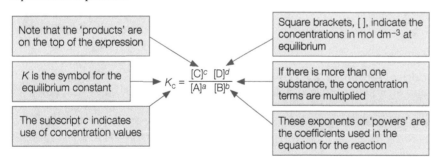

Note that the 'products' are on the top of the expression

K is the symbol for the equilibrium constant

The subscript c indicates use of concentration values

$$K_c = \frac{[C]^c [D]^d}{[A]^a [B]^b}$$

Square brackets, [ ], indicate the concentrations in mol dm⁻³ at equilibrium

If there is more than one substance, the concentration terms are multiplied

These exponents or 'powers' are the coefficients used in the equation for the reaction

Figure 7.5 The components of an equilibrium expression

If the concentrations of all the species in an equilibrium are known, the equilibrium expression can be used to calculate the value of the equilibrium constant. The changes in the concentrations of reactants and products on the way to reaching equilibrium are controlled by the stoichiometry of the reaction.

Expert tip

In order to calculate a standard value of the equilibrium constant, K_c, for a reaction the values of the concentrations entered in the equilibrium expression *must* numerically be in mol dm⁻³. Strictly speaking the values entered should be 'activity' values rather than concentrations. For gases and relatively dilute solutions the concentration values are sufficiently close to the 'activity' values that their use does not introduce any significant errors into the calculation. Since the 'activity' values for substances are simply a number – they do not have any units – it follows that *any K_c value will also simply be a number without units no matter which reaction you are studying.* Certainly in the IB examination you will not be asked for any units relating to K_c values.

Homogeneous and heterogeneous equilibria

An equilibrium in which all the substances are present in the same physical state is known as a *homogeneous equilibrium*. In contrast, one in which the substances involved are present in different phases is known as a *heterogeneous equilibrium*.

The concentration (more strictly, the activity) of any pure solid or pure liquid is constant (and has a value of 1) and so does not appear in an equilibrium expression for a heterogeneous system.

Expert tip

Note that you will only deal with examples and calculations relating to homogeneous equilibria in the context of this syllabus. That means that all the components of any equilibrium you must work with will either be in solution or in the gaseous phase.

Expert tip

The equilibrium constant is only meaningful if quoted with the stoichiometric (balanced) equation and state symbols.

■ QUICK CHECK QUESTION

2   Write the equilibrium expressions for the following reversible reactions. They are all examples of homogeneous equilibria.
   a $Fe^{3+}(aq) + SCN^-(aq) \rightarrow [Fe(SCN)]^{2+}(aq)$
   b $2SO_2(g) + O_2(g) \rightleftharpoons 2SO_3(g)$
   c $4NH_3(g) + 5O_2(g) \rightleftharpoons 4NO(g) + 6H_2O(g)$
   d $CH_4(g) + 2H_2O(g) \rightleftharpoons CO_2(g) + 4H_2(g)$

The law of mass action

The rate at which a substance reacts in homogeneous conditions is proportional to its concentration. Consider a general reversible equilibrium:

$$aA + bB \rightleftharpoons cC + dD$$

If the forward and backward reactions are reactions taking place in a single (elementary) step, then we expect the following:

the rate of forward reaction = $k_f[A]^a[B]^b$

the rate of reverse reaction = $k_r[C]^c[D]^d$

where k_f and k_r are rate constants and the square brackets represent the concentrations of the reactants and products.

At equilibrium, the rate of the forward reaction is equal to the rate of the reverse reaction:

$$k_f[A]^a[B]^b = k_r[C]^c[D]^d$$

$$\frac{k_f}{k_r} = \frac{[C]^c[D]^d}{[A]^a[B]^b} = K_c$$

The ratio of the two rate constants is equal to the equilibrium constant, K_c.

Expert tip

Dynamic is the opposite of static, meaning that the forward and backward reactions continue at equilibrium.

Equilibrium constant

The value, or size, of the equilibrium constant, K_c, indicates whether there are greater concentrations of reactants or products present in the system at equilibrium (Table 7.1). It is a measure of the extent to which the position of equilibrium favours the products over the reactants.

Table 7.1 The relationship between the value of K_c and the extent of a reaction

Reaction hardly goes	'Reactants' predominate at equilibrium	Equal amounts of reactants and products	'Products' predominate at equilibrium	Reaction goes virtually to completion
$K_c < 10^{-10}$	$K_c = 0.01$	$K_c = 1$	$K_c = 100$	$K_c > 10^{10}$

Equilibrium position

While the equilibrium constant gives a quantitative indication of the balance between reactants and products at equilibrium, the term 'equilibrium position' is used to describe this situation qualitatively (Figure 7.6). It gives an indication of whether the reactants or products are more abundant in the system. Remember, reactants are written on the left, products on the right in an equation.

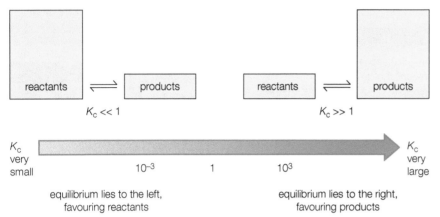

Figure 7.6 The significance of the equilibrium constant, K_c, in relation to the position of an equilibrium

Modifying equilibrium constants

The expression and value for an equilibrium constant vary depending on the direction from which the particular equilibrium is approached and also on how the chemical equation for the reaction is written.

For instance, the reaction

$$H_2(g) + I_2(g) \rightleftharpoons 2HI(g)$$

has the equilibrium expression:

$$K_c = \frac{[HI]^2}{[H_2][I_2]}$$

The above equilibrium constant can be referred to as K_{c_1}.

Table 7.2 shows the equilibrium expressions and how the equilibrium constants relate to the original equilibrium constant K_{c_1} for the following variations of the same reaction at the same temperature.

Table 7.2 How the values of the equilibrium constant are related for different forms of the equation for a reaction

Reverse the equation: $2HI(g) \rightleftharpoons H_2(g) + I_2(g)$	$K_{c_2} = \dfrac{[H_2][I_2]}{[HI]^2}$	$K_{c_2} = \dfrac{1}{K_{c_1}}$
Multiply the coefficients by 2: $2H_2(g) + 2I_2(g) \rightleftharpoons 4HI(g)$	$K_{c_3} = \dfrac{[HI]^4}{[H_2]^2 [I_2]^2}$	$K_{c_3} = (K_{c_1})^2$
Halve the coefficients: $\frac{1}{2}H_2(g) + \frac{1}{2}I_2(g) \rightleftharpoons HI(g)$	$K_{c_4} = \dfrac{[HI]}{[H_2]^{\frac{1}{2}}[I_2]^{\frac{1}{2}}}$	$K_{c_4} = \sqrt{K_{c_1}}$

Some reactions take place in a sequence of steps which may themselves be equilibria; the equilibrium constant for the overall reaction being the product of the constants for the individual steps making up the sequence (see Table 7.3). Table 7.3 summarizes generally how equilibrium constant expressions are modified by various changes to the reaction equation.

■ QUICK CHECK QUESTION

3   The figures given here are values for the equilibrium constant, K_c, for a reaction using the same starting concentrations but carried out at different temperatures. Which equilibrium mixture has the lowest concentration of products?

A 1.0

B 1.0×10^2

C 1.0×10^{-2}

D 0.1

Table 7.3 The equilibrium constant for the same reaction at the same temperature can be expressed in a number of ways

Change in reaction equation	Equilibrium constant expression	Equilibrium constant
Reverse the reaction	Inverse of expression	K_c^{-1}
Halve the stoichiometric coefficients	Square root of the expression	$\sqrt{K_c}$
Double the stoichiometric coefficients	Square of the expression	K_c^2
Sequence of reactions	Multiply the values for the individual steps	$K_c = K_{c_1} \times K_{c_2} \times K_{c_3}$

■ **QUICK CHECK QUESTIONS**

4 The following reaction is an esterification reaction producing ethyl ethanoate:

$CH_3CO_2H(l) + C_2H_5OH(l) \rightleftharpoons CH_3CO_2C_2H_5(l) + H_2O(l)$

The value of K_c for the above reaction at 25 °C is 4.0.

This equilibrium can be approached experimentally from the opposite direction. What is the value for K_c for this reaction, the hydrolysis of ethyl ethanoate, at 25 °C?

$CH_3CO_2C_2H_5(l) + H_2O(l) \rightleftharpoons CH_3CO_2H(l) + C_2H_5OH(l)$

5 At 25 °C the equilibrium constant K_c for the reaction

$CO(g) + 2H_2(g) \rightleftharpoons CH_3OH(g)$

is 1.7×10^4. What is the value of K_c at this temperature for the following reactions?

a $CH_3OH(g) \rightleftharpoons CO(g) + 2H_2(g)$

b $2CO(g) + 4H_2(g) \rightleftharpoons 2CH_3OH(g)$

■ **Reaction quotient**

The reaction quotient (Figure 7.7), Q, is determined by substituting reactant and product concentrations at any point during a reaction. If the system is at equilibrium, $Q = K_c$. If $Q \neq K_c$, however, the system is not at equilibrium. When $Q < K_c$, the reaction will move toward equilibrium by forming more products (the forward reaction is favoured); when $Q > K_c$, the reaction moves toward equilibrium by forming more reactants (the reverse reaction is favoured).

■ **QUICK CHECK QUESTION**

6 The value of the equilibrium constant, K_c, for a particular gas phase reaction is found to be 260 at 300 K. The pressure is changed at constant temperature and the value of the reaction quotient, Q, now becomes 120. Deduce the direction in which the system will move towards equilibrium; left towards reactants, or right towards products.

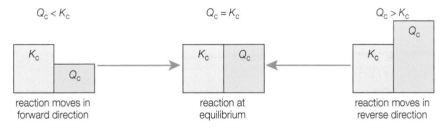

Figure 7.7 The relative sizes of the reaction quotient (Q) and equilibrium constant (K_c) indicate the direction in which a reaction mixture tends to change

The reaction quotient, Q, helps us to predict quantitatively the direction of the reaction:

■ if $Q < K_c$ then the reaction proceeds towards the products
■ if $Q > K_c$ then the reaction proceeds towards the reactants
■ if $Q = K_c$ then the reaction is at equilibrium and no net reaction occurs.

Factors affecting equilibria: Le Châtelier's principle

When the conditions under which a chemical equilibrium has been established are changed there is an effect on the position of the equilibrium. The direction of change can be predicted using Le Châtelier's principle. The principle states that when the conditions of a system at equilibrium change, the position of equilibrium shifts in the direction that tends to counteract the change. Put simply, the system responds to negate the change by responding in the opposite way.

Expert tip

Do note that for IB you do not need to learn a statement of the principle, as some published versions use quite complex language. However, it can be very useful to have a familiar version in mind when tackling questions.

NATURE OF SCIENCE

Le Châtelier's principle is a descriptive statement of what happens when a dynamic equilibrium is disturbed by a change in conditions; it is not an explanation as to why the change happens. In this sense it is similar to Markovnikov's rule in organic chemistry: through studying many examples we can use them to predict behaviour, but they offer no explanation of the phenomena considered.

This general principle is of importance industrially as it allows chemists to alter the reaction conditions to produce an increased amount of the product and, therefore, increase the profitability of a chemical process.

The possible changes in conditions that we need to consider in this context are:
- changes in the concentration of either reactants or products
- changes in pressure for gas phase reactions
- changes in temperature
- the presence of a catalyst.

Note that in all cases, once the equilibrium has been re-established after the change, the value of K_c will be unaltered except when there is a change in temperature (Table 7.4).

Table 7.4 A broad summary of the effects of changing conditions on the position of an equilibrium

Change made	Effect on 'position of equilibrium'	Value of K_c
Concentration of one of the components of the mixture	Changes	Remains unchanged
Pressure	Changes if the reaction involves a change in the total number of gas molecules	Remains unchanged
Temperature	Changes	Changes
Use of a catalyst	No change	Remains unchanged

Changes in concentration

The following statements indicate the change that occurs when the concentration of reactants or products are changed in a system at equilibrium:
- Increasing the concentration of a reactant will move the position of equilibrium to the right, favouring the forward reaction and increasing the equilibrium concentrations of the products.

Conversely, the opposite of this is also true:
- The addition of more product to an equilibrium mixture would shift the position of the equilibrium to the left; the reverse reaction would be favoured.

Concentration can be changed by either adding or removing one of the reactants or products.

Table 7.5 shows the effect of a sudden concentration change, i.e. a doubling of the hydrogen concentration from 0.01 mol dm^{-3} to 0.02 mol dm^{-3}, on the equilibrium position in the reaction between hydrogen and iodine as shown below:

$$H_2(g) + I_2(g) \rightleftharpoons 2HI(g)$$

Table 7.5 Data on the effect of doubling the [H$_2$] on the equilibrium position for the reaction

	Initial equilibrium mixture concentrations mol dm^{-3}	Doubling of [H$_2$(g)] mol dm^{-3}	Final equilibrium mixture concentrations mol dm^{-3}
[HI(g)]	0.07	0.07	0.076
[H$_2$(g)]	0.01	0.02	0.017
[I$_2$(g)]	0.01	0.01	0.007
	$K_c = 49$	$Q = 24.5$	$K_c = 49$

At equilibrium the K_c value is 49. When the concentration of hydrogen is doubled, as shown in the third column, the equilibrium law now produces a reaction quotient (Q) value of 24.5 which is below the K_c value of 49. This means the system is no longer at equilibrium and the system will respond.

In this example, after doubling the hydrogen concentration the value of reaction quotient Q is lower than the value of K_c. As a result, the equilibrium shifts to the right (the forward reaction is increased or favoured) to decrease the [reactant] and to increase [products] and thus increase the value of Q back to the value of the equilibrium constant at that temperature.

Change in pressure applied to a gas phase reaction

For gas phase reactions where there are differences in the total number of molecules on either side of the equation:
- increased pressure shifts the equilibrium position to the side of the equation with fewer molecules
- decreased pressure shifts the equilibrium position to the side with more molecules.

Pressure only affects the position of an equilibrium system that has a different number of gaseous molecules on the reactant side from that on the product side. Pressure has no effect on systems that have no gaseous molecules, or which has an equal number of gaseous moles on both sides.

The pressure in a closed system can be changed by changing the volume of the reacting vessel or by adding an unreactive gas like helium to the reacting mixture – this increases the pressure as it adds more particles to the same volume.

Change in temperature

None of the changes results mentioned above results in a change in the value of K_c. However, a change in temperature will alter K_c.

The key factor to be considered here is whether the forward reaction is exothermic (a negative ΔH value) or endothermic (a positive ΔH value). Remember that, in a reversible reaction, the reverse reaction has an enthalpy change that is equal and opposite to that of the forward reaction.
- When the temperature is increased, the equilibrium position will shift in the direction that will tend to lower the temperature, that is, the endothermic direction that absorbs heat.
- If the temperature is lowered, the equilibrium will shift in the exothermic direction so as to generate heat and raise the temperature. These effects are summarized in Table 7.6.

The enthalpy change for a reaction indicates how an increase in temperature affects the equilibrium: for an endothermic reaction, an increase in temperature shifts the equilibrium to the right; for an exothermic reaction, a temperature increases shifts the equilibrium to the left (Table 7.6).

Table 7.6 The effects of temperature changes on chemical equilibria

Nature of forward reaction (sign of Δ)	Change in temperature	Shift in the position of equilibrium	Effect on value of K_c
Endothermic (positive ΔH)	Increase	To the right	K_c increases
Endothermic (positive ΔH)	Decrease	To the left	K_c decreases
Exothermic (negative ΔH)	Increase	To the left	K_c decreases
Exothermic (negative ΔH)	Decrease	To the right	K_c increases

Expert tip

The effect on the position of equilibrium when an equilibrium is disturbed can be predicted by the application of Le Châtelier's principle. However, the rate at which an equilibrium is re-established is a kinetic issue.

If the temperature of an equilibrium system is increased, the rates of both the forward and reverse reactions increase. However, the rates do not increase by the same extent, because they will have different values of activation energy. Hence, the equilibrium position shifts. The larger the activation energy, the greater the change in rate. Smaller activation energies are less sensitive to changes in temperature.

Common mistake

A catalyst does not change the value of the equilibrium constant, K_c, but the equilibrium is reached more quickly. It reduces the forward and reverse activation energies equally.

Addition of a catalyst

The presence of a catalyst has no effect on the position of a chemical equilibrium. The effect of a catalyst is to provide an alternative pathway of lower activation energy. However, the effect is applicable to the E_a values of both the forward and reverse reaction: both values are reduced by the same amount. Consequently, *the presence of a catalyst increases the rate of the forward and reverse reactions equally*. There is no change in the position of the equilibrium or the value of K_c.

However, the advantage of using a catalyst is that its presence reduces the time required for the equilibrium to be established.

■ **QUICK CHECK QUESTIONS**

7   Using this reaction:

$Ce^{4+}(aq) + Fe^{2+}(aq) \rightleftharpoons Ce^{3+}(aq) + Fe^{3+}(aq)$

explain what happens to the position of equilibrium when:

a   the concentration of $Fe^{2+}(aq)$ ions is increased

b   water is added to the equilibrium mixture.

8   a   Predict the effect of the following changes, listed below, on the equilibrium position of the reaction

$CH_4(g) + H_2O(g) \rightleftharpoons CO(g) + 3H_2(g)$ $\Delta H = +206\,kJ\,mol^{-1}$

i   decreasing the temperature

ii   increasing the temperature

iii adding a catalyst

iv   adding hydrogen.

b   State how each of the above changes affects the value of the equilibrium constant, K_c.

9   a   Predict the effect on the position of equilibrium of increasing the pressure on the following gas phase reactions:

i $2NO_2(g) \rightleftharpoons 2NO(g) + O_2(g)$

ii $H_2(g) + I_2(g) \rightleftharpoons 2HI(g)$

b   Predict the effect on the position of equilibrium of decreasing the pressure on these two reactions:

i $N_2O_4(g) \rightleftharpoons 2NO_2(g)$

ii $CH_4(g) + H_2O(g) \rightleftharpoons CO(g) + 3H_2(g)$

10 Predict the effect on the position of equilibrium, and the value of K_c, of increasing the temperature on these reactions at equilibrium:

a $CO_2(g) + 2H_2(g) \rightleftharpoons CH_3OH(g)$ $\Delta H = -90\,kJ\,mol^{-1}$

b $H_2(g) + CO_2(g) \rightleftharpoons H_2O(g) + CO(g)$ $\Delta H = +41.2\,kJ\,mol^{-1}$

c $ClNO_2(g) + NO(g) \rightleftharpoons ClNO(g) + NO_2(g)$ $\Delta H = -18\,kJ\,mol^{-1}$

Topic **8** Acids and bases

8.1 Theories of acids and bases

Revised ☐

Essential idea: Many reactions involve the transfer of a proton from an acid to a base.

Theories of acids and bases

Revised ☐

- A Brønsted–Lowry acid is a proton/H^+ donor and a Brønsted–Lowry base is a proton/H^+ acceptor.
- Amphiprotic species can act as both Brønsted–Lowry acids and bases.
- A pair of species differing by a single proton is called a conjugate acid–base pair.

Arrhenius theory

Acids and bases were first recognized by the properties of their aqueous solution. Arrhenius recognized that the properties of acidic solutions are due to the presence of protons (H^+(aq)) and those of basic solutions are due to the presence of hydroxide ions (OH^-(aq)).

Brønsted–Lowry theory

Brønsted–Lowry theory defines acids as proton donors and bases as proton acceptors.

NATURE OF SCIENCE

In chemistry, theories are often falsified by experimental data and replaced by new theories. Early chemists originally believed that all acids contained oxygen, but HCl and HCN were shown to be acids that do not contain oxygen and a new theory of acidity had to be developed. When acids dissolve in water, the properties of the resulting solution led later chemists to conclude that it was ionized. They described the formation of the hydrochloric acid with the following equation:

$$HCl(g) + (aq) \rightarrow H^+(aq) + Cl^-(aq)$$

However, experiments showed this to be exothermic, but this equation implies it to be endothermic (bond breaking). Lowry and Brønsted simultaneously, and independently, proposed a theory to explain this discrepancy with the water molecules actively involved in the formation of hydrochloric acid.

Acids

According to this theory, hydrogen chloride molecules donate protons (hydrogen ions) to water molecules when they dissolve in water, producing oxonium ions. The water molecule accepts the proton and acts as a base (Figure 8.1).

The changes in the bonding during this reaction are described in Figure 8.2, with dots and crosses representing the valence electrons.

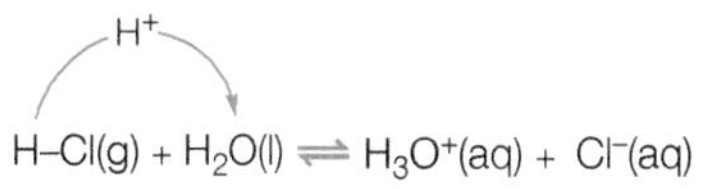

$$H\text{–}Cl(g) + H_2O(l) \rightleftharpoons H_3O^+(aq) + Cl^-(aq)$$

Figure 8.1 Hydrogen chloride is an acid and donates a proton when dissolved in water

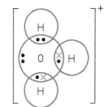

proton transfer
H
Cl
O
H
H
+
H
O
H
H
–
Cl

The proton, H^+, is donated by the hydrogen chloride molecule and accepted by the water molecule... ...forming the positive oxonium ion... ...and leaving a negative chloride ion

Figure 8.2 Hydrogen chloride is the acid (proton donor) and water is the base (proton acceptor)

An example of an acid–base reaction that does not involve water is the formation of ammonium chloride when ammonia gas mixes and reacts with hydrogen chloride gas. The ammonia molecule acts as a base and the hydrogen chloride molecule acts an acid.

A monoprotic acid is an acid that can donate one proton (hydrogen ion) per molecule in aqueous solution. Examples of monoprotic acids are hydrochloric acid, HCl, nitric(V) acid, HNO_3, and ethanoic acid, CH_3COOH, where the terminal hydrogen is donated. Sulfuric acid, H_2SO_4, is regarded as a strong diprotic acid. Figure 8.4 shows the dissociations or ionization of nitric, ethanoic and sulfuric acids.

+ H⁺
+ H⁺
+ 2H⁺

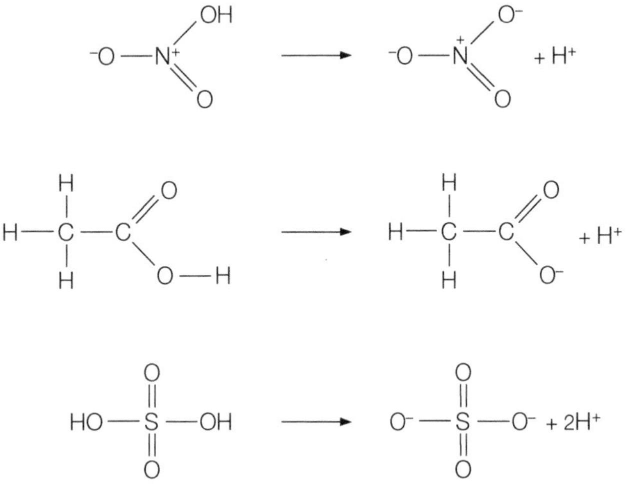

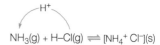

Figure 8.4 Structure of nitric(V), ethanoic and sulfuric(VI) acid molecules and their complete dissociation or ionization in water

Bases

According to this theory, ammonia molecules accept protons (hydrogen ions) from water molecules when they dissolve in water, producing ammonium ions. The water molecule donates the proton and acts as an acid (Figure 8.5). There is a lone pair on the nitrogen atom of ammonia that allows it to act as a base.

The changes in the bonding during this reaction are described in Figure 8.6 with dots and crosses representing the valence electrons.

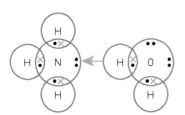

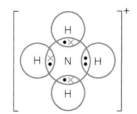

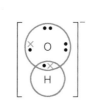

H
H
N
H
H
O
H
+
H
H
N
H
H
–
O
H

The proton, H^+, is donated by the water molecule and accepted by the ammonia molecule... ...forming the positive ammonium ion... ...and the negative hydroxide ion

Figure 8.6 Water acts as the acid (proton donor) and ammonia is the base (proton acceptor)

Expert tip

H^+ does not exist by itself in aqueous solution. It has to be solvated to be stable – i.e. for the acid to give up H^+ there must be a base to accept it.

H⁺
NH₃(g) + H–Cl(g) ⇌ [NH₄⁺ Cl⁻](s)

Figure 8.3 Ammonia acts as a base when reacting with hydrogen chloride gas

QUICK CHECK QUESTION

1 Explain why this reaction between ammonia and hydrogen chloride is not an acid–base reaction according to the Arrhenius theory.

Common mistake

Hydrogen chloride molecules only undergo dissociation or ionization in water to form oxonium ions, $H_3O^+(aq)$, and chloride ions, $Cl^-(aq)$. Hydrogen chloride molecules remain as undissociated molecules (H-Cl) in organic solvents.

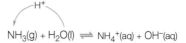

H⁺
NH₃(g) + H₂O(l) ⇌ NH₄⁺(aq) + OH⁻(aq)

Figure 8.5 Ammonia acts as a base when dissolved in water

Common Brønsted–Lowry bases are the oxide and hydroxide ions, ammonia, amines and the carbonate and hydrogen carbonate ions.

Expert tip

Note that in accepting a proton a Brønsted–Lowry base is forming a coordinate (dative covalent) bond with the proton (H^+ ion).

Amphiprotic substances

Water is an amphiprotic substance, one that can function as either a Brønsted–Lowry acid, or base depending on the substance which it reacts. The ions derived from polyprotic acids, such as sulfuric(VI) acid, H_2SO_4, and phosphoric(V) acid, H_3PO_4, are also amphiprotic species, for example:

$$H_2PO_4^-(aq) + OH^-(aq) \rightleftharpoons HPO_4^{2-}(aq) + H_2O(l)$$

$$H_2PO_4^-(aq) + H_3O^+(aq) \rightleftharpoons H_3PO_4(aq) + H_2O(l)$$

Examples of Brønsted–Lowry species

Brønsted–Lowry acids and bases may be molecules, anions or cations. Examples of such species are given in Table 8.1.

Table 8.1 Examples of Brønsted–Lowry acids and bases

Type of species	Acids	Bases
Molecular	HCl, HNO_3, $HClO_4$, H_3PO_4, H_2O	NH_3, N_2H_4, H_2O, and amines
Cations	NH_4^+, $[Fe(H_2O)_6]^{3+}$, H_3O^+	$[Fe(H_2O)_5OH]^{2+}$
Anions	HCO_3^-, HSO_4^-	CO_3^{2-}, SO_4^{2-}, H^-, NH_2^-, OH^-

Expert tip

The term amphoteric refers to any substance that can behave as both an acid (proton donor) or a base (proton acceptor). All amphiprotic species are amphoteric. The term amphoteric is a more general term and includes other definitions of acids and bases.

QUICK CHECK QUESTION

2 State what is meant by the term amphiprotic. Show that the hydrogen selenate ion, $HSeO_4^-$(aq), is amphiprotic.

Conjugate acid–base pairs

An acid–base equilibrium involves a competition for protons (H^+). An acid forms a conjugate base when it loses a proton and a base turns into a conjugate acid when it gains a proton. For every acid dissociation there has to be a base present to accept the proton.

In the equilibrium in Figure 8.7 the protons are held on the left hand side of the equation by dative bond formation by lone pair of the nitrogen atom on the ammonia molecule. On the right hand side they are held by dative bond formation by one of the lone pairs on the water molecule. This equilibrium involves two conjugate acid–base pairs that differ by H^+: NH_4^+ and NH_3 and H_3O^+ and H_2O.

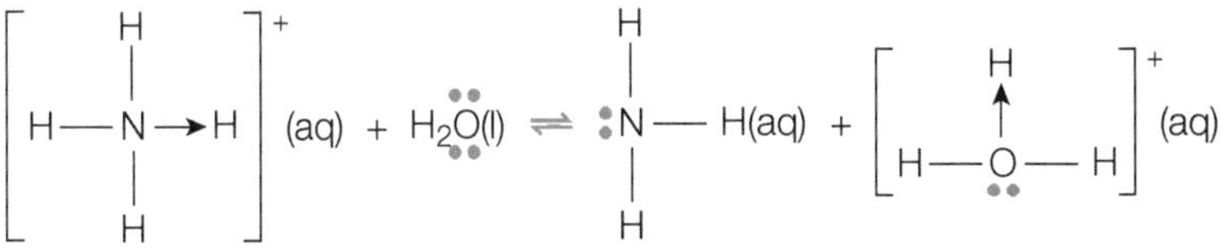

Figure 8.7 The equilibrium between ammonia and the ammonium ion, and water and the oxonium ion

Expert tip

A diprotic acid is termed a dibasic acid because it has two conjugate bases.

■ **QUICK CHECK QUESTIONS**

3 Identify which reactants are acids and which are bases in the following reaction:

$HCOOH(aq) + HClO_2(aq) \rightleftharpoons HCOOH_2^+(aq) + ClO_2^-(aq)$

4 Define the terms acids and bases in terms of Brønsted–Lowry theory. Consider the following equilibria in aqueous solutions.

(I) $NH_3(aq) + H_2O(l) \rightleftharpoons NH_4^+(aq) + OH^-(aq)$

$K_c = 1.8 \times 10^{-5}\,mol\,dm^{-3}$

(II) $C_6H_5O^-(aq) + HCOOH(aq) \rightleftharpoons C_6H_5OH(aq) + HCOO^-(aq)$

$K_c = 1.3 \times 10^6\,mol\,dm^{-3}$

Identify the conjugate acid–base pairs for each acid–base reaction. State with a reason, whether the base in reaction (I) or reaction (II) is stronger.

5 Use an equation to explain how methylamine, CH_3-NH_2, can act as a Brønsted–Lowry base.

6 Select the Brønsted–Lowry acid and conjugate base pairs for the following reaction:

$NH_3(aq) + H_2O(l) \rightleftharpoons NH_4^+(aq) + OH^-(aq)$

7 Write the conjugate bases for the following Brønsted–Lowry acids:

$HF, HNO_3, HSO_4^-, H_2S, N_2H_5^+$

8 Write the conjugate acids for the following Brønsted–Lowry bases:

$^-NH_2, H_2O, PH_3, C_2H_5OH, ^-OH, HCOO^-$

8.2 Properties of acids and bases

Revised ☐

Essential idea: The characterization of an acid depends on empirical evidence such as the production of gases in reactions with metals, the colour changes of indicators or the release of heat in reactions with metal oxides and hydroxides.

Properties of acids and bases

Revised ☐

■ Most acids have observable characteristic chemical reactions with reactive metals, metal oxides, metal hydroxides, hydrogen carbonates and carbonates.

■ Salt and water are produced in exothermic neutralization reactions.

■ Reactions of acids and bases

Acids and bases undergo a variety of reactions. With one exception all the reactions shown in Figure 8.8 are acid–base reactions and involve proton transfer: there are no changes in oxidation states. The reaction between a metal and an acid is a redox reaction and involves a change in oxidation states. Reactions that result in the formation of a salt and water only are termed neutralization reactions. These are always exothermic.

> **Expert tip**
>
> Specific equations (including ionic) can be deduced from the general equations, provided the formulas of the ions in the reactants or products are known.

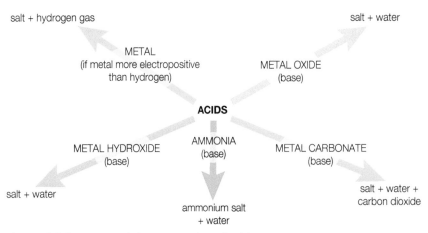

Figure 8.8 Summary of the properties of acids

Salts are ionic substances formed when the 'acidic' hydrogen of an acid is completely replaced by a metal (ion). Alkalis are soluble bases (Figure 8.9) and contain hydroxide ions. They can be used to neutralize acids to form salts. An acid–base indicator can be used to determine whether or not a solution is acidic or alkaline.

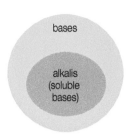

Figure 8.9 The relationship between bases and alkalis

■ **QUICK CHECK QUESTIONS**

9 State two observations if magnesium metal is reacted with aqueous HNO_3 under standard conditions. Write an ionic equation to describe this reaction (with state symbols).

10 The above reaction with magnesium is repeated using a solution of aqueous CH_3OOH (of the same concentration) instead of HNO_3. State and explain one difference observed during the reaction.

11 By means of balanced chemical equations, describe **three** different types of reaction of dilute aqueous ethanoic acid.

12 Identify the acids and bases required to make the salts: lithium nitrate and sodium sulfate. State the type of reaction that occurs and state the type of enthalpy change.

Expert tip

It takes just as much sodium hydroxide to neutralize 25.00 cm³ of 0.1 mol dm⁻³ of a weak monoprotic acid, such as ethanoic acid, as it does to neutralize 25.00 cm³ of 0.1 mol dm⁻³ of a strong monoprotic acid, such as nitric acid.

8.3 The pH scale

Revised ☐

Essential idea: The pH scale is an artificial scale used to distinguish between acid, neutral and basic/alkaline solutions.

The pH scale

Revised ☐

■ pH = –log [H⁺(aq)] and [H⁺(aq)] = 10⁻ᵖᴴ
■ A change of one pH unit represents a 10-fold change in the hydrogen ions, [H⁺].
■ pH values distinguish between acidic, neutral and alkaline solutions.
■ The ionic product constant, K_w = [H⁺] [OH⁻] = 10⁻¹⁴ at 298 K.

The pH scale

Revised ☐

The pH scale is a logarithmic scale (Table 8.2) for measuring the concentration of aqueous hydrogen (oxonium) ions in aqueous solutions. The possible range of [H⁺(aq)] is very large so it is just a convenience to use a log scale.

The pH of a solution can be measured using a pH probe and meter, or it can be estimated using universal indicator (a mixture of indicators).

Table 8.2 The relationship between pH and H⁺(aq) concentration

pH	Concentration of hydrogen ions, H⁺(aq)/mol dm⁻³
0	$1 \times 10^0 = 1.0$
1	$1 \times 10^{-1} = 0.1$
2	$1 \times 10^{-2} = 0.01$
3	$1 \times 10^{-3} = 0.001$
4	$1 \times 10^{-4} = 0.0001$
5	$1 \times 10^{-5} = 0.00001$
6	$1 \times 10^{-6} = 0.000001$
7	$1 \times 10^{-7} = 0.0000001$
8	$1 \times 10^{-8} = 0.00000001$
9	$1 \times 10^{-9} = 0.000000001$
10	$1 \times 10^{-10} = 0.0000000001$
11	$1 \times 10^{-11} = 0.00000000001$
12	$1 \times 10^{-12} = 0.000000000001$
13	$1 \times 10^{-13} = 0.0000000000001$
14	$1 \times 10^{-14} = 0.00000000000001$

The pH scale (Figure 8.10) runs from 0 to 14 and the smaller the number, the higher the concentration of hydrogen ions. A change in one pH unit corresponds to a ten-fold change ($10\times$) in $H^+(aq)$.

$$pH = -\log_{10}[H^+(aq)] \text{ or } [H^+(aq) = 10^{-pH}]$$

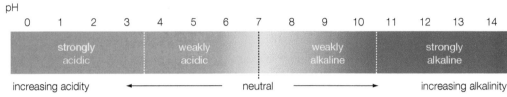

Figure 8.10 The pH scale and the colours of universal indicator

For a strong monoprotic acid, the pH of its aqueous solution is directly related to its concentration. However, this is not true for weak monoprotic acids as the pH depends on the acid dissociation constant, K_a.

Expert tip

Numbers can be expressed on a normal number line, where the gaps between the numbers are equal. On a log scale to the base 10, such as pH, the distance between 1 and 10 and between 10 and 100 (between powers of 10) is kept the same. The p-notation is a modified log scale, used by chemists to express small values with a large range simply (Figure 8.11).

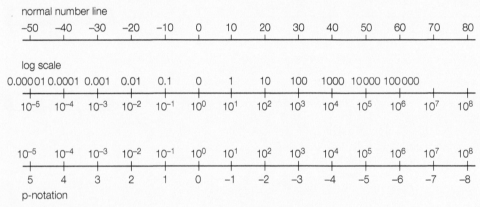

Figure 8.11 Normal number line, log scale and p-notation

■ QUICK CHECK QUESTIONS

13 Equal concentrations of aqueous HCN and HNO_3 solutions have pH values of 6.00 and 2.00, respectively. Determine the ratio of the hydrogen ion concentrations (H^+) in the two aqueous solutions.

14 Determine the pH of the solution when 50.00 cm³ of 0.50 mol dm⁻³ NaOH solution is mixed with 200.00 cm³ of 0.10 mol dm⁻³ HCl solution.

15 The pH of an aqueous solution is 7.00. If its pH is decreased to 3.00, deduce how much the hydrogen ion concentration changes.

16 A diluted hydrochloric acid solution is prepared by adding 25.00 cm³ of 15.00 mol dm⁻³ hydrochloric acid to water and making it up to 1.00 dm³ in a volumetric flask. Calculate the pH of the diluted hydrochloric acid solution.

17 Determine the pH of 25.00 cm³ of 0.125 mol dm⁻³ $HNO_3(aq)$ solution.

18 Ant stings contain methanoic acid. Sodium hydrogen carbonate is used to treat ant stings. Write the equation for the reaction between sodium hydrogen carbonate and methanoic acid. A methanoic acid solution has a pH of 2.42. Calculate its hydrogen ion concentration.

19 State how the pH changes when 100.00 cm³ of 0.10 mol dm⁻³ HCl(aq) is diluted with 900.00 cm³ of pure water.

20 Calculate the pH of an aqueous solution of barium hydroxide with 0.750 g in 2.000 dm³ of solution.

21 Calculate the mass (in grams) of sodium hydroxide that must be dissolved to make 1 dm³ of aqueous solution with a pH of 12.

Ionic product of water

There are low concentrations of hydrogen and hydroxide ions in pure water because of the transfer of hydrogen ions (protons) between water molecules. There is an equilibrium system, but the extent of ionization is very small.

$$H_2O(l) + H_2O(l) \rightleftharpoons H_3O^+(aq) + OH^-(aq)$$

which can be written more simply as:

$$H_2O(l) \rightleftharpoons H^+(aq) + OH^-(aq)$$

The equilibrium constant, $K_w = [H^+(aq)] \times [OH^-(aq)]$ since pure liquids are not included in equilibrium expressions. The value of K_w at 25 °C is 1.00×10^{-14} mol^2 dm^{-6}. K_w is the ionic product constant for water.

This expression can be used to calculate the pH of an alkaline aqueous solution since it allows the deduction of hydrogen ion concentration in aqueous solution from hydroxide ion concentration.

It is important to realize that $[H^+(aq)]$ only equals $[OH^-(aq)]$ in a neutral solution or pure water at 25 °C. If the solution is acidic $[H^+(aq)] > [OH^-(aq)]$ and if it is alkaline $[OH^-(aq)] > [H^+(aq)]$. This means that neutral $[H^+(aq)] = [OH^-(aq)]$ is only pH = 7 at 25 °C.

> **Expert tip**
>
> The electrical conductivity of even the purest form of water never falls to zero suggesting the presence of ions.

■ QUICK CHECK QUESTIONS

22 Calculate the pH of 0.0500 mol dm^{-3} caesium hydroxide solution and 0.02500 mol dm^{-3} barium hydroxide solution at 25 °C.

23 Complete the following table with the type of solution and the missing values.

H$^+$(aq)	[OH$^-$(aq)	Type of solution
	1×10^{-13}	
1×10^{-3}	1×10^{-11}	
1×10^{-5}	1×10^{-9}	
1×10^{-7}	1×10^{-7}	
1×10^{-9}	1×10^{-5}	
1×10^{-11}		
1×10^{-13}		

24 Calculate the concentration of OH$^-$(aq) in 0.05 mol dm^{-3} HCl(aq).

8.4 Strong and weak acids and bases

Revised ▢

Essential idea: The pH depends on the concentration of the solution. The strength of acids or bases depends on the extent to which they dissociate in aqueous solution.

Strong and weak acids and bases

Revised ▢

■ Strong and weak acids and bases differ in the extent of ionization.
■ Strong acids and bases of equal concentrations have higher conductivities than weak acids and bases.
■ A strong acid is a good proton donor and has a weak conjugate base.
■ A strong base is a good proton acceptor and has a weak conjugate acid.

■ Strong and weak acids and bases

Strong acids are strong electrolytes, ionizing completely in aqueous solution (Figure 8.12). The common strong acids are HCl, HNO_3 and H_2SO_4.

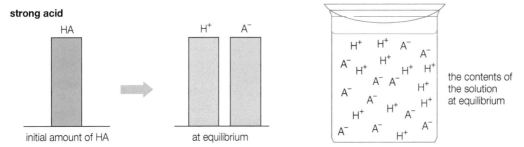

Figure 8.12 Graphical representation of the behaviour of a strong acid, HA, in aqueous solution

Weak acids are weak electrolytes and are only partially ionized in aqueous solution (Figure 8.13). The common weak acids are carbonic acid, H_2CO_3 and ethanoic acid, CH_3COOH.

> **Expert tip**
>
> All organic acids (unless otherwise stated) are weak in aqueous solution.

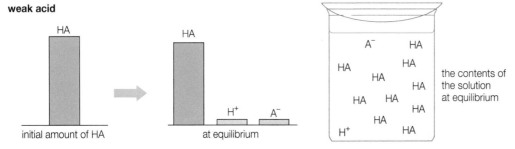

Figure 8.13 Graphical representation of the behaviour of a weak acid, HA, in aqueous solution

Strong bases (alkalis) are sodium, potassium and barium hydroxides. They are completely dissociated into ions in aqueous solution. Common weak bases include ammonia and the amines, for example, ethylamine. The molecules of weak bases will be in equilibrium with their ions.

The acid–base strengths of conjugate acid–base pairs are related (Figure 8.14). The stronger an acid, the weaker its conjugate base; the weaker an acid, the stronger its conjugate base. In every acid–base reaction, the position of the equilibrium favours the transfer of the proton (H^+) from the stronger acid to the stronger base.

The strengths of acids, bases and their respective conjugates can be measured and expressed in terms of K_a or pK_a. Table 8.3 shows selected acids and their conjugate bases arranged in order of their strength.

> **Expert tip**
>
> Magnesium and calcium hydroxides are classified as weak bases because of their low solubility in water.

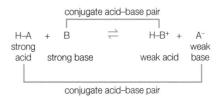

Figure 8.14 A conjugate acid–base pair

Table 8.3 Some common acids and conjugate bases in order of their strengths

Acid	Strength	Base	Strength
H_2SO_4	very strong	HSO_4^-	very weak
HCl		Cl^-	
HNO_3		NO_3^-	
H_3O^+	fairly strong	H_2O	weak
HSO_4^-		SO_4^{2-}	
CH_3COOH		CH_3COO^-	
H_2CO_3	weak	HCO_3^-	less weak
NH_4^+		NH_3	
HCO_3^-		CO_3^{2-}	
H_2O	very weak	OH^-	fairly strong

> **Common mistake**
>
> You must not include a one-way arrow when writing equations showing the dissociation or ionization of weak acids and weak bases; use a reversible reaction sign.

The terms 'strong' and 'weak' when applied to acids and bases are quite distinct from 'concentrated' and 'dilute', as illustrated in Table 8.4.

Table 8.4 Strong versus weak and concentrated versus dilute

	Concentrated solution	Dilute solution
Strong	5 mol dm^{-3} HCl	0.5 mo dm^{-3} HCl
Weak	10 mol dm^{-3} CH$_3$COOH	0.1 mol dm^{-3} CH$_3$COOH

> **Expert tip**
>
> In theory, there is no minimum pH as the concentration of the acid can increase indefinitely. In practice, the pH rarely falls below −0.5 as even strong acids do not dissociate fully at very high concentrations (there is not enough water). Strong acids can thus only be said to be fully dissociated if the solution is reasonably dilute (i.e. less than 0.1 mol dm^{-3}).

■ **QUICK CHECK QUESTIONS**

25 Hydrocyanic acid, HCN and chloric(VII) acid, HClO$_4$ are common acids used in the laboratory and have different strengths. HCN is a *weak acid* but HClO$_4$ is a *strong acid*. Both acids are *monoprotic*.

Define the terms *monoprotic*, *strong* and *weak acid* and write two equations to show the dissociation of HClO$_4$ and HCN in water.

26 Arrange the following in order of increasing pH:

2.0 mol dm^{-3} NH$_3$(aq), 0.020 mol dm^{-3} HCl(aq), 2.0 mol dm^{-3} NaOH(aq), 0.002 mol dm^{-3} CH$_3$CH$_2$COOH, 0.20 mol dm^{-3} H$_2$SO$_4$(aq), 0.020 mol dm^{-3} CH$_3$CH$_2$COOH, 0.0020 mol dm^{-3} NH$_3$(aq), 0.20 mol dm^{-3} HCl(aq)

Experiments to distinguish strong and weak acids and bases

Revised ☐

■ pH measurement (by a pH probe and meter)

A strong acid produces a higher concentration of hydrogen ions (H$^+$(aq)) in aqueous solution than a weak acid, with the same concentration at the same temperature. The pH of a strong acid will be lower than a weak acid. A strong base will have a higher pH in aqueous solution than a weak base with the same concentration at the same temperature.

■ Electrical conductivity measurement

Strong acids and strong bases in solution will give much higher readings on a conductivity meter than equimolar (equal concentrations) solution of weak acids or bases because of a much higher concentration of ions in solution. All the ions present in the solution contribute to the conductivity (which is measured as reciprocal of resistance).

■ Relative rates of reaction assessment

Since the concentration of hydrogen ions is much greater, the rate of reaction of strong acids with reactive metals, metal oxides, metal hydroxides, metal hydrogen carbonates and metal carbonates is greater than that of weak acids with the same concentration and temperature.

■ **QUICK CHECK QUESTIONS**

27 The pH values of three acidic solutions, **A**, **B** and **C**, are shown in the following table:

Solution	Acid	pH
A	HCl(aq)	2.00
B	HCl(aq)	4.00
C	CH$_3$COOH(aq)	4.00

Solutions **A** and **C** have the same acid concentration. Explain, by reference to both acids, why they have different pH values. Deduce by what factor the values of [H$^+$(aq)] in solutions **A** and **C** differ.

28 Potassium carbonate (of equal mass) is added to two separate solutions of sulfuric acid and chloropropanoic acid of the same concentration and temperature. State **one** similarity and **one** difference in the observations you could make.

> **Common mistake**
>
> Do not forget that the lower the pH, the higher the concentration of hydrogen ions (protons).

Factors controlling acid strength

The tendency of a substance to show acidic or basic characteristics in water can be correlated with its chemical structure. Acid character requires the presence of a highly polar H–X bond. Acidity is also favoured when the H–X bond is weak (Figure 8.15) and when the X^- ion is very stable (Figure 8.16).

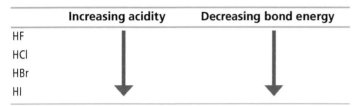

	Increasing acidity	**Decreasing bond energy**
HF		
HCl		
HBr		
HI		

Figure 8.15 Effect of the size of the halogen atom on the acidity of hydrides from group 17

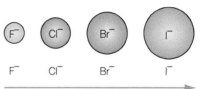

electron density decreasing; attraction for a proton decreasing; basicity decreasing.

Figure 8.16 Trends in properties of halide ions

For oxyacids with the same number of –OH groups and the same number of oxygen atoms, acid strength increases with increasing electronegativity of the central atom. For oxyacids with the same central atom, acid strength increases as the number of oxygen atoms attached to the central atom increases (Figure 8.17).

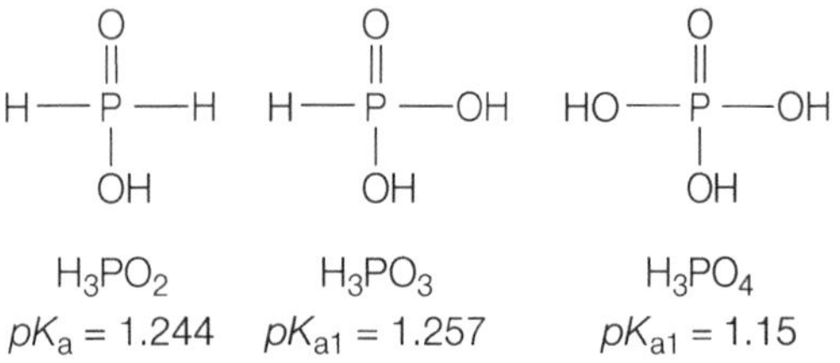

H_3PO_2 $pK_a = 1.244$
H_3PO_3 $pK_{a1} = 1.257$
H_3PO_4 $pK_{a1} = 1.15$

Figure 8.17 Acid strength increases from left to right

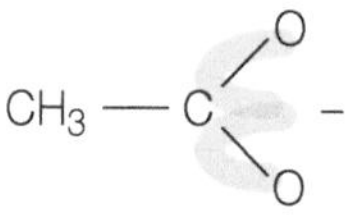

Figure 8.18 The delocalization of charge in the ethanoate ion

Carboxylic acid which are organic acids containing the –COOH group are the most important class of organic acids. The presence of delocalized pi bonding (resonance) in the conjugate base (Figure 8.18) is partially responsible for the acidity of these compounds.

8.5 Acid deposition

Essential idea: Increased industrialization has led to greater production of nitrogen and sulfur oxides, resulting in acid rain, which is damaging our environment. These problems can be reduced through collaboration with national and intergovernmental organizations.

Acid deposition

- Rain is naturally acidic because of dissolved CO_2 and has a pH of 5.6. **Acid deposition** has a pH below 5.6.
- Acid deposition is formed when nitrogen or sulfur oxides dissolve and react in water to form HNO_3, HNO_2, H_2SO_4 and H_2SO_3.
- Sources of the oxides of sulfur and nitrogen and the effects of acid deposition should be covered.

Rain is naturally slightly acidic because of the carbon dioxide (an acidic oxide) dissolving and reacting in the rain water to form carbonic acid according to the equation:

$$CO_2(g) + H_2O(l) \rightleftharpoons H_2CO_3(aq)$$

Carbonic acid is a weak acid that dissociates partially to produce H^+ ions which make the rain water slightly acidic with a pH of 5.6.

$$H_2CO_3(aq) \rightleftharpoons HCO_3^-(aq) + H^+(aq)$$

Key definition

Acid deposition – any process in which acidic substances (particles, gases and precipitation) leave the atmosphere and are deposited on the Earth.

Expert tip

H_2SO_3 is a much stronger acid than H_2CO_3 so the dissociation 'pulls' the equilibria to the right so making SO_2 more soluble than CO_2.

In addition to the carbonic acid present, rain water also contains naturally produced low levels of sulfur dioxide and nitrogen oxides.

Sulfur dioxide (SO_2) occurs naturally in the air from volcanic emissions and through a series of reactions with water produces sulfurous and sulfuric acids.

Atmospheric nitrogen(II) oxide (NO) is formed during lightning storms by the reaction of nitrogen and oxygen, the two predominant atmospheric gases. The energy for the reaction is provided by the electrical discharge inherent in the lightning. In air, NO is oxidized to nitrogen dioxide (NO_2), which in turn reacts with water to give nitric acid (HNO_3).

Acidic rain is rain with a pH of less than 5.6 as it contains nitric acid and sulfuric acid in addition to the carbonic acid.

Acid deposition (a form of secondary pollution) refers to the various processes by which acidic substances such as acidic gases (acidic oxides such as SO_2) or acidic precipitates (e.g. ammonium salts) leave the atmosphere.

There are two types of acid deposition:
- wet deposition: acid rain, fog, sleet and snow
- dry deposition: acidic gas molecules (e.g. SO_2) or acidic particles attached onto small airborne particles (e.g. ammonium salts) such as dust; this can include sulfur dioxide or nitrogen oxides.

Sources of nitrogen oxides:
- Combustion of fossil fuels inside internal combustion engines in cars and in furnaces of fossil fuel burning power stations as both produce high temperatures that cause the oxidation of nitrogen to nitrogen(II) oxide (nitrogen monoxide) and nitrogen dioxide. This conversion takes place in the both vehicle petrol and diesel engines, though the level of pollution is greater from diesel engines as their operating temperature is higher. This reaction also takes place in the jet engines of aircraft.
- Excess use of nitrogen containing fertilizers; excess fertilizer is decomposed by denitrifying bacteria, releasing nitrogen oxide into atmosphere.

Equations for the production of nitrogen oxides:

$$N_2(g) + O_2(g) \rightarrow 2NO(g)$$

$$2NO(g) + O_2(g) \rightarrow 2NO_2(g)$$

Two ways in which nitric(V) acid is formed in the atmosphere:
- $2NO_2(g) + H_2O(l) \rightarrow HNO_3(aq) + HNO_2(aq)$

  ($HNO_2(aq)$ = nitrous acid or nitric(III) acid)

- $4NO_2(g) + 2H_2O(l) + O_2(g) \rightarrow 4HNO_3(aq)$

Sources of sulfur oxides:
- Coal-burning power stations: coal is formed from decayed animal and plant material which contains protein which contain sulfur; also coal contains sulfur in the form of FeS_2, iron pyrites. When coal is burned, the sulfur in it oxidizes.
- Roasting of metal sulfides such as ZnS and Cu_2S:

  $$Cu_2S(s) + 2O_2(g) \rightarrow 2CuO(s) + SO_2(g)$$

Equations for the production of the oxides:

$$S(g) + O_2(g) \rightarrow SO_2(g)$$

$$2SO_2(g) + O_2(g) \rightarrow 2SO_3(g)$$

The formation of atmospheric sulfurous acid (sulfuric(IV) acid) and sulfuric(VI) acid:

$$SO_2(g) + H_2O(l) \rightarrow H_2SO_3(aq)$$

$$SO_3(g) + H_2O(l) \rightarrow H_2SO_4(aq)$$

Expert tip

You should make sure that you know that rain is naturally acidic due to the presence of CO_2, NO and SO_2. The predominant reason to give in an exam is the acidity arising from carbon dioxide in the air.

You should distinguish this natural acidity from that caused by the presence of nitrogen oxides and sulfur dioxide from human sources.

Expert tip

Be careful of the technically correct names for *nitrous acid* (HNO_2) and *sulfurous acid* (H_2SO_3). They are *nitric(III) acid* and *sulfuric(IV) acid*, respectively.

Look carefully at the oxidation number of nitrogen and sulfur in each case.

■ **QUICK CHECK QUESTION**

29 Explain why rainwater with a pH of 5.8 is not classified as 'acid deposition' even though its pH is less than 7.0.

Environmental effects of acid deposition

Revised ☐

Vegetation	• Increased soil acidity leaches important nutrients out of the top soil, (Ca^{2+}/K^+/ Mg^{2+} for example). • Mg^{2+} is necessary to make chlorophyll, so removal of this ion results in lowering rate of photosynthesis and reducing growth of plants and crop yields. • Yellowing and loss of leaves (linked to loss of Mg^{2+}). • Increased concentration of Al^{3+} in the soil which damages roots. • Stunted growth of crops. • Acidic fog is a particular problem for high-altitude forests. • Dry deposition (e.g. sulfur dioxide) blocks the pores in the leaves. • Thinning of tree tops.
Aquatic locations and life	• Increased levels of aluminum ions, Al^{3+}, dissolved from the soil by the acidic water kills fish as it reduces the effectiveness of gills. • A lot of fish, algae, insects, larvae and even plankton cannot survive in water below certain pH. • Eutrophication as a result of a high concentration of nitrate ions in the water.
Materials	• Corrosion of materials such as limestone, marble and dolomite ($CaCO_3.MgCO_3$). Wet deposition: $CaCO_3(s) + H_2SO_4(aq) \rightarrow CaSO_4(s) + H_2O(l) + CO_2(g)$ Wet deposition: $CaCO_3(s) + 2HNO_3(aq) \rightarrow Ca(NO_3)_2(aq) + H_2O(l) + CO_2(g)$ Dry deposition: $2CaCO_3(s) + 2SO_2(g) + O_2(g) \rightarrow 2CaSO_4(s) + H_2O(l) + CO_2(g)$ • Faster corrosion of iron and steel structures in buildings or bridges (remember rusting is an electrochemical process). Wet deposition: $Fe(s) + 2HNO_3(aq) \rightarrow Fe(NO_3)_2(aq) + H_2(g)$ Dry deposition: $Fe(s) + SO_2(g) + O_2(g) \rightarrow FeSO_4(s)$ • Removes protective oxide layer on aluminium: $Al_2O_3(s) + 6HNO_3(aq) \rightarrow 2Al(NO_3)_3(aq) + 3H_2O(l)$
Human health	• Irritation of mucuos membranes and lung tissue when breathing in fine droplets of acid rain. • Increased risk of respiratory illnesses such as asthma and bronchitis caused by, for instance, the fine sulfate and nitrate particles formed during reactions with acid rain which are able to penetrate the lungs. • Acidic water also dissolves and leaches potentially toxic ions such as Al^{3+} and Pb^{2+} that end up in water supplies and the food chain.

Methods to counteract the environmental effects of acid rain

Revised ☐

▪ Counteracting the effects of sulfur oxides

▪ Pre-combustion methods

Removal of sulfur from the fossil fuel; both from oil and coal.

This removal of sulfur is what was done so effectively in the refining of gasoline (petrol) and diesel for use in road vehicles.

- Hydrodesulfurization (HDS): catalytic removal of sulfur using hydrogen to form hydrogen sulfide which is then removed.
- If sulfur is present as a metal sulfide then the coal is crushed and mixed with water making the denser metal sulfide sink and easier to be removed.

■ Post-combustion methods

Removal of sulfur dioxide from fumes before they are released into atmosphere – flue-gas desulfurization (FGD):

■ *Fluidized bed combustion* – uses a new type of burner:
  □ Coal and calcium carbonate are pulverized and mixed together.
  □ Mixture is placed on a bed where the coal is combusted.
  □ Heat from the coal burning decomposes the limestone:

$$CaCO_3(s) \rightarrow CaO(s) + CO_2(g)$$

  □ Calcium oxide then combines with sulfur dioxide released from the burning coal.
  $$CaO(s) + SO_2(g) \rightarrow CaSO_3(s)$$

■ *Alkaline scrubbing* – here sulfur dioxide is passed through an alkaline solution (scrubbing = the gases are passed through a stream of a liquid containing a chemical that will react with the pollutant – an aqueous slurry of calcium carbonate, for instance).

Chemicals used in FGD are calcium carbonate, magnesium oxide, magnesium hydroxide. When calcium carbonate is used the equations are the same as those for the fluidized combustion bed.

$$CaCO_3(s) + SO_2(g) \rightarrow CaSO_3(s) + CO_2(g)$$

$$also\ MgO(s) + SO_2(g) \rightarrow MgSO_3(s)$$

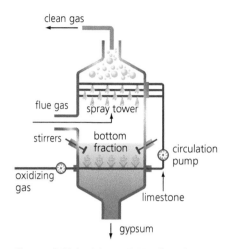

Figure 8.19 Inside an FGD cleaning tower; showing the counterflow of the gases and reactant spray

■ Counteracting the effects of nitrogen oxides

■ *Catalytic converters:*
$$2NO(g) + 2CO(g) \rightarrow 2CO_2(g) + N_2(g)$$

■ *Lean burn engines:* which are important in reducing emissions of not only CO and HC but also of NO; these engines ensure that the fuel : air ratio in the carburetor (where the fuel and air are mixed before combustion occurs) is close to 1:18 as opposed to 1:16 so that more complete combustion occurs.

■ *Recirculation of exhaust gases:* cooler exhaust gases from the car engine are recirculated around the engine to cool the engine. As production of nitrogen oxide is temperature dependent this cooling reduces NO_x production.

■ General methods of counteracting acid deposition in the environment

■ *Liming of lakes:* Adding powdered calcium carbonate (limestone), or calcium hydroxide (lime) to lakes can neutralize the acidity. It introduces calcium ions into the water and precipitates aluminium ions from solution.

■ *Switch to alternative methods for energy generation:* The move away from fossil fuels to alternative energy sources will help to reduce the levels of acidic gases in the atmosphere.

NATURE OF SCIENCE

The origins of the environmental problems surrounding acid deposition lie in our technological development – the use of fossil fuels in power generation and the extensive development of modern transport systems. However, technology can also be used to limit and remedy the problems that arise. The efforts and agreements needed to counteract the impact of acid deposition include a local element but must also be based on international collaboration.

■ QUICK CHECK QUESTIONS

30 Explain why limestone buildings and statues become eroded in regions of high acid deposition.

31 State the equations for the formation of nitrogen(II) oxide, and then nitrogen(IV) oxide, in an internal combustion engine. Describe, again with equations, how this oxide is converted to nitric(V) acid in the atmosphere.

9.1 Oxidation and reduction

Essential idea: Redox (reduction–oxidation) reactions play a key role in many chemical and biochemical processes.

Oxidation and reduction

- Oxidation and reduction can be considered in terms of oxygen gain/hydrogen loss, electron transfer or change in oxidation number.
- An oxidizing agent is reduced and a reducing agent is oxidized.
- Variable oxidation numbers exist for transition metals and for most main-group non-metals.
- The activity series ranks metals according to the ease with which they undergo oxidation.
- The Winkler Method can be used to measure biochemical oxygen demand (BOD), used as a measure of the degree of pollution in a water sample.

Definitions of oxidation and reduction

Oxidation is defined as gain of oxygen, loss of hydrogen, loss of electrons or increase in oxidation state. *Reduction* is defined as loss of oxygen, gain of hydrogen, gain of electrons or decrease in oxidation state. Remember OILRIG: **o**xidation **i**s **l**oss (of electrons); **r**eduction **i**s **g**ain (of electrons). Table 9.1 summarizes these different definitions.

Table 9.1 The focus of the different definitions of oxidation and reduction

Focus of definition	Oxidation	Reduction
Oxygen	Gain of oxygen	Loss of oxygen
Hydrogen	Loss of hydrogen	Gain of hydrogen
Electron transfer	Loss of electrons	Gain of electrons
Change in oxidation number	Increase in oxidation number	Decrease in oxidation number

Redox reactions involve the processes of reduction and oxidation, which must occur together. Redox reactions involve the transfer of electrons. Many chemical and biochemical reactions, such as respiration and photosynthesis, involve enzyme-controlled redox reactions.

Oxidizing and reducing agents

Oxidizing agents gain electrons; *reducing agents* lose electrons. Oxidizing agents cause a substance to undergo oxidation and reducing agents cause a substance to undergo reduction. An oxidizing agent undergoes reduction (decrease in oxidation state); a reducing agent undergoes oxidation (an increase in oxidation state). The actions of an oxidizing agent and a reducing agent during a redox reaction are summed up in Table 9.2.

Table 9.2 The roles of an oxidizing agent and a reducing agent in a redox reaction

Oxidizing agent	Reducing agent
Gains electrons	Loses electrons
Oxidizes another reactant	Reduces another reactant
Is itself reduced during the reaction	Is itself oxidized during the reaction

Common oxidizing agents are the halogens, ozone, manganate(VII) ions, dichromate(VI) ions, oxyacids (such as concentrated nitric(V) acid) and hydrogen peroxide. Common reducing agents are hydrogen, carbon, carbon monoxide, hydrogen sulfide, reactive metals (high up in the activity series) and sulfur dioxide.

Oxidizing agents may be in equilibrium with reducing agents. This is known as a redox pair and is analogous to an acid–base conjugate pair (transfer of protons) (Figure 9.1). *Strong* oxidizing agents, e.g. F_2, have a *weak* conjugate reducing agent, F^-. *Strong* reducing agents, e.g. Na, have a *weak* conjugate oxidizing agent, Na^+. This relationship is shown in the activity series of metals and non-metals.

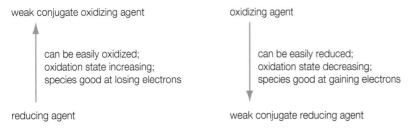

Figure 9.1 The relationship between oxidizing and reducing agents

A *disproportionation* reaction is a redox reaction in which both oxidation and reduction of the same species occurs. Copper(I) compounds often undergo disproportionation, for example:

$$\text{oxidation}$$
$$Cu_2O(s) + H_2SO_4(aq) \longrightarrow Cu(s) + CuSO_4(aq) + H_2O(l)$$
$$\text{reduction}$$

■ Redox equations

All redox reactions can be described by two half-equations. One half-equation describes the oxidation process and the other half-equation describes reduction. Hydrogen peroxide can act as both an oxidizing agent and a reducing agent. It acts as a reducing agent when in the presence of a more powerful oxidizing agent. (This can be predicted by the use of standard electrode potentials which are measures of oxidizing and reducing strength or power). Table 9.3 shows some important half-equations for reactions in aqueous solution.

Table 9.3 Important half-equations for selected redox reactions in aqueous solution

Reduction half-equations (oxidizing agents)	Oxidation half-equations (reducing agents)
$MnO_4^- + 8H^+ + 5e^- \rightarrow Mn^{2+} + 4H_2O$	$C_2O_4^{2-} \rightarrow 2CO_2 + 2e^-$
$Cr_2O_7^{2-} + 14H^+ + 6e^- \rightarrow 2Cr^{3+} + 7H_2O$	$2S_2O_3^{2-} \rightarrow S_4O_6^{2-} + 2e^-$
$I_2 + 2e^- \rightarrow 2I^-$	$H_2O_2 \rightarrow O_2 + 2H^+ + 2e^-$
$Fe^{3+} + e^- \rightarrow Fe^{2+}$	$2I^- \rightarrow I_2 + 2e^-$
$H_2O_2 + 2H^+ + 2e^- \rightarrow 2H_2O$	$Fe^{2+} \rightarrow Fe^{3+} + e^-$

The two half-equations are then added to form the overall redox or ionic equation. The number of electrons has to be adjusted so the number of electrons produced equals the number of electrons consumed, for example:

Oxidation half-equation:
$$C_2O_4^{2-} \rightarrow 2CO_2 + 2e^-$$
Reduction half-equation:
$$2Fe^{3+} + 2e^- \rightarrow 2Fe^{2+}$$
Ionic equation:
$$2Fe^{3+} + C_2O_4^{2-} \rightarrow 2Fe^{2+} + 2CO_2$$

■ **QUICK CHECK QUESTIONS**

1 Hydrogen peroxide spontaneously decomposes into water and oxygen. Write a balanced equation and deduce the oxidation numbers of oxygen. Define the term disproportionation.

2 Hydrazine (N_2H_4) and dinitrogen tetroxide (N_2O_4) form a self-igniting mixture that has been used as a rocket propellant. The reaction products are nitrogen and water. Write a balanced chemical equation for this reaction and identify the substances being oxidized and reduced.

3 Ammonia burns in oxygen to form nitrogen monoxide and water. Write a balanced chemical equation for this reaction and identify the substances being oxidized and reduced.

Expert tip

The terms oxidizing and reducing agent are relative terms. A chemical normally called an oxidizing agent will itself be oxidized if it meets a more powerful oxidizing agent.

■ **QUICK CHECK QUESTION**

4   Insert electrons on the appropriate side of the following half-equations to balance and complete them, so that the electrical charges on both sides are equal and cancel. For each completed half-equation describe the process as oxidation or reduction.

$Li \rightarrow Li^+$

$H_2 \rightarrow 2H^+$

$O \rightarrow O^{2-}$

$Cu^+ \rightarrow Cu^{2+}$

$Cr^{3+} \rightarrow Cr^{2+}$

$Fe^{3+} \rightarrow Fe^{2+}$

$Al \rightarrow Al^{3+}$

$Cl^- \rightarrow Cl_2$

▨ Oxidation numbers

Oxidation numbers (or oxidation states) are a method developed by chemists to keep track of the electrons transferred during a redox reaction. The 'rules' are based on the assumption that all compounds are ionic and that the more electronegative atom has 'control' over the electrons in a bond (Figure 9.2).

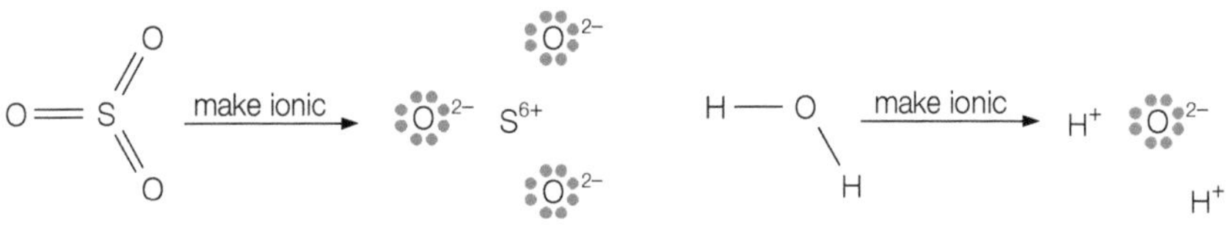

Figure 9.2 The ionic formulations of the sulfur trioxide and water molecules

■ Rules for working out oxidation numbers

■   The oxidation number of an atom in an element is zero.

■   Element–same element (homopolar) bonds are ignored for the purpose of calculating oxidation numbers.

■   In simple ions the oxidation number of the element is equal to the charge on the ion.

■   The sum of the oxidation numbers in a compound is zero.

■   The sum of the oxidation numbers for all the elements an oxyanion is equal to the charge on the ion.

■   Some elements have fixed oxidation numbers in all (or most) of their compounds.

Metals		Non-metals	
Group 1 metals	+1	Hydrogen (except in metal hydrides), H^-	+1
Group 2 metals	+2	Fluorine	−1
Aluminium	+3	Oxygen (except in peroxides and compounds with fluorine)	−2
		Chlorine (except in compounds with oxygen and fluorine)	−1

An increase in oxidation number is defined as oxidation; a decrease in oxidation number is defined as reduction. Figure 9.3 shows that movement up the number line involves the loss of electrons and a shift to a more positive oxidation number. Movement down the diagram involves a gain of electrons and a shift to a less positive or more negative oxidation number. This is reduction.

Expert tip

An oxidation number is written sign first; charge is written number first, then sign.

Expert tip

The terms 'oxidation number' and 'oxidation state' may be used interchangeably in the IB Chemistry examination.

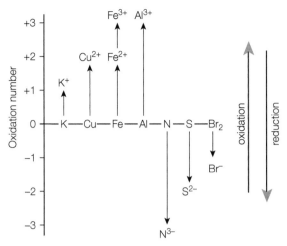

Figure 9.3 Oxidation numbers of atoms and simple ions. Note that the oxidation number for the atoms in the element is zero

---

■ **QUICK CHECK QUESTIONS**

5   Deduce the oxidation numbers of nitrogen in the following species:

NH_2^-, N_2, NH_3, N_2H_4, $N_2H_5^+$, NO, N_2O, N_2O_3, N_2O_5, NO_2^-, NO_3^-, N^{3-}, HNO_3, HNO_2, HN_3, N_2H_2

6   The structure of the thiosulfate ion is shown below. Explain why the oxidation number of the central sulfur atom is +4 and the terminal sulfur is zero.

$$:\overset{\displaystyle :S:}{\underset{\displaystyle}{\|}}$$

7   Outline the difference between 'Cr^{6+}' and the +6 oxidation state of chromium in CrO_4^{2-}.

8   Deduce the oxidation numbers of all the elements in the following compounds:

$POCl_3$, $NaHC_2O_4$, Na_2O, Na_2O_2, Cl_2O_7, ICl, Ca_3N_2, KH

NATURE OF SCIENCE

Historically, oxidation was defined as gain of oxygen or loss of hydrogen. Reduction was defined as loss of oxygen or gain of hydrogen. These two definitions show patterns among reactions, but they are rather restrictive. Only reactions involving hydrogen or oxygen are covered. The definitions were broadened later (after chemical bonding theories were developed) to include more reactions. Oxidation is the loss of electrons and reduction is the gain of electrons. The definitions of redox were further broadened with the development of oxidation numbers to include reactions involving covalent compounds.

■ **QUICK CHECK QUESTION**

9   a Deduce the balanced equation for the reaction between potassium and molecular oxygen.

   b State the electron configurations of the reactants and product particles.

   c Explain whether oxygen is oxidized or reduced in terms of electron transfer and oxidation numbers.

■ Rules for common inorganic names

Oxidation numbers are also used in the naming of inorganic compounds, especially of transition metals, and compounds of elements, such as sulfur or chlorine, that can be oxidized or reduced to varying degrees. This system is known as Stock notation.

For example, sulfur has different oxidation numbers depending on which other elements it is bonded to. Sulfur is given the negative oxidation state when it is the more electronegative element.

H_2S	S	SCl_2	SO_2	H_2SO_4
(−2)	(0)	(+2)	(+4)	(+6)

■ The ending *–ide* shows that a compound contains two elements. The more electronegative element is placed at the end.
For example, sulfur(II) chloride, SCl_2 and sulfur(IV) oxide, SO_2.
■ The Roman numbers in the name indicates the oxidation numbers of the element which has variable oxidation states.
For example, iron(II) sulfate, $FeSO_4$, and iron(III) sulfate, $Fe_2(SO_4)_3$.
■ The traditional names of oxyacids end in *–ic* or *–ous*, for example, sulfurous acid, H_2SO_3, and sulfuric acid, H_2SO_4. However, the more systematic names are sulfuric(IV) and sulfuric(VI) acids, where +4 and +6 are the oxidation states of sulfur.

■ QUICK CHECK QUESTION

10 a State the Stock names of the following compounds:

K_2SO_3, Na_2SO_4, $Fe(NO_3)_3$, $Fe(NO_3)_2$, $CuSO_4$, Cu_2O, Cr_2O_3, Cu_2SO_4, Mn_2O_7, MnO_2

b State the formulas of sodium chlorate(I), iron(III) oxide, caesium nitrate(V), phosphorus(III) chloride, sulfur(IV) chloride and sodium chlorate(V).

Expert tip

Many chemical reactions are redox reactions and involve electron transfer. Acid–base reactions involve the transfer of protons (Brønsted–Lowry theory) or the formation of dative bonds (Lewis theory). They are *not* redox reactions, and hence do not involve a change in oxidation number in any of the species in the reaction.

Transition metals also form complex ions. The oxidation number of the metal can be determined by taking all the charges of the surrounding small molecules or ions (ligands) and the overall charge of the complex ions.

The oxidation number of copper in $[CuCl_4]^{2-}$ is +2 since the complex consists of 4 chloride ions bonded to a central copper(II) ion: Cu^{2+} $4Cl^-$. The oxidation number of copper in $[Cu(H_2O)_6]^{2+}$ is also +2 since the six water molecules are neutral and the charge on the complex is due to the copper(II) ion.

■ QUICK CHECK QUESTIONS

11 Deduce the oxidation states of transition metals in the following complexes and complex ions: $[Cr(H_2O)_6]^{3+}$, $[Cu(NH_3)_4(H_2O)_2]^{2+}$, $[Ni(NH_3)_2Cl_2]$ and $[Ag(NH_3)_2Cl]$.

12 The conversion of iron(II) iodide to iron(III) iodide can be used to show the definitions of the terms oxidation and reduction. Write an equation for this reaction and deduce what is oxidized and reduced, by reference to both electron transfer and oxidation numbers.

13 Explain why the following reaction is not a redox reaction:

$BaCl_2(aq) + H_2SO_4(aq) \rightarrow BaSO_4(s) + 2HCl(aq)$

State the type of reaction.

Activity series

Metals act as reducing agents and their reducing powers can be compared by replacement (displacement) reactions, where a more reactive metal replaces a less reactive metal in the form of its ion. The resulting order is called the activity series (Table 9.4). Those metals that lose electrons readily to form positive ions are powerful reducing agents and placed high in the activity series.

Table 9.4 The activity series for selected metals

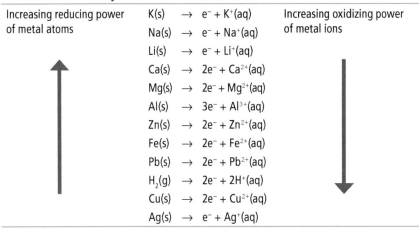

Increasing reducing power of metal atoms		Increasing oxidizing power of metal ions
	$K(s) \rightarrow e^- + K^+(aq)$	
	$Na(s) \rightarrow e^- + Na^+(aq)$	
	$Li(s) \rightarrow e^- + Li^+(aq)$	
	$Ca(s) \rightarrow 2e^- + Ca^{2+}(aq)$	
	$Mg(s) \rightarrow 2e^- + Mg^{2+}(aq)$	
	$Al(s) \rightarrow 3e^- + Al^{3+}(aq)$	
	$Zn(s) \rightarrow 2e^- + Zn^{2+}(aq)$	
	$Fe(s) \rightarrow 2e^- + Fe^{2+}(aq)$	
	$Pb(s) \rightarrow 2e^- + Pb^{2+}(aq)$	
	$H_2(g) \rightarrow 2e^- + 2H^+(aq)$	
	$Cu(s) \rightarrow 2e^- + Cu^{2+}(aq)$	
	$Ag(s) \rightarrow e^- + Ag^+(aq)$	

The ions of metals low in the activity series are good oxidizing agents because they readily gain electrons. Therefore, any metal will displace from aqueous solution ions of a metal below it in the activity series, for example,

$Zn(s) + Al^{3+}(aq) \rightarrow$ No reaction

$Zn(s) + Cu^{2+}(aq) \rightarrow Zn^{2+}(aq) + Cu(s)$

During this reaction the blue solution turns colourless (provided the zinc is in excess) and a brown deposit is formed.

All metals above hydrogen in the activity series can displace hydrogen ions from dilute acids to form hydrogen gas, for example,

$Mg(s) + 2H^+(aq) \rightarrow Mg^{2+}(aq) + H_2(g)$

Metals below hydrogen in the activity series cannot displace hydrogen ions from dilute acids to form hydrogen gas:

$Cu(s) + 2H^+(aq) \rightarrow$ No reaction

■ QUICK CHECK QUESTION

14 A part of the activity series of metals, in order of decreasing reactivity, is shown below.

calcium, magnesium, zinc, iron, lead, copper, silver, gold

a If a piece of lead metal were placed in separate solutions of silver(I) nitrate and magnesium nitrate determine which solution would undergo reaction.

b State the type of chemical change taking place in the lead and write the half-equation for this change.

c State, giving a reason, what visible change would take place in the aqueous solutions.

Common mistake

It is wrong to think that all metals react with dilute acids. Metals below hydrogen in the activity series do not react. These metals include copper, gold and silver. Aluminium should react with dilute acid but is covered in a thin and unreactive oxide layer, which protects the reactive metal underneath.

Replacement reactions can occur with non-metals like the halogens. Here, a more reactive halogen will displace a less reactive halogen from its simple ions. The electronegativity and oxidizing power of the halogens both decrease down the group as the atomic radius increases and attraction for electrons decreases. This means a higher halogen will displace a lower halogen from its salts, e.g. chlorine will oxidize iodide ions to iodine molecules and this may be detected by a colour change in the solution.

e.g. $Cl_2(aq) + 2I^-(aq) \rightarrow 2Cl^-(aq) + I_2(aq)$

     pale   brown
   yellow-green

Balancing redox reactions

Half-equation method

Redox reactions involve the transfer of electrons from the oxidizing agent to the reducing agent. For example, the reaction between manganate(VII) ions and ethanedioate ions in acidic aqueous solution involves the transfer of five electrons to each manganate(VII) ion (the oxidizing agent) from ethanedioate ions (the reducing agent).

$$16H^+(aq) + 2MnO_4^-(aq) + 5C_2O_4^{2-}(aq) \rightarrow 2Mn^{2+}(aq) + 8H_2O(l) + 10CO_2(g)$$

Balancing redox equations by inspection ('trial and error') is quite difficult, as you must take into account not only the mass balance but also the charge balance in the equation.

For some redox reactions this is relatively simple and involves ensuring the number of electrons in each half-equation are equal. Consider the oxidation of magnesium by gold(I) ions. Write the two half-equations, then double the gold half-equation so that both half-equations involve two electrons and then simply add them together:

$$2Au^+(aq) + 2e^- \rightarrow 2Au(s)$$

$$Mg(s) \rightarrow Mg^{2+}(aq) + 2e^-$$

to give

$$2Au^+(aq) + Mg(s) \rightarrow 2Au(s) + Mg^{2+}(aq)$$

For more complicated redox reactions a set of rules, called the half-reaction method, has been developed. The following rules work for reactions performed in acidic or in neutral solution.

1   Write down the two incomplete half-reactions.

   $$MnO_4^-(aq) \rightarrow Mn^{2+}(aq)$$
   $$C_2O_4^{2-}(aq) \rightarrow CO_2(g)$$

2   Balance each half reaction:

   a   First, balance elements other than hydrogen and oxygen.

   $$MnO_4^-(aq) \rightarrow Mn^{2+}(aq)$$
   $$C_2O_4^{2-}(aq) \rightarrow 2CO_2(g)$$

   b   Then balance oxygen atoms by adding water molecules on the side lacking in oxygen atoms.

   $$MnO_4^-(aq) \rightarrow Mn^{2+}(aq) + 4H_2O(l)$$
   $$C_2O_4^{2-}(aq) \rightarrow 2CO_2(g)$$

   c   Then balance the hydrogen by adding hydrogen ions for acidic solutions.

   $$8H^+(aq) + MnO_4^-(aq) \rightarrow Mn^{2+}(aq) + 4H_2O(l)$$
   $$C_2O_4^{2-}(aq) \rightarrow 2CO_2(g)$$

   d   Finish by balancing the charge by adding electrons.

3   For the manganate(VII) half-reaction, note that there is a charge of 7+ on the left and 2+ on the right. Hence 5 electrons need to be added to the left:

   $$5e^- + 8H^+(aq) + MnO_4^-(aq) \rightarrow Mn^{2+}(aq) + 4H_2O(l)$$

4   In the ethaneidoate half-reaction, there is a 2 – charge on the left and zero charge on the right, so we need to add two electrons to the products side:

   $$C_2O_4^{2-}(aq) \rightarrow 2CO_2(g) + 2e^-$$

5   Multiply each half-reaction by the appropriate factors so that the number of electrons gained equals electrons lost.

6   To balance the 5 electrons for manganate(VII) and 2 electrons for ethanedioate, we need 10 electrons for both.

■ QUICK CHECK QUESTION

15   Sea water contains approximately 65 000 tonnes of bromine. This bromine is removed from the sea water by displacement of the bromide ions using chlorine gas. Write an overall equation and explain why it is a redox reaction and what is observed during the reaction.

7 Multiplying gives:

$$10e^- + 16H^+(aq) + 2MnO_4^-(aq) \rightarrow 2Mn^{2+}(aq) + 8H_2O(l)$$

$$5C_2O_4^{2-}(aq) \rightarrow 10CO_2(g) + 10e^-$$

Now add the half-equations and simplify.

$$16H^+(aq) + 2MnO_4^-(aq) + 5C_2O_4^{2-}(aq) \rightarrow 2Mn^{2+}(aq) + 8H_2O(l) + 10CO_2(g)$$

8 The equation is now balanced. Note that all of the electrons have cancelled out.

Expert tip

These steps for balancing half-equations in acidic aqueous solution must be followed in sequence.

■ QUICK CHECK QUESTIONS

16 Deduce the ionic equation for the reaction between acidified dichromate(VI) ions and sulfur dioxide solution, $SO_2(aq)$. The products are chromium(III) ions and sulfate(VI) ions.

17 Titanium(III) ions, $Ti^{3+}(aq)$ can be reduced to titanium(II) ions, by silver which forms silver(I) ions. Deduce the half-equations and state the ionic equation.

18 $2Cu^{2+}(aq) + 4I^-(aq) \rightarrow 2CuI(s) + I_2(aq)$

Explain why this is a redox reaction. Identify the oxidizing and reducing agents.

19 Balance the following disproportionation reaction which occurs in acidic aqueous solution:

$I_3^- \rightarrow IO_3^- + I^-$.

■ Oxidation number method

The oxidation number method for balancing redox equations is based on the principle that the total increase in oxidation state equals the total decrease in oxidation state.

$$HNO_3(aq) + H_3AsO_3(aq) \rightarrow NO(g) + H_3AsO_4(aq) + H_2O(l)$$

The nitrogen atoms change from +5 to +2, so they are reduced. The arsenic atoms, which change from +3 to +5, are oxidized.

Determine the net increase in oxidation number for the element that is oxidized and the net decrease in oxidation number for the element that is reduced.

As +3 to +5 net change = +2

N +5 to +2 net change = −3

Determine a ratio of oxidized to reduced atoms that would yield a net increase in oxidation number equal to the net decrease in oxidation number.

Arsenic atoms have a net increase in oxidation number of +6. (Six electrons would be lost by three arsenic atoms.) Two nitrogen atoms would have a net decrease of −6. (Two nitrogen atoms would gain six electrons.) Hence the ratio of arsenic atoms to nitrogen atoms is 3:2.

Add coefficients (numbers) to the formulas which contain the elements whose oxidation number is changing.

$$2HNO_3(aq) + 3H_3AsO_3(aq) \rightarrow NO(g) + H_3AsO_4(aq) + H_2O(l)$$

Balance the rest of the equation by inspection (trial and error):

$$2HNO_3(aq) + 3H_3AsO_3(aq) \rightarrow 2NO(g) + 3H_3AsO_4(aq) + H_2O(l)$$

■ Redox titrations

Redox titrations involve calculations similar to those described for acid–base titrations (Topic 1 Stoichiometric relationships). Indicators are sometimes not needed since there may be a colour change involving the oxidizing and reducing agents. Tables 9.5 and 9.6 summarize the reactions of selected oxidizing and reducing agents.

Table 9.5 Common oxidizing agents

Oxidizing agents	Reduced to
MnO_4^- manganate(VII) ions purple	Mn^{2+} (under acidic conditions) manganese(II) ions pale pink (at high concentration)/colourless (at low concentration)
MnO_4^- manganate(VII) ions purple	MnO_2 (under neutral/alkaline conditions) manganese(IV) oxide brown precipitate
$Cr_2O_7^{2-}$ dichromate(VI) ions orange	Cr^{3+} chromium(III) ions green
IO_3^- iodate(V) ions colourless	I_2 iodine brown (in aqueous solution)
$I_2(aq)$ iodine brown	I^- iodide ions colourless
Fe^{3+} iron(III) ions pale yellow/brown	Fe^{2+} iron(II) ions pale green (but oxidized by oxygen in air)
H_2O_2 hydrogen peroxide colourless	H_2O water colourless
NO_2^- nitrate(III) ions colourless	$NO(g)$ nitrogen(II) oxide (colourless) NO can be easily oxidized to nitrogen(IV) oxide, NO_2 (brown gas)

Table 9.6 Common reducing agents

Reducing agents	Oxidized to
$C_2O_4^{2-}$ ethanedioate ions	CO_2 carbon dioxide gas
$S_2O_3^{2-}$ thiosulfate ions	$S_4O_6^{2-}$ tetrathionate ions
H_2O_2 hydrogen peroxide	O_2 oxygen gas
NO_2^- nitrate(III) ions	NO_3^- nitrate(V) ions
I^- iodide ions (colourless)	$I_2(aq)$ iodine (brown)
Fe^{2+} iron(II) ions (light green)	Fe^{3+} iron(III) ions (pale yellow/brown)

Redox titrations can be used for two main reasons:
- to find the concentration of a solution
- to establish the stoichiometry of a redox equation and hence to suggest an equation for the reaction.

If the concentration of a solution B is unknown and is to be found by a redox titration, then a standard solution of A is required and we need to know the amount (in mol) of solution B that reacts with one mole of solution A. The balanced equation allows us to calculate the amount of B from the amount of A which allows the concentration of solution B to be calculated.

If the stoichiometry of the reaction is to be found, then the concentration of both solution A and solution B must be known. The volumes from the titration allow us to calculate the amount of solution B that reacts with one mole of solution A.

Many redox titrations use acidified aqueous potassium dichromate(VI), $K_2Cr_2O_7$, and potassium manganate(VII), $KMnO_4$, as oxidizing agents. These oxidizing agents can be used to oxidize iron(II) ions to iron(III) ions, or ethanedioate ions to carbon dioxide. The solution is acidified using sulfuric acid: protons are required for the redox reactions.

Iodine is a weak oxidizing agent but is often used to oxidize thiosulfate ions to tetrathionate ions. Small amounts of iodine can be detected by using starch as an 'indicator'. The colour change is from colourless to dark blue-black.

$$I_2(aq) + 2S_2O_3^{2-}(aq) \rightarrow 2I^-(aq) + S_4O_6^{2-}(aq)$$

■ QUICK CHECK QUESTIONS

20 In an experiment, iron ore was converted to an iron(II) aqueous solution by reaction with dilute acid. The iron(II) ion solution was made up to 250.00 cm³ in a volumetric flask. Iron(II) ions react with dichromate(VI) ions:

$$6Fe^{2+}(aq) + Cr_2O_7^{2-}(aq) + 14H^+(aq) \rightarrow 6Fe^{3+}(aq) + 2Cr^{3+}(aq) + 7H_2O(l)$$

25.00 cm³ of the $Fe^{2+}(aq)$ solution required 21.50 cm³ of aqueous dichromate(VI) ions for complete reaction. The concentration of the dichromate(VI) solution used was 0.100 mol dm⁻³.

Calculate the amount of $Fe^{2+}(aq)$ present in the volumetric flask and hence determine the percentage by mass of iron present in the 12.25 g of the sample of iron ore.

21 In an experiment to determine the concentration of sodium chlorate(I), NaOCl, in household bleach, 25.00 cm³ of the bleach was diluted to 250.00 cm³ in a volumetric flask. 20.00 cm³ of the diluted solution was pipetted into a conical flask containing excess hydrochloric acid and potassium iodide solution.

Sodium chlorate(I) reacts with the hydrochloric acid to form chloric(I) acid, HOCl:

$$NaOCl(aq) + HCl(aq) \rightarrow HOCl(aq) + NaCl(aq)$$

The chloric(I) acid then reacts with the iodide ions in the presence of acid to form iodine and chloride ions. The iodine produced required 16.85 cm³ of 0.152 mol dm⁻³ sodium thiosulfate solution (using starch as an indicator) for complete reaction as shown in the following equation:

$$I_2(aq) + 2Na_2S_2O_3(aq) \rightarrow 2NaI(aq) + Na_2S_4O_6(aq)$$

a Construct a balanced equation for the reaction between chloric(I) acid and iodide ions.

b Calculate the amount (mol) of iodine produced in 20.00 cm³ of the diluted bleach solution.

c Calculate the concentration of sodium hypochlorite, in g per dm³, in the original bleach solution.

▨ Winkler method

Biochemical oxygen demand (BOD) is the amount of oxygen that micro-organisms consume while decomposing organic matter:

$$\text{organic matter} + O_2 \rightarrow H_2O + CO_2$$

The Winkler titration method is an iodometric titration, in which the amount of dissolved oxygen (related to the BOD) in the sample is determined indirectly via iodine.

A solution of manganese(II) ions is added to the sample, followed by strong alkali in glass-stoppered bottles. Any dissolved oxygen present in the water sample rapidly oxidizes an equivalent amount of the manganese(II) hydroxide precipitate to manganese(IV) oxide. In the presence of iodide ions in an acidic solution, the oxidized manganese (+4) is converted back to the +2 oxidation state, with the production of iodine equivalent to the original dissolved content.

The reactions are as follows:

Manganese(II) sulfate reacts with potassium hydroxide to give a white precipitate of manganese(II) hydroxide. In the presence of oxygen, brown manganese(IV) basic oxide is formed.

$$MnSO_4(aq) + 2KOH(aq) \rightarrow Mn(OH)_2(s) + K_2SO_4(aq)$$

$$2Mn(OH)_2(s) + O_2(aq) \rightarrow 2MnO(OH)_2(s)$$

Addition of sulfuric acid dissolves the brown manganese(IV) basic oxide which reacts instantly with iodide ions to produce iodine molecules.

$$2MnO(OH)_2(s) + 4H_2SO_4(aq) \rightarrow 2Mn(SO_4)_2(aq) + 6H_2O(l)$$

$$2Mn(SO_4)_2(aq) + 4KI(aq) \rightarrow 2MnSO_4(aq) + 2K_2SO_4(aq) + 2I_2(aq)$$

In effect, oxygen oxidizes Mn^{2+} to Mn^{4+} and the Mn^{4+} oxidizes I^- to I_2. Iodine is then determined via titration with sodium thiosulfate with starch as an end point 'indicator' (blue-black to colourless).

$$4Na_2S_2O_3(aq) + 2I_2(aq) \rightarrow 2Na_2S_4O_6(aq) + 4NaI(aq)$$

From the above stoichiometric equations we can find that four moles of thiosulfate are titrated for each mole of molecular oxygen (O_2).

■ **QUICK CHECK QUESTION**

22 In an experiment, the amount of dissolved oxygen from a sample of river water was tested using the Winkler method. The water was collected in two BOD bottles. One BOD bottle was incubated in a BOD incubator for 5 days at 25 °C while the concentration of dissolved oxygen in the second BOD bottle was tested on the first day. After 5 days, the water from the incubated bottle was then tested using the Winkler method.

It was found that 24.50 cm³ of the water from the incubated bottle at 25 °C required 12.50 cm³ of 0.00200 mol dm⁻³ sodium thiosulfate solution to react with the iodine produced. Using the following equations calculate the concentration of dissolved oxygen in parts per million (ppm) in the incubated bottle (ppm is equivalent to mg dm⁻³).

$$4Mn(OH)_2(aq) + O_2(g) + 2H_2O\ (l) \rightarrow 4Mn(OH)_3(s)$$
$$2KI(aq) + 2Mn(OH)_3(s) \rightarrow I_2(aq) + 2Mn(OH)_2(aq) + 2KOH(aq)$$
$$2S_2O_3{}^{2-}(aq) + I_2(aq) \rightarrow S_4O_6{}^{2-}(aq) + 2I^-(aq)$$

9.2 Electrochemical cells

Essential idea: Voltaic cells convert chemical energy to electrical energy and electrolytic cells convert electrical energy to chemical energy.

Electrochemical cells

- Voltaic cells convert energy from spontaneous, exothermic chemical processes to electrical energy.
- Oxidation occurs at the **anode** (negative electrode) and reduction occurs at the **cathode** (positive electrode) in a voltaic cell.
- Electrolytic cells convert electrical energy to chemical energy, by bringing about non-spontaneous processes.
- Oxidation occurs at the anode (positive electrode) and reduction occurs at the cathode (negative electrode) in an electrolytic cell.

Key definitions

Anode – the half-cell/electrode where *oxidation* occurs.

Cathode – the electrode/half-cell where *reduction* occurs. Electrons always flow from the anode to the cathode.

■ Voltaic (galvanic) cells

Voltaic or galvanic cells use the electron flow in a spontaneous redox reaction to perform useful work. Such cells are used as batteries, pH meters and fuel cells. The cell requires that the oxidation and reduction half-reactions are kept separate but are connected by a wire and by a salt bridge.

Electrons will flow through that wire creating an electrical current. The salt bridge (Figure 9.4) allows the movement of unreactive spectator ions in solution to maintain charge neutrality (electroneutrality) in each half-cell.

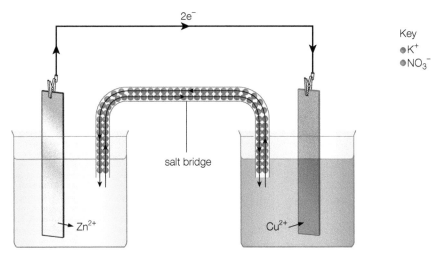

Figure 9.4 Flow of ions in a salt bridge containing potassium and nitrate(V) ions

The anode is where oxidation takes place (electrons are released on to the electrode surface, so it is the negative electrode). Electrons move from the anode to the cathode (where reduction takes place – electrons are attracted to the 'positive' electrode).

If a strip of zinc metal is placed in an aqueous solution of copper(II) sulfate solution, copper is deposited on the zinc and the reacts and dissolves by forming zinc ions. This cell is known as the Daniell cell (Figure 9.5).

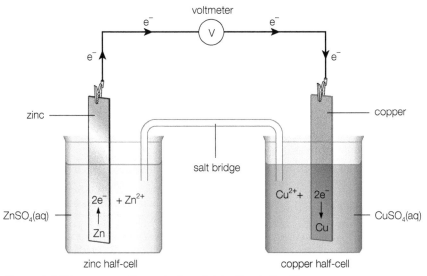

Figure 9.5 The copper/zinc voltaic cell (the Daniell cell)

Zinc atoms are spontaneously oxidized to zinc ions (Figure 9.6a) by copper(II) ions (oxidizing agent). The copper(II) ions are spontaneously reduced to copper atoms by zinc atoms (reducing agent) (Figure 9.6b). The entire process is exothermic and spontaneous (Figure 9.6).

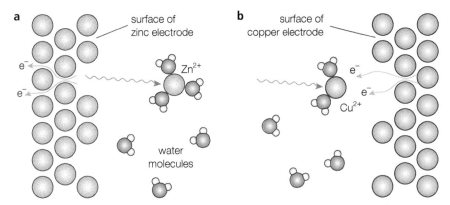

Figures 9.6 a Zinc atoms forming hydrated zinc ions at the surface of the zinc electrode of a Daniell cell; **b** hydrated copper(II) ions converted to copper atoms at the surface of the copper electrode

This voltaic cell consists of an oxidation half-reaction at the anode:

$$Zn(s) \rightarrow Zn^{2+}(aq) + 2e^-$$

and a reduction half-reaction at the cathode:

$$Cu^{2+}(aq) + 2e^- \rightarrow Cu(s)$$

A salt bridge allows cations to move from anode to cathode (to replace the loss of positive charge) and anions to move from cathode to anode (to neutralize the excess positive charge).

Cell diagrams

There is a shorthand description for describing the electrodes and electrolytes present in a voltaic cell. The standard EMF ('voltage') of a cell, $E^\ominus_{cell}$, is written next to the diagram. In a cell diagram the single line | represents a boundary between two phases (solid and solution), while the double lines || represent a salt bridge.

The cell diagram for the spontaneous reaction that occurs when a copper half-cell is connected to a zinc half-cell (Daniell cell) is:

$$Zn(s) \mid Zn^{2+}(aq) \mid\mid Cu^{2+}(aq) \mid Cu(s) \quad E^\ominus_{cell} = +1.10\,V$$

The 'rules' are that the oxidized forms are written in the middle of the cell diagram and the reduced forms are written on the outside of the cell diagram. The half-cell with the more negative electrode potential (more reactive metal) is placed on the left and the half cell with the more positive electrode potential (less reactive metal) is placed on the right. The anode is on the left and the cathode is on the right of the cell diagram. The cell diagram traces the path of the electrons from anode to the cathode.

Electrolytic cells

Ionic substances become conductors when molten or if dissolved in water to form an aqueous solution. These liquids and solutions are known as electrolytes and include acids, alkalis and salts.

When electricity is passed through an electrolyte a chemical reaction occurs (Figure 9.7) and products are formed at the electrodes. The product at the cathode (negative electrode where electrons enter the electrolyte) is often a metal which plates the electrode, causing an increase in mass. At the anode (the positive electrode where electrons leave the electrolyte) the product is always a non-metal.

■ QUICK CHECK QUESTION

23 Voltaic cells are made of two half-cells and involve a spontaneous redox reaction.

a Draw a diagram of a voltaic cell made from a magnesium half-cell and a gold half-cell. Label the positive electrode (cathode) and the negative electrode (anode) and show the direction of electron flow.

b Write the half-equations of the reaction that occur at the two electrodes.

c Describe the purpose of the salt bridge and identify one substance that might be used in it.

Conduction in electrolytes is due to the presence of mobile ions. During electrolysis, ions gain or lose one or more electrons (discharge) to form atoms. These either deposit as a metal or bond together to form molecules.

Electrolysis is a redox reaction because it involves the gain and loss of electrons in a chemical reaction. The number of electrons entering the electrolyte from the cathode must be equal to the number of electrons leaving the electrolyte at the anode.

Molten salts are binary compounds that contain one type of positive ion (cation) and one type of negative ion (anion). For example, sodium bromide, NaBr, is composed of sodium ions, Na^+, and bromide ions, Br^-.

During electrolysis of molten sodium bromide the cations move to the cathode and gain electrons, and the anions move to the anode and lose electrons, for example:

Cathode: $Na^+(l) + e^- \rightarrow Na(l)$

Anode: $2Br^-(l) \rightarrow Br_2(g) + 2e^-$

The ratio of substances discharged during electrolysis can be determined by comparing the number of electrons entering and leaving the electrolyte.

For example, during the electrolysis of molten sodium fluoride:

At the cathode:

$Na^+(l) + e^- \rightarrow Na(l)$

The formation of 1 mole of sodium atoms requires 1 mole of electrons

At the anode:

$2F^-(l) \rightarrow F_2(g) + 2e^-$

The formation of 1 mole of fluorine molecules results in the formation of two moles of electrons.

Since the numbers of electrons entering the cathode and leaving the anode are equal, the two half-equations can be modified to show that the formation of one fluorine molecule results in the formation of two sodium atoms.

$2Na^+(l) + 2e^- \rightarrow 2Na(l)$

and

$2F^-(l) \rightarrow F_2(g) + 2e^-$

Table 9.7 compares a voltaic cell (with two half-cells with metal electrodes connected by a salt bridge) with an electrolytic cell (of a molten binary ionic compound).

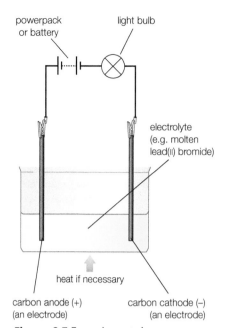

Figure 9.7 Experimental apparatus for electrolysis

Table 9.7 Comparing a voltaic cell with an electrolytic cell

A voltaic cell	An electrolytic cell
Oxidation occurs at the anode (negative)	Oxidation occurs at the anode (positive)
Reduction occurs at the cathode (positive)	Reduction occurs at the cathode (negative)
It uses a redox reaction to produce a voltage	It uses electricity to carry out a redox reaction
A voltaic cell involves a spontaneous redox reaction (exothermic)	An electrolytic cell involves a non-spontaneous redox reaction (endothermic)
The voltaic cell converts chemical energy to electrical energy	The electrolytic cell converts electrical energy to chemical energy
The cathode is the positive electrode and the anode is the negative electrode (during discharge)	The cathode is the negative electrode and the anode is the positive electrode
There are two separate aqueous solutions connected by a salt bridge and an external circuit	There is one molten liquid

■ QUICK CHECK QUESTIONS

24 a Draw a diagram of an electrolytic cell suitable for the electrolysis of molten lead(II) chloride, $PbCl_2(l)$. Label the essential components of the electrolytic cell clearly.

  b Describe the two different ways in which charge flows during electrolysis.

  c Write an equation to show the formation of product at each of the electrodes.

  d Explain why electrolysis is described as 'non-spontaneous'.

25 The diagram below shows the industrial approach used to obtain pure strontium metal by the electrolysis of molten strontium bromide, $SrBr_2$.

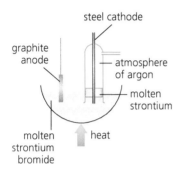

Figure 9.8 Industrial production of strontium

  a State the energy conversion in an electrolytic cell.

  b Explain why the electrolysis only occurs in the molten state.

  c Write equations including state symbols for the reactions at the graphite anode and the steel cathode. Identify the reactions as reduction or oxidation.

  d Explain why an inert atmosphere of argon is used.

10.1 Fundamentals of organic chemistry

Revised ▢

Essential idea: Organic chemistry focuses on the chemistry of compounds containing carbon.

Fundamentals of organic chemistry

Revised ▢

- A homologous series is a series of compounds of the same family, with the same general formula, which differ from each other by a common structural unit.
- Structural formulas can be represented in full and condensed formats.
- Structural isomers are compounds with the same molecular formula but different arrangements of atoms.
- Functional groups are the reactive parts of molecules.
- Saturated compounds contain single bonds only and unsaturated compounds contain double or triple bonds.
- Benzene is an aromatic, unsaturated hydrocarbon.

▨ Basic concepts

All organic compounds contain carbon and hydrogen, with the simplest being the hydrocarbons that contain only these two elements. The complexity of carbon chemistry begins with different types of hydrocarbon that can exist (Table 10.1). Most organic compounds also contain other elements, such as oxygen, nitrogen and the halogens.

Table 10.1 The different types of hydrocarbon compound

Saturated Compounds containing only carbon–carbon single bonds	Unsaturated Compounds containing carbon–carbon multiple bonds		
Alkanes	Alkenes	Alkynes	Arenes
Aliphatic Compounds with structures that do not involve a benzene ring or related structure		Aromatic Compounds with a benzene ring, or similar ring structure involving a delocalized pi system	

Functional groups (Table 10.2), which have their own characteristic reactions, are attached to the hydrocarbon 'backbone' of an organic molecule. Functional groups also give organic molecules specific physical properties. Functional groups are the reactive parts of organic molecules; reacting by characteristic reaction mechanisms.

Expert tip

The terms *saturated* and *unsaturated* just refer to the carbon skeleton ('backbone') of the molecule and the presence, or otherwise, of multiple bonds somewhere in the structure. They do not usually include the other groups (e.g. carbonyl, >C=O) in a molecule. Thus it is possible to have saturated and unsaturated carboxylic acid or nitrile, for example.

Key definition

Functional group – an atom / group of atoms in an organic molecule that gives the compound its characteristic properties; it is the reactive part of the molecule.

Table 10.2 Some functional groups

Homologous series	Functional group and condensed structural formula	Suffix in name of compound	General formula	Structure of the functional group
Alkanes	$-CH_2-CH_2-$	-ane	C_nH_{2n+2}	*
Alkenes	$-CH=CH-$ alkenyl	-ene	C_nH_{2n}	$\diagdown C = C \diagup$
Alkynes	$-C \equiv C-$ alkynyl	-yne	C_nH_{2n-2}	$-C \equiv C-$
Halogenoalkanes	$-X$ (where X = F, Cl, Br, I)	Name uses a prefix (chloro-, bromo-, etc)	$C_nH_{2n+1}X$	$-X$ (where X = F, Cl, Br, I)
Alcohols	$-OH$ hydroxyl	-ol	$C_nH_{2n+1}OH$ or ROH	$-O-H$
Aldehydes	$-CHO$ aldehyde (carbonyl)	-al	$C_nH_{2n+1}CHO$ or RCHO	$-C \diagup ^O _{\diagdown H}$
Ketones**	$-CO-$ carbonyl	-one	$C_nH_{2n+1}COC_mH_{2m+1}$ or RCOR′	$^R_{R'} \diagdown C = O$
Carboxylic acids	$-COOH$ or $-CO_2H$ carboxyl	-oic acid	$C_nH_{2n+1}COOH$ or RCOOH	$-C \diagup ^O _{\diagdown O-H}$

*The alkane structure is the basic backbone into which the functional groups are introduced.
**R and R′ represent hydrocarbon chains (alkyl groups) attached to the group. These chains can be identical or different (as represented here).

A **homologous series** is a series of compounds of the same family, containing the same functional group, and having the same general formula. The members of a series differ from each other by a common structural unit, usually $-CH_2-$.

Physical properties, such as boiling points, show a steady trend in values down a homologous series; the alkanes are shown here (Table 10.3). This trend correlates with an increase in the strength of London (dispersion) forces. Molecules with hydrogen bonds will have higher boiling and melting point than hydrocarbons of similar molar mass. The presence of the capacity for hydrogen bonding in the smaller molecules of a series allows them to dissolve in water.

> **Key definition**
>
> **Homologous series** – a series of compounds with the same functional group. Members of the series have the same general formula and differ from each other by a common structural unit, usually $-CH_2-$.

Table 10.3 Some details of the early members of a homologous series – the alkanes

Alkane	Molecular formula C_nH_{2n+2}	Number of carbon atoms	Melting point/K	Boiling point/K		Physical state at room temperature and pressure
Methane	CH_4	1	91	109		gas
Ethane	C_2H_6	2	90	186	b.p. increasing	gas
Propane	C_3H_8	3	83	231		gas
Butane	C_4H_{10}	4	135	273		gas
Pentane	C_5H_{12}	5	144	309		liquid
Hexane	C_6H_{14}	6	178	342		liquid

The members of a homologous series have very similar chemical properties because they all have the same functional group.

Table 10.4 shows additional homologous series and their functional groups. Large molecules may contain two or more functional groups.

Table 10.4 Further functional groups

Homologous series	Functional group and condensed structural formula	Suffix in name of compound	General formula	Structure of the functional group
Ethers	R–O–R′ ether	alkoxy- (prefix)	$C_nH_{2n+1}OC_mH_{2m+1}$	
Esters	R–OO–R′ ester	-oate	$C_nH_{2n+1}COOC_mH_{2m+1}$	
Amines	R–NH₂ amino	-amine (or prefix amino-)	$C_nH_{2n+1}NH_2$	
Amides	R–CONH₂ carboxamide	-amide	$C_nH_{2n+1}CONH_2$	
Nitriles	R–CN nitrile	-nitrile	$C_nH_{2n+1}CN$	
Arenes	phenyl-	-benzene (or prefix phenyl-)	$C_6H_5–$	

Degree of substitution

Alcohols, halogenoalkanes and amines are classified as *primary*, *secondary* and *tertiary* (Figure 10.1) depending on the structure of the molecule around the functional group.

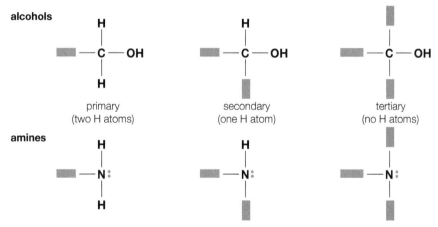

Figure 10.1 A comparison of the use of the terms primary, secondary and tertiary as applied to alcohols and amines – the shaded blocks represent hydrocarbon chains or rings (alkyl or aryl groups)

Note that in the case of alcohols or halogenoalkanes it is a question of how many alkyl groups or hydrogen atoms are attached to the carbon atom bonded to the functional group. In the case of amines, the terms apply to the number of alkyl groups attached to the nitrogen atom. Amides can also be classified as primary, secondary or tertiary on a similar basis to amines.

Expert tip

Sometimes the name of the functional group is different from the homologous series. For example, hydroxyl, –OH, is the functional group, but alcohol is the homologous series.

Common mistake

The amide functional group is distinct from the carbonyl and amine functional groups. There is interaction (resonance or delocalization) between the different regions of the group and therefore amide group has its own unique properties.

Expert tip

Methanol has only hydrogen atoms, and no alkyl group, attached to the relevant carbon atom but is still classified as a primary alcohol.

■ Formulas

Chemists use a wide variety of formulas to represent organic molecules (Table 10.5). These include general (for homologous series), empirical, molecular, structural, displayed structural, skeletal and stereochemical (three-dimensional displayed structural) formulas. Ball-and-stick and space-filling models (Figure 10.2) are used to visualize organic molecules.

Table 10.5 Examples of the different types of formula used in depicting organic molecules

Type of formula	Description	Example
Empirical formula	Shows most simple whole number ratio of all the atoms present in a molecule.	CH_2O
Molecular formula	Shows the actual number of the different atoms present in a molecule of the compound; no information on how the atoms are arranged.	$C_6H_{12}O_6$
Full structural formula	Structural formula show how atoms are arranged together in the molecule; a full structural formula (sometimes called a graphic formula or displayed formula) shows every atom and bond (must include all the Hs and show all the bonds).	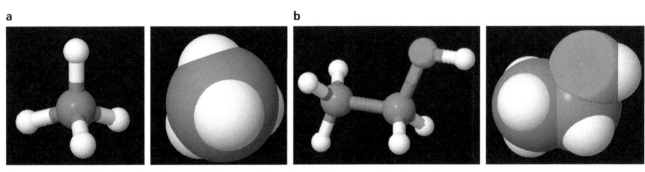
Condensed structural formula	Structural formula which shows order in which all atoms (or groups of atoms) are arranged, but which omits to show the bonds.	$CH_3CH_2CH_2CH_2CH_2CH_3$ or $CH_3(CH_2)_4CH_3$
Skeletal structural formula	Shows the simplified form of the displayed structural formula by removing Cs and Hs which are bonded to carbon atoms.	
Stereochemical formula (three-dimensional displayed structural formula)	Shows the spatial arrangement of bonds, atoms and functional groups in three dimensions.	

a   **b**

Figure 10.2 Ball and stick and space filling models of **a** methane; and **b** ethanol

■ Isomers

Structural isomerism may exist because:
- the backbone (chain) of carbon atoms is branched in different ways
- the functional group is in a different position or
- the functional groups are different (Figure 10.3).

Note that the branched chain hydrocarbons have side-chains that are referred to as alkyl groups (Table 10.6).

Table 10.6 The names of the most common alkyl groups

Alkyl group	Formula
methyl	CH_3-
ethyl	CH_3CH_2-
propyl	$CH_3CH_2CH_2-$
butyl	$CH_3CH_2CH_2CH_3-$

> **Common mistake**
>
> When asked to draw full structural (displayed) formulas, make sure that you show all the bonds between the atoms present in the structure – do not forget the –O–H bond, just writing –OH.

> **Key definition**
>
> **Structural isomers** – compounds with the same molecular formula but different arrangement of atoms.

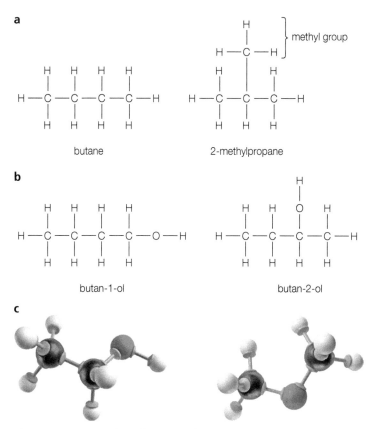

a

butane

2-methylpropane

methyl group

b

butan-1-ol

butan-2-ol

c

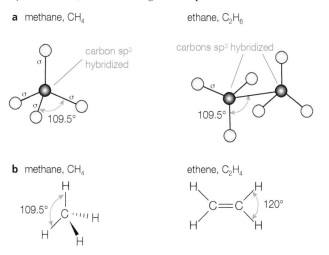

Figure 10.3 Examples of **a** chain; **b** position; and **c** functional group structural isomerism

■ Shapes of organic molecules

The geometry and structure (Figure 10.4) of alkanes, alkenes and arenes (benzene and its derivatives) may be described in terms of sigma and pi bonds and hybridized carbon atoms. Hydrocarbons with sigma bonds only are saturated hydrocarbons, those with sigma and pi bonds are unsaturated hydrocarbons.

a methane, CH_4 ethane, C_2H_6

carbon sp^3 hybridized

carbons sp^3 hybridized

109.5°

109.5°

b methane, CH_4 ethene, C_2H_4

109.5°

120°

Figure 10.4 a The hybridization of each carbon atom in any alkane is sp^3 hybridization. **b** Diagrams showing the bond angles in methane and ethene where the carbon atoms are sp^2 hybridized

The 2s and 2p orbitals of carbon may be hybridized to give the tetrahedral sp^3 (alkanes) planar sp^2 (alkenes) and linear sp (alkynes) arrangements. Two or more pi bonds may be conjugated and a benzene ring (sp^2 hybridized) with 6 pi electrons has a special stability (known as aromaticity). Compounds that do not contain a benzene ring are described as aliphatic.

◼ Organic nomenclature

Organic compounds are named in a systematic way, directly related to their molecular structure. Organic compounds with more than one functional group are named as follows (Figure 10.5).

1   Find the longest carbon chain and name this as the parent alkane.

2   Identify the major functional group and replace *-ane* with a suffix.

3   Number the chain starting nearest the major functional group. Note that the carbon atom of an aldehyde group or a carboxyl group is counted as the first in the chain.

4   Identify any substituents, including minor functional groups on the chain and their number.

5   The names and numbers are given in the prefix in alphabetical order.

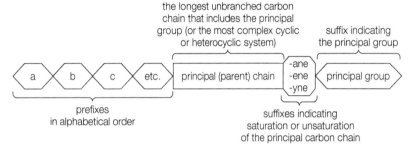

Figure 10.5 The structure of the name of an organic compound

◼ **QUICK CHECK QUESTIONS**

1   Which of the following represents the general formula of the alkyne homologous series?

    A C_nH_{2n} **C** C_nH_n

    B C_nH_{2n+2} **D** C_nH_{2n-2}

2   From the list of compounds below select the one which is:

    a an ester   **d** a tertiary amine

    b an alkene   **e** a ketone

    c a secondary alcohol   **f** an amide

    $CH_3CH_2CHCHCH_3$, $CH_3CH(CH_3)CHO$, $(CH_3)_3COH$, $CH_3CH_2COCH_3$, $CH_3CH_2COOCH_3$, $CH_3CH_2CONH_2$, $CH_3CH_2CH(OH)CH_3$, $CH_3CH_2OCH_3$, $(CH_3CH_2)_3N$

3   Draw the full structural formulae for:

    a butan-2-ol   **d** 1-chloro-2-methylbutane

    b 2-methylpentanal   **e** ethanenitrile

    c pentan-2-one

4   State the number of structural isomers there are of the following, and give their condensed structural formulas:

    a C_4H_8 that are alkenes

    b C_4H_9Br

    c $C_4H_{10}O$

5   Name the following compounds:

    a $CH_3CH_2CH_2CONH_2$ **e** $HCOOCH_2CH_3$

    b $CH_3CH_2COOCH_2CH_3$ **f** $CH_3(CH_2)_2COOCH_2CH_3$

    c $(CH_3CH_2)_2NH$ **g** $CH_3C(CH_3)_2CN$

    d $CH_3COOCH_2CH_3$ **h** $CH_3(CH_2)_3CN$

■ Practical techniques

Practical techniques (Figure 10.6) used in the preparation of organic compounds include reflux, distillation (to separate miscible liquids with different boiling points), suction (vacuum) filtration, separation of immiscible liquids in a separating funnel and recrystallization (to purify solids).

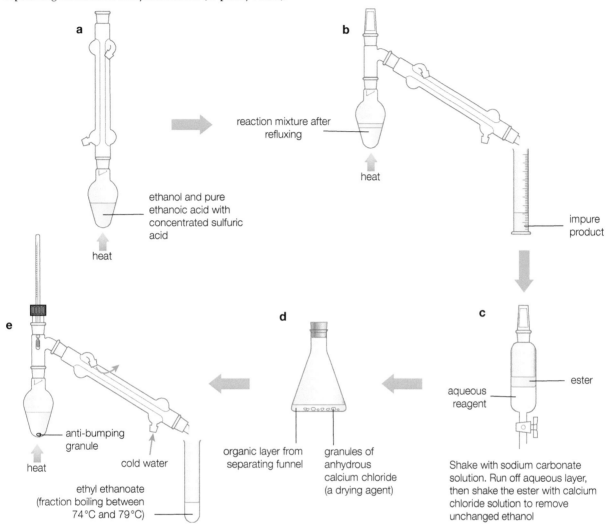

Figure 10.6 The various stages in the practical method of purifying the ethyl ethanoate produced by an esterification reaction; used here to illustrate a range of the techniques used in organic chemistry

Solvent extraction is a method used to separate and purify substances using a solvent which dissolves the desired product but leaves other compounds dissolved in the original solvent. It is essential that the two solvents do not mix.

Filtration under reduced pressure is the process of separating a solid from a liquid rapidly. A partial vacuum is created inside a Buchner flask using a suction pump. This causes the liquid component of the mixture to be drawn rapidly through the apparatus, while the solid sample collects on the filter paper seated in the Buchner funnel.

Recrystallization is a technique used for purification of a solid crystalline product. The required product must readily dissolve in the hot solvent but must be insoluble in the cold solvent. The impurities must either be insoluble in solution in both the cold and hot solvent so they can be filtered off before crystallization, or soluble in the cold solvent so they remain in solution.

Many organic reactions involve losses of product during separation and purification, and some are reversible reactions. The percentage yield indicates the proportion of the theoretical maximum yield that has been obtained.

10.2 Functional group chemistry

Revised ☐

Essential idea: Structure, bonding and chemical reactions involving functional group interconversions are key strands in organic chemistry.

Functional group chemistry

Revised ☐

- *Alkanes*: alkanes have low reactivity and undergo free-radical substitution reactions.
- *Alkenes*: alkenes are more reactive than alkanes and undergo addition reactions. Bromine water can be used to distinguish between alkenes and alkanes.
- *Alcohols*: alcohols undergo nucleophilic substitution reactions with acids (also called esterification or condensation) and some undergo oxidation reactions.
- *Halogenoalkanes*: halogenoalkanes are more reactive than alkanes. They can undergo (nucleophilic) substitution reactions. A nucleophile is an electron-rich species containing a lone pair that it donates to an electron-deficient carbon.
- *Polymers*: addition polymers consist of a wide range of monomers and form the basis of the plastics industry.
- *Benzene*: benzene does not readily undergo addition reactions but does undergo electrophilic substitution reactions.

Important concepts in organic chemistry

Revised ☐

The study of organic reactions is traditionally organized by functional group. Each functional group has its own characteristic reactions.

Reactions may also be studied by type or by mechanism, and Table 10.7 lists some of the most important types of reaction.

Table 10.7 Different categories of organic reaction

Substitution	Reaction in which one atom or functional group replaces another. Usually involves saturated or aromatic compounds.
Addition	Reaction where two reactants combine to make a single product. Usually involves a reactant molecule with a multiple bond, such as C=C or C=O.
Condensation (addition–elimination)	Reaction in which two molecules join to make a larger molecule, accompanied by the loss of a small molecule such as H_2O or HCl.
	Reaction involves functional groups on both molecules: for example, an alcohol reacting with a carboxylic acid to give an ester.
	Initial step is often nucleophilic attack of one molecule on the other.

More generally organic compounds may participate in acid–base, redox and hydrolysis reactions, but substitution and addition remain of particular significance in organic chemistry (Figure 10.7).

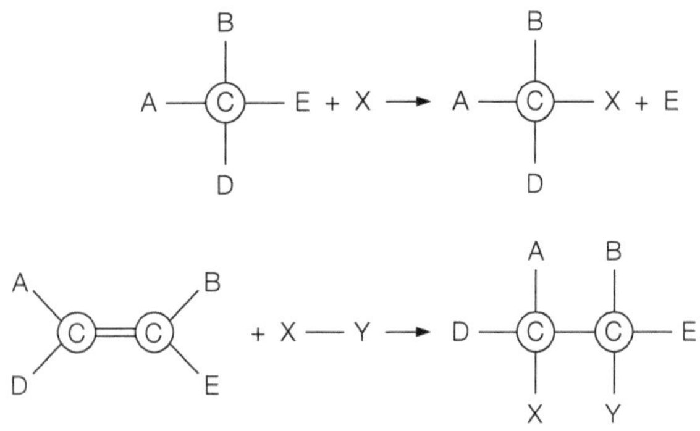

Figure 10.7 Generalized representations of substitution and addition reactions

> **Expert tip**
>
> Hydrolysis reactions involve the breaking of covalent bonds by reaction with water. The water acts as a reactant, though the reaction is often carried out in acidic or alkaline conditions. Acids and bases can play an important role in the catalysis of organic reactions. Acids and bases are useful catalysts. They can alter the reactivity of some organic molecules and increase the rate of reaction.

Reactive organic species

Organic *reaction mechanisms* may involve free radicals, nucleophiles or electrophiles. Each of these reactive species is capable of forming a new covalent bond to an atom.

Radicals contain an unpaired electron and are highly reactive intermediates, attacking any atom with which they are capable of forming a single covalent bond. *Nucleophiles* are electron pair donors (Lewis bases) and attack atoms with a low electron density; while *electrophiles* are electron pair acceptors (Lewis acids), usually cations, and attack atoms with high electron density. The features of these reactive species are summarized, with examples, in Table 10.8.

Expert tip

Electrophiles are Lewis acids and nucleophiles are Lewis bases. The terms electrophile and nucleophile are generally used within the context of bond forming reactions to carbon.

Table 10.8 Key types of reactive species in organic chemistry

Free radicals	Neutral reactive species that contains a lone electron. The product of homolytic fission; participating in chain reactions. Examples are Cl•, CH_3•, •$CClF_2$.
Nucleophiles	Electron-rich species having at least one lone pair of electrons – can also have a negative charge (–) or a partial negative charge (δ–). Attracted to electron-deficient regions of target molecules – examples are negatively charged or have a at least one lone pair of electrons. Examples are. Cl^-, OH^-, H_2O: and :NH_3.
Electrophiles	Electron-deficient species having a positive charge (+) or a partial positive charge (δ+). Attracted to electron-rich regions of target molecules. Examples are $H^{\delta+}$ from HBr, $Br^{\delta+}$ from Br_2, NO_2^+, CH_3^+ and CH_3CO^+.

Expert tip

Note that in modern terminology 'free radicals' are now simply referred to as 'radicals'. Do be aware that these terms are referring to the same reactive species.

Depicting bond breakage

Covalent bonds are broken during reactions by the movement of electrons. Bonds may be broken by homolytic or heterolytic fission.

Table 10.9 The types of bond breaking involved in organic reactions

Homolytic fission (homolysis)	Heterolytic fission (heterolysis)
Is when a covalent bond breaks by splitting the shared pair of bonding electrons between the two products	Is when a covalent bond breaks with both the shared electrons going to one of the products
Produces two free radicals, each having an unpaired electron	Produces two oppositely charged ions

Curly arrows are used to show the movement of electrons in the bond breaking that takes place during a reaction mechanism. A half-headed curly arrow ('fish hook') is used to represent one electrons; a full or double-headed curly arrow is used to represent the movement an electron pair. Figure 10.8 shows the cleavage of a chlorine molecule to produce two chlorine radicals (homolysis), and the breaking of a carbon–halogen bond during substitution (heterolysis).

a Cl:Cl ⟶ Cl• + Cl•
b HO:
C–X
δ+
δ–

Figure 10.8 Depicting the movement of electrons and reaction mechanisms using curly arrows

Expert tip

In drawing a mechanism, the tail of the curved arrow shows where the electrons are coming from, and the head of the arrow shows where the electrons are going. The arrow ends on a nucleus (electrons become a lone pair) or *between two nuclei* (electrons move into position to become a bond). The start of the arrow must be placed either on a bond or lone pair.

Alkanes

Alkanes are saturated hydrocarbons with the general formula C_nH_{2n+2}. The carbon atoms in alkane molecules may be bonded in straight chains or branched chains, but all the carbon–carbon bonds are single bonds.

Physical properties of alkanes

Alkane molecules do not dissolve in polar solvents, such as water. The molecules are held together by weak London (dispersion) forces. The longer the molecules, the greater the attraction between them (Figure 10.9). The molecules of branched alkanes are more compact (smaller surface area), so the London (dispersion) forces are weaker than the equivalent straight chain isomer (of the same molecular formula).

Expert tip

Melting points of organic compounds may not show such smooth trends as boiling points. This is because the shape of the molecule and its packing into a lattice also determines melting point.

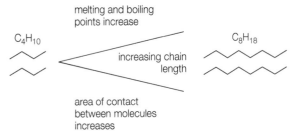

Figure 10.9 The greater the surface area of contact, the greater the strength of London (dispersion) forces between alkanes (and other similar molecules)

The boiling point of alkanes rises as the number of carbon atoms per alkane molecule increases (Figure 10.10). There is a sharp increase initially, as the percentage increase in mass is high, but as successive methylene, $-CH_2-$ groups are added the rate of increase in boiling point decreases.

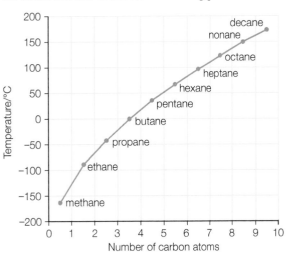

Figure 10.10 Graph of the boiling points of alkanes against the number of carbon atoms present

■ Chemical properties of alkanes

Alkanes are chemically unreactive with reagents in aqueous solution, such as acids and alkalis, as well as oxidizing and reducing agents because of the relatively high bond enthalpies for C–C and C–H bonds and the low polarity of the C–H bond. Alkane molecules will be non-polar or have very low polarity.

The important reactions of alkanes are: combustion, halogenation and cracking (Option C Energy) (Figure 10.11).

Alkanes are widely used as fuels. During complete combustion they produce carbon dioxide and water. However, they produce toxic carbon monoxide gas and/or carbon (soot) in a limited supply of oxygen.

A general equation for the complete combustion of a hydrocarbon is:

$$C_xH_y + \left(x + \frac{y}{4}\right)O_2 \rightarrow xCO_2 + \frac{y}{2}H_2O$$

The halogenation of alkanes is an example of a *radical chain reaction*. Radical chain reactions involve three types of elementary step, illustrated here by the reaction of methane and chlorine in the gaseous phase in the presence of ultraviolet radiation.

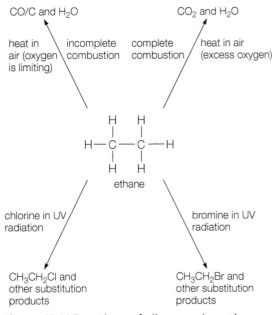

Figure 10.11 Reactions of alkanes using ethane as an example

initiation $Cl-Cl(g)$ $\xrightarrow{\text{ultraviolet}}$ $2Cl\bullet(g)$

propagation
- $Cl\bullet(g) + CH_4(g) \longrightarrow \bullet CH_3(g) + HCl(g)$
- $\bullet CH_3(g) + Cl_2(g) \longrightarrow CH_3Cl(g) + Cl\bullet(g)$

possible termination steps
- $Cl\bullet(g) + Cl\bullet(g) \longrightarrow Cl_2(g)$
- $Cl\bullet(g) + \bullet CH_3(g) \longrightarrow CH_3Cl(g)$
- $\bullet CH_3(g) + \bullet CH_3(g) \longrightarrow C_2H_6(g)$

overall reaction $Cl_2(g) + CH_4(g)$ $\xrightarrow{\text{ultraviolet}}$ $CH_3Cl(g) + HCl(g)$

Figure 10.12 The reaction mechanism for the free-radical substitution between methane and chlorine in ultraviolet light

This is a photochemical reaction with the initiation step producing chlorine radicals (atoms) by homolytic fission. The initiation step is then followed by propagation steps involving a chain reaction which regenerates the chlorine radicals. Termination steps occur when two radicals combine to form a molecule.

In the presence of excess chlorine (and ultraviolet radiation), the chloromethane product can re-enter into a chain reaction and undergo further substitution.

Expert tip

When propane reacts with chlorine in the presence of UV light, two initial products are possible depending on which hydrogen is substituted: 1-chloropropane and 2-chloropropane. The position and extent of the substitution cannot be controlled for radical substitution. It is a very rapid and non-specific reaction (Figure 10.13).

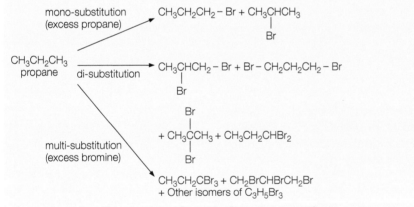

Figure 10.13 Possible products from the bromination of propane

Crude oil (petroleum) and natural gas (mainly methane) are important sources of hydrocarbons, which provide fuels and the source of many other compounds, such as plastics and pharmaceuticals. Fractional distillation of crude oil is followed by cracking and other reactions to form more useful alkanes and alkenes from high molar mass alkanes (Option C).

■ Alkenes

Alkenes are more reactive than alkanes because they contain a pi bond (a region of high electron density) located between a pair of carbon atoms. This is weaker than the sigma bond and is broken during addition reactions of alkenes. These reactive molecules add across the carbon atoms forming the double bond (Figure 10.14).

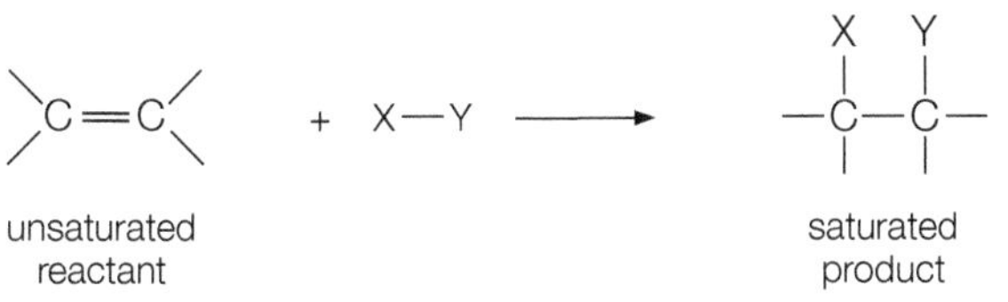

$$\underset{\substack{\text{unsaturated}\\\text{reactant}}}{\diagdown C = C \diagup} \quad + \quad X-Y \quad \longrightarrow \quad \underset{\substack{\text{saturated}\\\text{product}}}{-\overset{\displaystyle X}{\underset{\displaystyle |}{C}}-\overset{\displaystyle Y}{\underset{\displaystyle |}{C}}-}$$

Figure 10.14 Generalized addition reaction to an alkene

■ Addition reactions

Under suitable conditions, addition reactions occur with hydrogen (hydrogenation), water (in the form of steam) (hydration), halogens (halogenation) and hydrogen halides (hydrohalogenation) (Figure 10.15). Alkene molecules may also bond to each other via addition reactions to form polymers. Alkenes are unsaturated and their addition products are saturated.

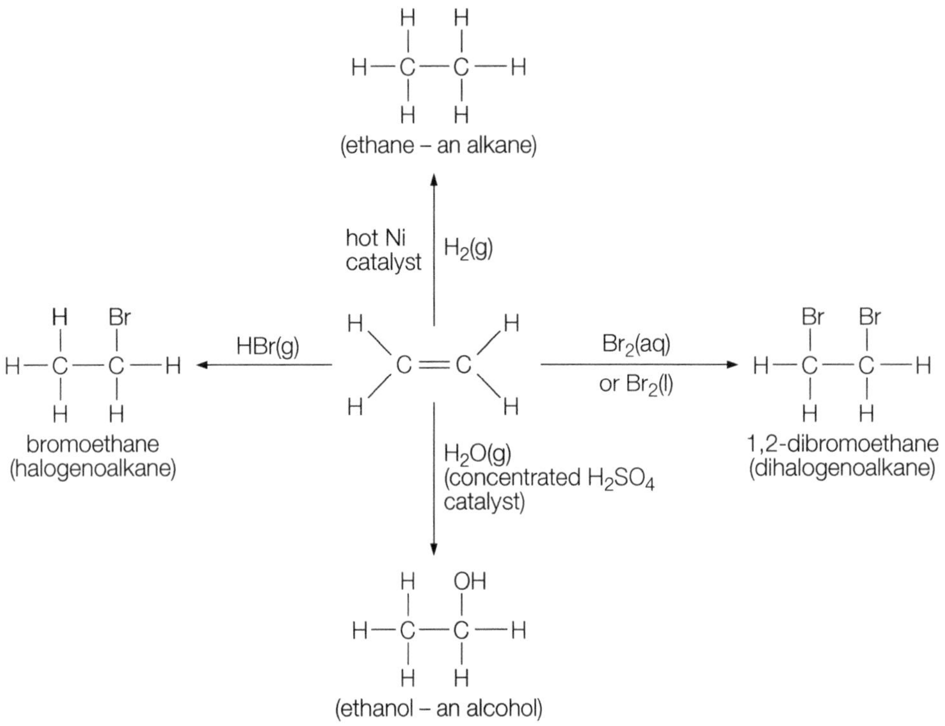

Figure 10.15 Summary of addition reactions of alkenes

■ Use of addition reactions

■ Bromination

Bromine forms an orange molecular solution in water or non-polar solvents. When a solution of bromine is added and shaken to an alkene the product is colourless. The decolourization of bromine solution provides a useful test for unsaturation. If the reactant is unsaturated the orange colour will remain.

■ Hydration

Ethene is an industrially important product formed during the cracking of crude oil. Ethanol for industrial use is formed by the hydration of ethene by steam.

■ Hydrogenation

The addition of molecular hydrogen to unsaturated oils is used on the industrial scale to make margarine. Hydrogenation reduces the number of carbon–carbon double bonds in polyunsaturated oils with the formation of a semi-solid saturated or less unsaturated product.

■ Alcohols

■ Physical properties of alcohols

Alcohols have significantly higher boiling points than hydrocarbons of similar molar mass. There is intermolecular hydrogen bonding due to the presence of hydroxyl groups, –OH. Boiling point increases with molar mass since the increasing number of electrons leads to stronger London (dispersion) forces. Boiling point decreases with branching due to the reduction in surface area.

The smaller alcohols are soluble in water due to the formation of hydrogen bonds with water molecules. The solubility of alcohols decreases as the size of the alkyl group increases which increases the hydrophobic nature and reduces the molecular polarity.

Common mistake

Ethanol is used as a fuel as it gives out considerable energy when burnt. When writing the equation for the combustion of any alcohol, do not forget the oxygen (O) present in the alcohol itself or you will balance the equation incorrectly.

■ Chemical properties of alcohols

Chemical oxidation of a primary alcohol occurs in two steps: an aldehyde is formed first (this can be removed by distillation) and this is oxidized further (by refluxing) to a carboxylic acid (Figure 10.16). Secondary alcohols are oxidized to ketones. Tertiary alcohols are not oxidized under mild conditions.

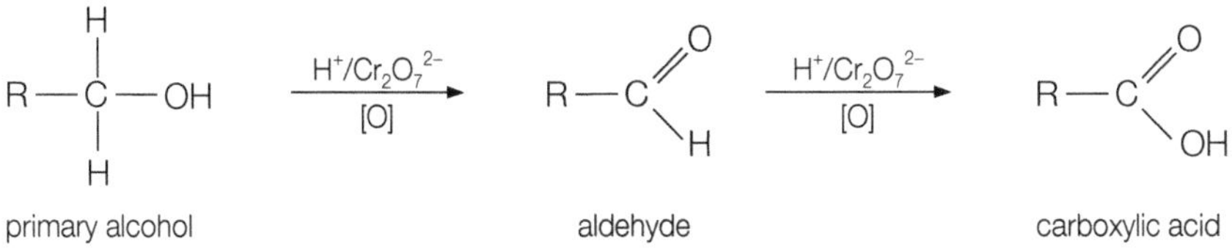

Figure 10.16 The oxidation of a primary alcohol

Mild oxidation is achieved by heating the alcohol with acidified dichromate(VI) ions or manganate(VII) ions. Acidified potassium dichromate(VI) changes colour from orange to green when a primary alcohol (Figure 10.16) or secondary alcohol is oxidized by it. Acidified potassium manganate(VII) changes colour from purple to brown when a primary or secondary alcohol is oxidized by it.

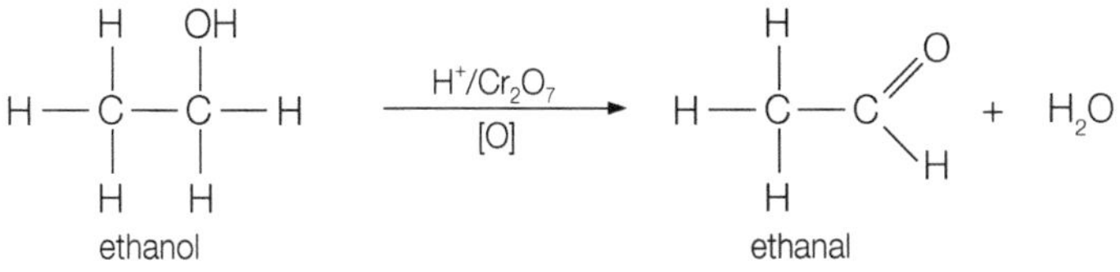

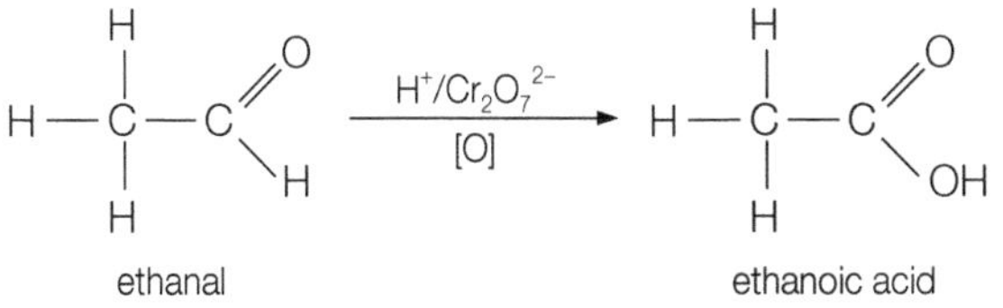

Figure 10.17 Oxidation of ethanol (primary alcohol) to ethanal (aldehyde) and then ethanoic acid (carboxylic acid)

Complete oxidation of alcohols occurs on combustion in the presence of excess oxygen to form carbon dioxide and water. Alcohols may be used as fuels.

Common mistake

Do not forget that the alcohol contains oxygen when balancing a combustion equation.

Esters are formed when an alcohol and a carboxylic acid are warmed in the presence of concentrated sulfuric acid (Figure 10.18). This acts as a dehydrating agent and supplies protons which act as a catalyst. The formation of an ester is known as esterification and is an example of a condensation reaction. Esters are used as flavours, solvents and fragrances.

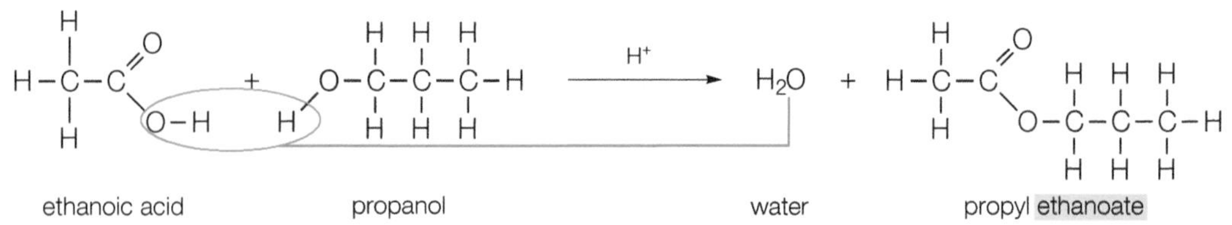

H–C–C (with H above and H below on left carbon; =O on top, O–H below) + O–C–C–C–H (with H's) $\xrightarrow{H^+}$ H_2O + H–C–C (=O on top, O–C–C–C–H below)

| ethanoic acid | propanol | water | propyl ethanoate |

Figure 10.18 The esterification reaction to form propyl ethanoate involves the elimination of water

The first part of the name of an ester is taken from the alkyl group of the alcohol from which it was synthesized. The second part of its name (highlighted in Figure 10.18) denotes the acid from which the compound is derived (see Figure 10.19).

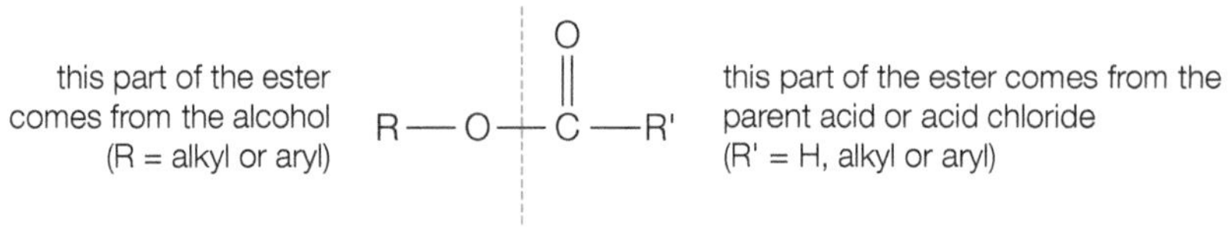

this part of the ester comes from the alcohol (R = alkyl or aryl)   R—O—C—R' (with O double bonded above C)   this part of the ester comes from the parent acid or acid chloride (R' = H, alkyl or aryl)

Figure 10.19 The naming of esters

■ Halogenoalkanes

Halogen alkanes react with a wide range of nucleophiles. The lone pair of electrons is attracted to the partially positive (electron-deficient) carbon atom in the C–X bond of a halogen. The halogen is substituted by the nucleophile which forms a new covalent bond to the carbon atom attacked.

Halogenoalkanes (R–X) undergo substitution when reacted with warm dilute alkali (hydroxide ions). The reaction with hydroxide ions is known as hydrolysis and an alcohol is formed.

$$R–X + OH^- \rightarrow R–OH + X^-$$

The presence of halide ions and relative rate of reaction can be confirmed by the addition of silver nitrate solution (Figure 10.20) (after acidifying after the addition of dilute nitric acid). The slower the reaction the longer time is needed before a visible precipitate is formed.

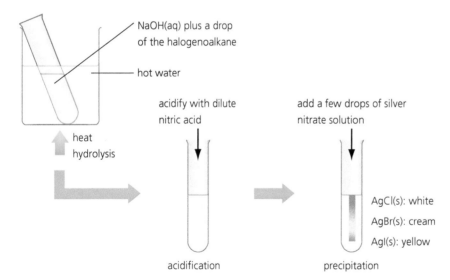

Figure 10.20 Experimental method for estimating the rate of substitution of a halogenoalkane

■ Addition polymers

Polymers are macromolecules that are formed from a large number of small molecules known as monomers. The properties of polymers depend on chain length, strength of intermolecular forces between chains, degree of chain branching (if present) and crystallinity.

Addition polymerization occurs when a monomer joins to another (usually identical) monomer by an addition reaction. This type of polymerization occurs with alkenes, such as ethene (Figure 10.21), and other molecules containing a carbon–carbon double bond, such as chloroethene and tetrafluoroethene (Figure 10.22).

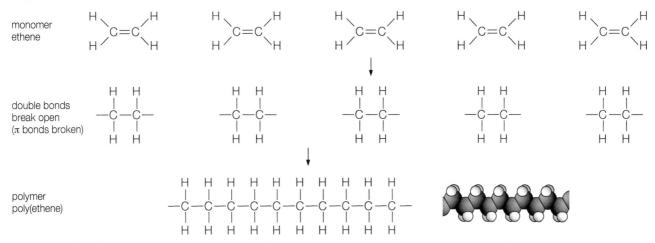

Figure 10.21 A diagrammatic representation of the polymerization of ethene to form poly(ethene)

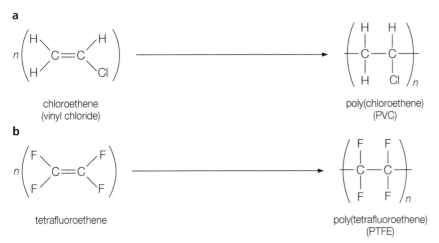

Figure 10.22 The formation of PVC and PTFE

Arenes

The benzene molecule, C_6H_6, is symmetrical, with a planar (flat) hexagon shape. Its carbon–carbon bonds are of a length and bond energy intermediate between single and double bonds. The structure is often represented by a regular hexagon with a circle inside it (Figure 10.23).

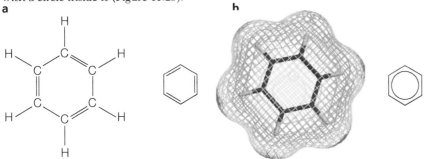

Figure 10.23 Representations of the structure of the benzene molecule (C_6H_6). The three ring structures are all valid structural formulas

Arenes, such as benzene, have considerable energetic stability because of the presence of delocalized pi electrons. A benzene molecule is stabilized by the presence of six pi electrons located in a delocalized molecular orbital that extends over all the six carbon atoms of the ring.

Evidence for this greater energetic stability comes from a comparison of the experimental enthalpy change on adding three moles of hydrogen molecules to one mole of benzene (hydrogenation), to that calculated for three times the enthalpy change on adding a mole of hydrogen molecules to a mole of cyclohexene (Figure 10.24). Benzene is more stable by about $152\,kJ\,mol^{-1}$.

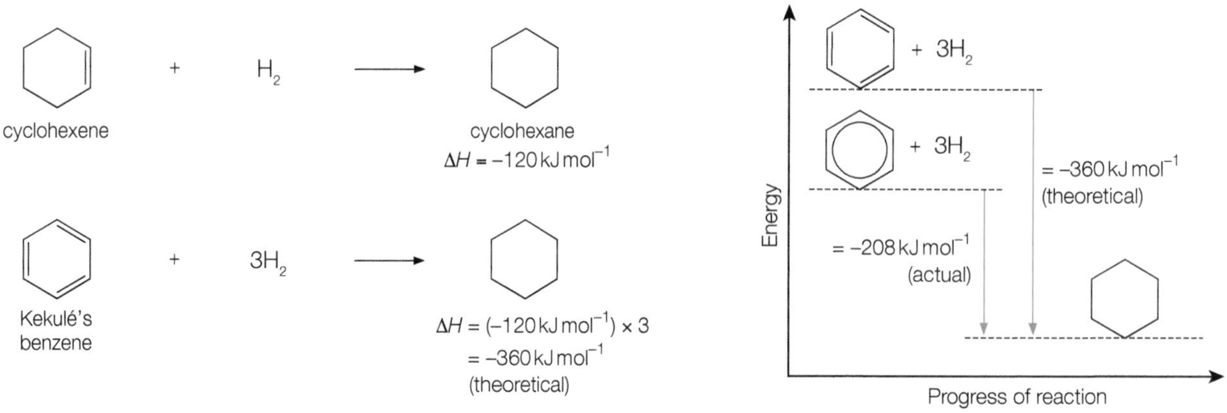

cyclohexene
$+$ H_2 $\longrightarrow$
cyclohexane
$\Delta H = -120\,kJ\,mol^{-1}$

Kekulé's benzene
$+$ $3H_2$ $\longrightarrow$
$\Delta H = (-120\,kJ\,mol^{-1}) \times 3$
$= -360\,kJ\,mol^{-1}$
(theoretical)

$+$ $3H_2$
$+$ $3H_2$
$= -360\,kJ\,mol^{-1}$ (theoretical)
$= -208\,kJ\,mol^{-1}$ (actual)

Energy

Progress of reaction

Figure 10.24 Comparison of the enthalpy of hydrogenation of the localized Kekulé structure and the actual delocalized structure of benzene

Additional supporting evidence for a delocalized structure for benzene is that only one isomer exists for 1,2-disubstituted benzene compounds. If there were a localized system of alternating single and double bonds then two isomers would exist (Figure 10.25).

Arene chemistry is dominated by substitution reactions that allow arenes to retain the benzene ring structure; the delocalized pi electron cloud. Hydrogen atoms in the benzene ring may be replaced (substituted) by a wide range of atoms or groups of atoms, including halogen atoms, nitro- ($-NO_2$) groups, or alkyl and acyl groups (Figure 10.26). These are *electrophilic substitution* reactions, with the electrophiles being NO_2^+ and Cl^+, respectively.

Figure 10.25 The hypothetical (Kekulé) isomers of 1,2-dichlorobenzene – in fact the compound has one single, unique structure

$C_6H_6 + HNO_3 \xrightarrow{\text{conc } HNO_3/H_2SO_4} C_6H_5NO_2 + H_2O$

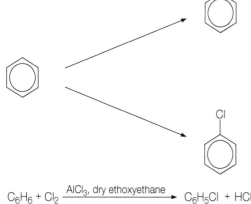

NO$_2$

Cl

Figure 10.26 The substitution of benzene to form nitrobenzene or chlorobenzene

$C_6H_6 + Cl_2 \xrightarrow{\text{AlCl}_3,\ \text{dry ethoxyethane}} C_6H_5Cl + HCl$

The variety of electrophilic substitution reactions on benzene provides access to many useful compounds including medicines, dyes, explosives and polymers, such as PET.

■ QUICK CHECK QUESTIONS

6 a Write equations for the complete and incomplete combustion (giving CO) of propane and butane.

b Write the equation for the complete combustion of ethanol.

7 Write chemical equations, using condensed structural formulas, for the reactions between each of the combinations of reactants and name the product in each case:

a propan-1-ol and ethanoic acid

b butan-1-ol and methanoic acid

c propanoic acid and ethanol

8 Write the equations for the two propagation steps involved in the free radical mechanism for the chain reaction between ethane and chlorine.

9 Give the names and condensed structural formulas of the products of complete and partial oxidation (if any) of the following alcohols:

a propan-1-ol

b propan-2-ol

c 2-methylpropan-2-ol

d butan-1-ol

10 What is the composition of the 'nitration mixture' used to provide the electrophile for the nitration of benzene? What is the name and formula of this electrophile?

11.1 Uncertainties and errors in measurements and results

Essential idea: All measurement has a limit of precision and accuracy, and this must be taken into account when evaluating experimental results.

Uncertainties and errors in measurements and results

- Qualitative data includes all non-numerical information obtained from observations, not from measurement.
- Quantitative data are obtained from measurements, and are always associated with random errors/uncertainties, determined by the apparatus, and by human limitations such as reaction times.
- Propagation of random errors in data processing shows the impact of the uncertainties on the final result.
- Experimental design and procedure usually lead to systematic errors in measurement, which cause a deviation in a particular direction.
- Repeat trials and measurements will reduce random errors but not systematic errors.

Qualitative and quantitative data

Qualitative data include all non-numerical information obtained from observations rather than from measurement. For example, sulfur is a bright yellow crystalline solid.

Quantitative chemistry involves the recording of measurements with associated units. Quantitative data about sulfur would include melting and boiling points (under standard thermodynamic conditions), density and electrical and thermal conductivities.

The instruments used to record the data may have a digital display or use a scale (analogue).

Uncertainty in measurements is unavoidable and it estimates the range within which the true value will lie. This is usually expressed as an absolute value, but can be given as a percentage.

For example, a volume of solution in a burette may be recorded as $20.00\,cm^3 \pm 0.05\,cm^3$ or $20.00\,cm^3 \pm 0.25\%$. This implies the true volume is between $19.95\,cm^3$ and $20.05\,cm^3$. For apparatus with a scale the general rule is that the absolute uncertainty or random error is half of the smallest division that a reading can be recorded to.

On the ammeter (Figure 11.1), the needle is between $1.6\,A$ and $1.7\,A$ so we might estimate the current to be $1.65\,A$. The smallest division is $0.1\,A$. Half of this is $0.05\,A$ so this is the random uncertainty or absolute error in the reading.

For digital instruments the absolute uncertainty or random error is quoted as plus-or-minus the smallest reading. This is because of the error in the reading when the instrument is zeroed. For example, a mass reading with an electronic balance that measures to three decimal places should be quoted as $2.356\,g \pm 0.001\,g$.

> ### Expert tip
>
> Linearity is a design feature of many instruments (for example, a mercury-in-glass thermometer) and it means that the readings are directly proportional to the magnitude of the variable being measured.

Figure 11.1 An analogue ammeter showing a current of $1.65\,A \pm 0.05\,A$

As general rules the uncertainty ranges due to readability from analogue scales and digital displays are summarized in Table 11.1.

Table 11.1 Estimating uncertainties from analogue scales and digital displays

Instrument or apparatus	Example	Random uncertainty
Analogue scale	Rulers, voltmeters, colorimeters, volumetric glassware	± (half the smallest scale division (least count))
Digital display	Top pan balances, spectrophotometers, stopwatches, pH meters	± (1 in the least significant digit)

Random uncertainties

Random uncertainties are present in all measurements. They occur if there is an equal probability of the reading being too high or too low for each measurement. The effects of random uncertainties should mean that a large number of repeated measurements will be distributed in a normal or Gaussian distribution (Figure 11.2). Random errors are errors with no pattern or bias.

They occur when reading a scale and recording a digital readout. Judging when an indicator changes colour or the time taken for sulfur to obscure a cross are also examples where the measurements will have random errors.

The uncertainty of each reading cannot be reduced by repeat measurement but the more measurements which are taken, the closer the mean value of the measurements is likely to be to the true value of the quantity. Taking repeat readings is therefore a way of reducing the effect of random uncertainties.

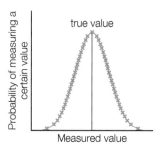

Figure 11.2 Normal distribution of repeated measurements with random errors

> ## ■ QUICK CHECK QUESTION
>
> 1 Deduce whether the following measurements are within the expected range of values:
>
> a 3.48 J and (3.42 ± 0.04) J
>
> b 13.206 g cm^{-3} and (13.106 ± 0.014) g cm^{-3}

Expert tip

Random uncertainties are also known as random errors. However, the term 'error' has the everyday meaning of mistake. Random uncertainties are *not* due to mistakes and cannot be avoided.

Systematic errors

These can be due to a fault in the equipment or apparatus, or design of the experiment, for example, large heat losses from a calorimeter during a determination of an enthalpy change.

Sometimes, a systematic error can also arise from incorrect use of apparatus or from a slow human reaction time. For example, consistently reading a measuring cylinder from too high or too low introduces another systematic error (parallax error) (Figure 11.3).

Common mistake

Digital readings have a random error which should be included in any readings from a digital instrument. Digital readings are not error free.

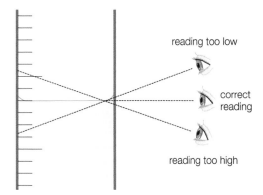

Figure 11.3 Parallax error with a measuring cylinder

A zero error is another example of a systematic error (Figure 11.4). A zero error arises when an instrument gives a non-zero reading for a true zero value of the quantity it measures. A zero error is a systematic error which must be added to (or subtracted from) all readings obtained with that particular instrument.

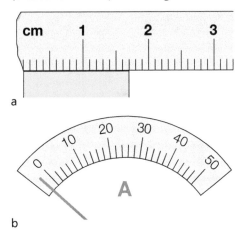

a

b

Figure 11.4 a A zero error with a metre rule; **b** an ammeter reading with a zero error of about –2 A

The presence of systematic errors can often be identified by comparison with accepted literature values for quantities, especially constants. Systematic and random uncertainty errors can often be recognized from a graph of the results (Figure 11.5).

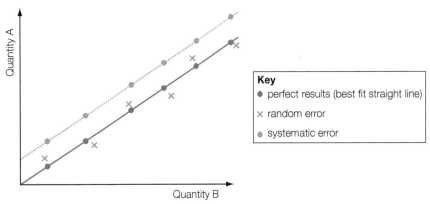

Key
● perfect results (best fit straight line)
× random error
● systematic error

Figure 11.5 Perfect results (no errors), random uncertainties and systematic errors (positive bias) of two proportional quantities

The effect of systematic error cannot be reduced by taking repeated readings. If a systematic error is suspected, it must be solved either by a redesign of the experimental technique, correct use of the apparatus or instrument, or theoretical analysis.

■ Experimental error

Experimental error is defined as the difference between an experimental value and the actual value of a quantity. This difference indicates the accuracy of the measurement. The accuracy is a measure of the degree of closeness of a measured or calculated value to its actual value. The percentage error is the ratio of the error to the actual value multiplied by 100:

$$\text{Percentage error} = \frac{\text{experimental value} - \text{accepted value}}{\text{accepted value}} \times 100$$

If the percentage error is greater than the percentage uncertainty due to random uncertainties (errors) the experiment has systematic error, but if the percentage error is smaller than the percentage uncertainty due to random uncertainties (random errors), any deviation from the literature value is due to random errors alone, usually the limitations of the instrumentation and apparatus. Figure 11.6 compares random and systematic errors.

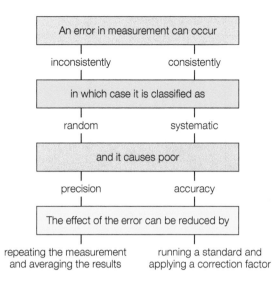

Figure 11.6 Concept chart for contrasting random and systematic errors

Precision and accuracy

Precision refers to how close repeated experimental measurements of the same quantity are to each other. A precise measurement is one where independent measurements of the same quantity closely cluster about a single value that may or may not be the correct value.

Accuracy refers to how close the readings are to the accepted value. The accepted value of a measurement is the value of the most accurate measurement available. It is sometimes referred to as the 'true' value. A measurement which can be described as accurate is one that has been obtained using accurately calibrated instruments correctly and where no systematic errors arise.

The difference between accuracy and precision is shown in Figure 11.7. The centre of the innermost circle represents a completely accurate value.

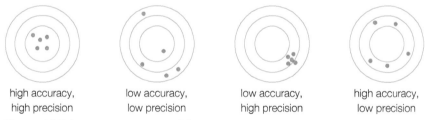

| high accuracy, high precision | low accuracy, low precision | low accuracy, high precision | high accuracy, low precision |

Figure 11.7 Accuracy versus precision

■ QUICK CHECK QUESTION

2   Distinguish between the terms accuracy and precision.

Expert tip

Class A glassware provides the highest accuracy in volumetric glassware. Class B volumetric glassware has random uncertainties twice those of Class A (with the exception of measuring cylinders).

Systematic errors are not random and therefore can never cancel out. They affect the accuracy but not the precision of a measurement. Figure 11.8 shows the effect of random and systematic error on the accuracy and precision of temperature measurements.

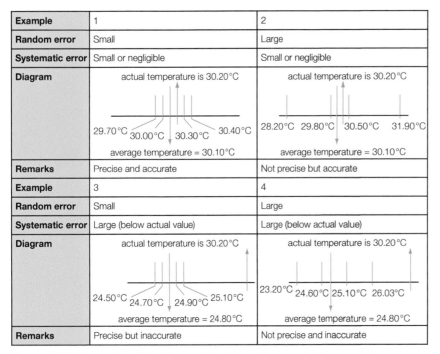

Example	1	2
Random error	Small	Large
Systematic error	Small or negligible	Small or negligible
Diagram	actual temperature is 30.20°C 29.70°C 30.00°C 30.30°C 30.40°C average temperature = 30.10°C	actual temperature is 30.20°C 28.20°C 29.80°C 30.50°C 31.90°C average temperature = 30.10°C
Remarks	Precise and accurate	Not precise but accurate
Example	3	4
Random error	Small	Large
Systematic error	Large (below actual value)	Large (below actual value)
Diagram	actual temperature is 30.20°C 24.50°C 24.70°C 24.90°C 25.10°C average temperature = 24.80°C	actual temperature is 30.20°C 23.20°C 24.60°C 25.10°C 26.03°C average temperature = 24.80°C
Remarks	Precise but inaccurate	Not precise and inaccurate

Figure 11.8 Examples of random and systematic error in temperature measurements

■ Significant figures

A convenient method of expressing the uncertainty in a measurement is to express it in terms of significant figures. It is assumed that all the digits are known with certainty except the last digit which is uncertain to the extent of ±1 in that decimal place (Figure 11.9).

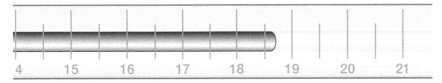

Figure 11.9 A magnified thermometer scale showing a temperature of 18.7°C: the last digit is uncertain

Hence, a measured quantity is expressed as a number which includes all digits that are certain and a last digit which is uncertain. The total number of digits in the number is called the number of significant digits. In order to determine the number of significant figures:

- all non-zero digits are significant, for example, 165 cm has 3 s.f. and 0.165 cm also has 3 s.f.
- zeros to the left of the first non-zero digit are not significant, for example, 0.005 g only has 1 s.f. but 0.026 g has two s.f.
- zeros to the right of the decimal point are significant, for example, 2.05 g has 3 s.f. and 2.500 g has 4 s.f.
- exact numbers, for example 2, and irrational numbers, for example pi and e, have an infinite number of significant figures
- if a number ends in zeros that are not to the right of a decimal, the zeros may or may not be significant, for example, 1500 g may have 2, 3 or 4 s.f.
- the ambiguity can be removed by expressing the number in standard form: 1.5×10^3 g (2 s.f.); 1.50×10^3 g (3 s.f.) and 1.500×10^3 g (4 s.f.).

Expert tip

When a calculation involves two or more steps and you write answers for intermediate steps, keep at least one non-significant digit for the intermediate answers. This ensures that small errors from rounding at each step do not combine to affect the final result.

■ QUICK CHECK QUESTION

3   State how many figures are present in each of the following measured values:

246.33, 1.0009, 700000, 107.854, 0.00340, 350.670, 0.0001, -154.090×10^{27}.

Rounding off

When carrying out calculations with experimental measurements, the final calculated answer will often contain figures that are not significant. The final answer then needs to be rounded off to the appropriate number of significant figures.

If the digit following the last digit to be kept is less than 5, the last digit is left unchanged, for example, 46.32 rounded to 2 s.f. is 46.

If the digit following the last digit to be kept is more than 5, the last digit retained is increased by 1, for example, 52.87 rounded to 3.s.f. is 52.9.

■ Calculations involving addition and subtraction

In calculations involving addition and subtraction, the final answer should be reported to the same number of decimal places as the number with the minimum number of decimal places. For example, 35.52 + 10.3 = 45.82 rounded to 45.8; 3.56 − 0.021 = 3.539 rounded to 3.54.

■ Calculations involving multiplication and division

In calculations involving multiplication and division, the final result should be reported as having the same number of significant figures as the number with the least number of significant digits. For example, 6.26 × 5.8 = 36.308 rounded to 36; $\frac{5.27}{12}$ = 0.439 rounded to 0.44.

If, in a calculation, some exact number is involved, it itself is regarded as having an infinite number of significant digits and the number of digits is limited by the other number. For example, 11 × 2.55 = 28.05 rounded to 28.0.

■ QUICK CHECK QUESTION

4 Give the answers to each of the following calculations with physical quantities to the correct level of significant figures or decimal places.

   a $35.50 \, \text{m}^3 - 22.0 \, \text{m}^3$

   b $46 \, \text{g} \times 4.21 \, \text{J mol}^{-1} \text{K}^{-1} \times 10 \, \text{K}$

   c $\dfrac{10.75681 \, \text{g}}{10.5 \, \text{cm}^3}$

   d $1.05 \, \text{J mol}^{-1} + 945.1 \, \text{J mol}^{-1}$

Propagation of errors

Random uncertainties (random errors) in raw data combine through a calculation to give an estimation of the overall uncertainty (or error) in the final calculated result.

There are some simple rules for propagating errors.

■ Adding and subtracting measurements

When adding or subtracting measurements, the maximum uncertainty (random error) is the sum of the uncertainties associated with each individual measurement.

For example, if two temperatures are measured as 21.40 °C ± 0.05 °C and 31.70 °C ± 0.05 °C, the difference in temperature is 10.3 °C ± 0.1 °C.

For example, if two masses weighed on electronic balances with different absolute random errors are added together to give a combined mass. 0.343 g ± 0.001 g + 0.277 g ± 0.002 g = 0.620 g ± 0.003 g.

■ Multiplying and dividing measurements

The percentage uncertainty in a quantity, formed when two or more quantities are combined by either multiplication or division, is the sum of the uncertainties in the quantities which are combined.

Common mistake

The rules for add/subtract are different from multiply/divide. A very common student error is to swap the two sets of rules. Another common error is to use just one rule for both types of operations.

Expert tip

The Internet has on-line sites that will report the number of significant figures in a number, or round a number to a chosen number of significant figures; there are even on-line calculators that will perform arithmetic according to the rules of significant figures.

Worked example

The following results were obtained when measuring the surface area of a glass block with a 30 cm rule with an absolute error (random uncertainty) of ± 0.1 cm

length = 9.7 cm ± 0.1 cm

width = 4.4 cm ± 0.1 cm

This gives the following percentage errors:

$$\text{length} = \frac{0.1}{9.7} \times 100\% = 1.0\%$$

$$\text{width} = \frac{0.1}{4.4} \times 100\% = 2.2\%$$

So the percentage error in the volume = 1.0% + 2.2% = 3.2%

Hence

surface area = 9.7 cm × 4.4 cm = 42.68 cm² ± 3.2%

The absolute error in the surface area is now 3.2% of 42.68 cm² = 1.37 cm²

Quoted to 1 significant figure the uncertainty becomes 1 cm². The correct result, then, is 43 ± 1 cm². Note that the surface area is expressed to a number of significant figures which is consistent with the estimated uncertainty.

Expert tip

A calculated absolute uncertainty usually has only one significant figure. If necessary the calculated value is adjusted so it has the same number of decimal places as the calculated uncertainty.

Expert tip

Repeated measurements can generate an average for a set or repeated measurements. The average should be stated to the propagated error of the values in the measurements.

For example, the mass of a nickel cathode is measured three times (weighing by difference) using an electronic balance. The values are (1.321 ± 0.002) g, (1.318 ± 0.002) g and (1.322 ± 0.002) g. The average is therefore [(1.321 + 1.318 + 1.322)/3 ± 0.002] g = (1.320 ± 0.002) g.

Worked example

20.00 ± 0.03 cm³ dilute nitric(V) acid was titrated with a 0.100 mol dm⁻³ sodium hydroxide solution to determine its concentration.

Trial	1	2	3
Final burette reading/ ± 0.05 cm³	21.30	40.35	20.35
Initial burette reading/ ± 0.05 cm³	1.00	20.00	0.00
Volume NaOH used/ ± 0.10 cm³	20.30	20.35	20.35

Calculate the volume of sodium hydroxide needed to neutralize the acid.

$$\text{Volume of NaOH} = \frac{20.30 + 20.35 + 20.35}{3}$$

$$= 20.33 \text{ cm}^3 \pm 0.10 \text{ cm}^3$$

$$= 20.33 \text{ cm}^3 \pm 0.49\%$$

$$\text{Random uncertainty} = \frac{0.01}{20.33} \times 100\% = \pm 0.49\%$$

Calculate the concentration of the dilute nitric acid

$$\text{Amount of NaOH reacted} = \frac{20.33}{1000 \text{ dm}^3} \times 0.100 \text{ mol dm}^{-3}$$

$$= 2.033 \times 10^{-3} \text{ mol}$$

Amount of HNO₃ in 20.00 cm³ = 2.033 × 10⁻³ mol

$$\text{Concentration of HNO}_3 = \frac{20.33 \times 10^{-3}}{0.02000}$$

$$= 0.1017 \pm 0.64\%$$

$$= 0.102 \pm 0.7\% \text{ mol dm}^{-3}$$

$$= 0.102 \pm 0.001 \text{ mol dm}^{-3}$$

$$\text{Random uncertainty} = \pm 0.49\% + \frac{0.03 \times 100\%}{20.00} = 0.579\%$$

Allow more significant figures in reporting the answer during data processing. The final answer must adhere to the number of the significant figure of the least accurate instrument.

■ **QUICK CHECK QUESTIONS**

5 The Gibbs relationship is $\Delta G = \Delta H - T\Delta S$. Consider the following experimental data:

$$\Delta H = (-766 \pm 20)\,kJ\,mol^{-1}$$
$$T = (373.0 \pm 0.5)\,K$$
$$\Delta S = (19.6 \pm 0.7)\,J\,mol^{-1}\,K^{-1}$$

   a Calculate the value of ΔG and its absolute uncertainty, expressing both values together with the appropriate number of significant figures and units.

   b The literature value of ΔG is $-785\,kJ\,mol^{-1}$. Calculate the percentage error in the experimental results.

6 $20.00\,cm^3$ of potassium hydroxide, KOH(aq), was measured with a pipette into a conical flask. $(24.40 \pm 0.05)\,cm^3$ of $(0.140 \pm 0.002)\,mol\,dm^{-3}$ hydrochloric acid, HCl(aq), was required to neutralize it. Assume that there was a 0.2% random uncertainty in the pipette reading.

   a Calculate the amounts of hydrochloric acid used and potassium hydroxide in $20.00\,cm^3$ of pipetted solution.

   b Calculate the concentration of potassium hydroxide in $mol\,dm^{-3}$.

   c Calculate the percentage uncertainties in the volume and concentration of hydrochloric acid and the absolute uncertainty in the concentration of the potassium hydroxide solution.

7 $(50.00 \pm 0.10)\,cm^3$ of a transition metal nitrate solution was added a plastic cup using a $50.00\,cm^3$ pipette. Excess magnesium powder was added and the temperature rose from $(26.0 \pm 0.5)\,°C$ to $(36.0 \pm 0.5)\,°C$. The heat released the reaction was then calculated using the equation: $q = mc\,\Delta T$, and was calculated to be $2080\,J$.

   Calculate the error in q. Identify two systematic errors present in this approach and suggest how random error in the investigation can be reduced further.

8 The enthalpy of combustion of ethanol (C_2H_5OH) was determined experimentally by burning a known mass of ethanol in a spirit burner and transferring the heat to water in a open copper calorimeter. The experimental value was $-855\,kJ\,mol^{-1}$; the literature value was $-1367\,kJ\,mol^{-1}$.

   Calculate the percentage error and state one source of systematic error in this experiment (other than heat loss).

11.2 Graphical techniques

Revised ▢

Essential idea: Graphs are a visual representation of trends in data.

Graphical techniques

Revised ▢

■ Graphical techniques are an effective means of communicating the effect of an independent variable on a dependent variable, and can lead to determination of physical quantities.

■ Sketched graphs have labelled but unscaled axes, and are used to show qualitative trends, such as variables that are proportional or inversely proportional.

■ Drawn graphs have labelled and scaled axes, and are used in quantitative measurements.

▨ Drawing and interpreting graphs

Relationships between an independent and a dependent variable can be shown by a suitable line graph (Figure 11.10). The independent variable is changed during an experiment and the dependent variable is what is measured during the experiment.

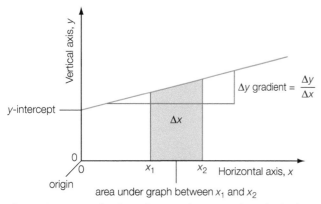

Figure 11.10 Terminology for graphs. Note that the independent variable is plotted on the horizontal axis, and the dependent variable on the vertical axis

In a series of experiments to investigate the effect of temperature on reaction time between sodium thiosulfate and dilute hydrochloric acid, temperature is the independent variable and time (to form a sulfur precipitate).

Figure 11.11 shows raw data for the reaction between sodium thiosulfate and acid warmed and cooled together at different temperatures of sodium thiosulfate. The independent variable is plotted along the horizontal (*x*-axis) and the dependent variable (or processed dependent variable) is plotted up the vertical axis (*y*-axis).

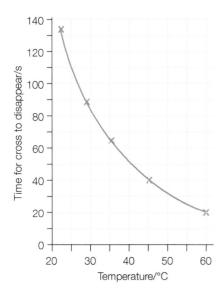

Figure 11.11 A graph of time (for cross to disappear) against temperature for the reaction between sodium thiosulfate and acid

The dependent variable is often transformed into a processed variable. In this example, the times can be processed to give reciprocal of time (s^{-1}) which is proportional to the rate. A linear and directly proportional graph is obtained (Figure 11.12).

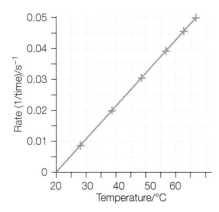

Figure 11.12 A graph of 'rate' (1/time) against temperature for the reaction between sodium thiosulfate and acid

A chemical relationship may be linear (straight line graph) or some type of curved graph (inverse (Figure 11.13), such as logarithmic (Figure 11.14) or exponential relationship (Figure 11.15)). Often the data for a graph with a non-linear relationship is transformed to a linear plot.

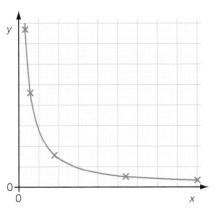

Figure 11.13 Inverse relationship

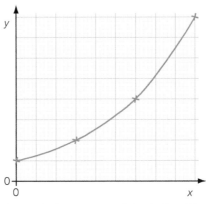

Figure 11.14 Exponential relationship

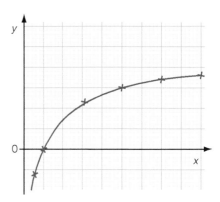

Figure 11.15 Logarithmic relationship

Line graphs can be used to obtain unknown values by one or more of the following: measuring the intercept on the *x* or *y*-axis, measuring the gradient at a data point, extrapolation or interpolation.

Interpolation involves determining an unknown value within the range of the data points. Extrapolation requires extending the graph to determine an unknown value which lies outside the experimental range of values. Extrapolation can be done with a curve (Figure 11.16), but it is more accurate when performed with a line of best fit (linear relationship).

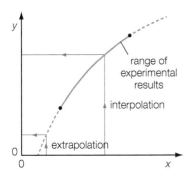

Figure 11.16 Interpolating and extrapolating a non-linear relationship to find intercepts on the *y*-axis (interpolation) and *x*-axis (extrapolation) axis

The line of best fit must be a straight line (Figure 11.17) for experimental data from a directly proportional relationship. It must pass through the origin for a proportional or directly proportional relationship (Figure 11.18).

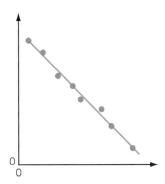

Figure 11.17 Line of best fit for experimental data for a linear relationship

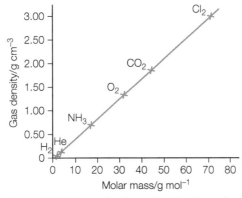

Figure 11.18 A graph of gas density (g cm^{-3}) against molar mass (g mol^{-1}) for selected common gases (a linear and directly proportional relationship)

■ **QUICK CHECK QUESTION**

9   A series of solutions of a coloured salt were prepared with different
    concentrations and their absorbance values measured in a colorimeter using
    a cell of 1 cm path length.

Concentration/mol dm⁻¹	Absorbance
0.2	0.27
0.3	0.41
0.4	0.55
0.5	0.69

Draw a suitable graph and use it to graphically deduce the approximate
concentration value of a solution that has an absorbance of 0.60.

Graphs may be sketched graphs (Figures 11.19 and 11.20), where the axes are
labelled, but there is no scale. Graphs may be drawn graphs, where the axes
are labelled (with units) and have a scale. Sketched graphs are used to show
qualitative trends and drawn graphs are used in quantitative measurements.

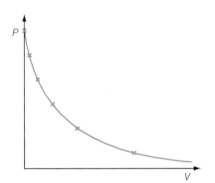

Figure 11.19 Sketch of gas pressure
against gas volume (for a fixed
mass of ideal gas at constant
temperature)

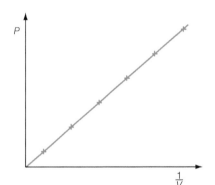

Figure 11.20 Sketch of gas pressure
against reciprocal of gas volume (for
a fixed mass of ideal gas at constant
temperature)

■ **QUICK CHECK QUESTION**

10 Assuming ideal gas behaviour,
    sketch two graphs of volume,
    V, against temperature, T, in the
    labelled axis below for a fixed
    mass of gas at varying pressures,
    P_1 and P_2 where $P_1 > P_2$.

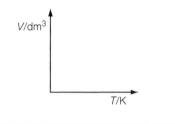

11.3 Spectroscopic identification of organic compounds

Revised ▢

Essential idea: Analytical techniques can be used to determine the structure of
a compound, analyse the composition of a substance or determine the purity of a
compound. Spectroscopic techniques are used in the structural identification of
organic and inorganic compounds.

Spectroscopic identification of organic compounds

Revised ▢

■   The degree of unsaturation or index of hydrogen deficiency (IHD) can be used
    to determine from a molecular formula the number of rings or multiple bonds
    in a molecule.
■   Mass spectrometry (MS), proton nuclear magnetic resonance spectroscopy
    (^{1}H NMR) and infrared spectroscopy (IR) are techniques that can be used to
    help identify compounds and to determine their structure.

▨ Introduction

Spectroscopic methods, such as IR and NMR, depend on the interaction
of electromagnetic radiation with matter. The different regions of the

electromagnetic spectrum all have different frequency (energy) ranges, which have specific applications in the study of ions, atoms and molecules.

Quantum theory is applied to all spectroscopic techniques and assumes that the particles of a substance emit or absorb electromagnetic radiation in multiples of small amounts (quanta) of energy. Some examples of the types of electromagnetic radiation absorbed are shown in Table 11.2.

Table 11.2 Summary of spectroscopic techniques

Electromagnetic radiation absorbed	What the energy is used for	Spectroscopic technique
Ultraviolet/visible	Movement of electrons to higher energy levels	Ultraviolet/visible spectroscopy
Infrared	To vibrate bonds	Infrared spectroscopy
Microwaves	To rotate molecules	Microwave spectroscopy
Radio waves	To change nuclear spin	NMR spectroscopy

Index of hydrogen deficiency (IHD)

The IHD indicates the number of double bond equivalents (the number of double bonds and/or rings) in a compound. A triple bond counts as two double bond equivalents.

The IHD for a hydrocarbon with x carbon atoms and y hydrogen atoms, C_xH_y, is given by the following relationship:

$$IHD = \frac{(2x + 2 - y)}{2}$$

For example, the IHD for C_2H_4 is $\frac{(2(2) + 2 - 4)}{2} = 1$. This means it can have either one double bond or one ring, but it cannot have a triple bond. Since you cannot form a ring with only two C's, it must have a double bond.

If the organic compound is not a hydrocarbon then the following should be noted:
- Oxygen and sulfur atoms do not alter the value of the IHD.
- Halogens are treated like hydrogen atoms, for example, CH_2Cl_2 has the same IHD value as CH_4.
- For each nitrogen atom, add one to the number of carbons and one to the number of hydrogen.

For example, the IHD of $C_5H_7OBr = \frac{(12 - 8)}{2} = 2$. It may have two rings; two π bonds or a ring and a π bond.

Expert tip

Determining the IHD for organic molecules can be useful for the following reasons: seeing what types of structural units may be possible and quickly checking structures to see if they fit the molecular formula rather than simply counting hydrogens (when a mistake is possible).

■ QUICK CHECK QUESTIONS

11 An unknown compound has a molecular formula $C_6H_{10}Cl_2$. Deduce the IHD and draw two skeletal formulas that are consistent with the IHD.

12 Deduce the IHD for the following molecules:

a $CH_3CHCHCH_2CHCH_2$

b CH_3OCOCH_2Cl

c $CH_3C{\equiv}CCOCH_3$

d C_5H_7OBr

e $C_6H_6Br_2$

f C_4H_6

◼ Mass spectrometry (MS)

The mass spectrum of a compound enables the relative molecular mass of the compound to be determined using the molecular ion peak. The molecular ion peak is the peak produced by the unipositive ion formed by the loss of the loss of one electron from a molecule of the compound (Figure 11.21).

Figure 11.21 Ionization of a propanone molecule to form a molecular ion

The molecular ion is often very unstable and susceptible to fragmentation. The fragmentation peaks (Table 11.3) provide invaluable information on the possible structure of the molecular ion. A peak at M-15 indicates the loss of a methyl group; a peak at M-29 indicates the loss of an ethyl group.

Table 11.3 Common peaks in mass spectra

Mass lost	Fragment lost	Mass lost	Fragment lost
15	CH_3^+	29	$CH_2CH_3^+$, CHO^+
17	OH^+	31	CH_3O^+
18	H_2O	45	$COOH^+$
28	$CH_2{=}CH_2$, $C{=}O^+$		

A fragmentation process that leads to a more stable ion is favoured over another that produces a less stable ion. The release of small highly charged ions, such as H^+, is not favoured and double bonds are usually not broken during electron bombardment. Some fragment ions containing oxygen will be stabilized by resonance.

Useful information can be compared by the relative heights of the $(M + 2)^+$ peak and the $(M)^+$ peak. The structure probably contains a chlorine atom if the latter peak is approximately one-third as tall as the former peak. The presence of a bromine atom is indicated when these two peaks are similar in height. These observations can be rationalized in terms of the isotope ratios of chlorine and bromine atoms.

Worked example

Examine the mass spectrum for benzene carboxylic acid and identify the ions responsible for each of the major peaks.

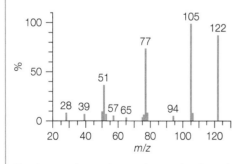

The four major peaks are due to the detection and measurement of the following cations:

m/z	Cation responsible
122	$[C_6H_5COOH]^+$
105	$[C_6H_5CO]^+$
77	$[C_6H_5]^+$
51	$[C_4H_3]^+$

■ **QUICK CHECK QUESTIONS**

13 The diagram below shows the mass spectrum of propanal.

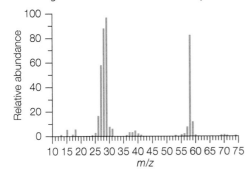

Write equations to explain the strong peaks observed at *m/z* 29 and 57.

14 The mass spectra of halogenoalkanes show more than one line corresponding to the molecular ion. This is due to the presence of isotopes such as ^{35}Cl, ^{37}Cl (present in a ratio of 3:1) ^{79}Br and ^{81}Br (present in a ratio of 1:1).

Analyse the following spectra of halogenoalkanes A and B and deduce the formula of all the molecular ion species and identify species responsible for the peak at 29 in mass spectrum A.

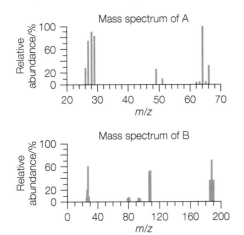

Worked example

A compound gives a mass spectrum with peaks at *m/z* = 77 (40%), 112 (100%), 114 (33%), and essentially no other peaks. Identify the compound.

The molecular ion peak is 112 and there is a M + 2 peak at 114. Therefore, it is a halogen-containing molecule. This molecular ion peak and M + 2 peak are in a 3 to 1 ratio; this implies chlorine. So, 112 − 35 = 77; number of carbon atoms = $\frac{77}{12}$ = 6, hence it is chlorobenzene, C_6H_5–Cl.

■ **QUICK CHECK QUESTION**

15 Propose a structure that is consistent with a compound that only contains sp^3 hybridized carbon atoms and gives a molecular ion at *m/z* = 84.

▨ Infrared spectroscopy (IR)

IR spectroscopy is based on the principle that when a polar covalent bond between atoms in a molecule vibrates (Figure 11.22), it may change the dipole moment of the molecule. If it does, IR radiation (of a specific energy) will be absorbed as a result and an absorption spectrum can be recorded.

$$\overset{\longleftarrow}{Cl} \quad \overset{\longrightarrow}{Cl}$$

Figure 11.22 Bond stretching in the hydrogen chloride molecule

Polyatomic molecules, such as water (Figure 11.23), carbon dioxide (Figure 11.24) and methane, vibrate in a number of different ways, known as modes of vibration. Only the modes that are associated with a changing dipole moment for the molecules will give rise to absorption bands in the IR spectrum.

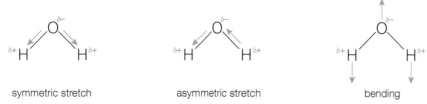

symmetric stretch   asymmetric stretch   bending

Figure 11.23 Stretching and bending (rocking) vibrations in the water molecule

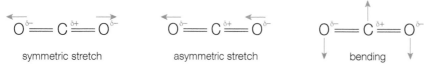

symmetric stretch   asymmetric stretch   bending

Figure 11.24 Stretching and bending (rocking) vibrations in the carbon dioxide molecule

Functional groups in organic molecules (Figure 11.25) are identified by their position of their absorption bands (measured by a wavenumber range) on the infrared absorption spectrum.

■ **QUICK CHECK QUESTIONS**

16 Describe what happens on a molecular level when gaseous phosphine molecules, $PH_3(g)$, strongly absorb IR radiation.

17 Three modes of vibrations of the carbon dioxide molecule are shown below. Complete the table by indicating if each mode would be *IR active* or *IR inactive*.

Mode of vibration	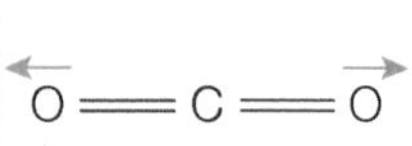 O=C=O	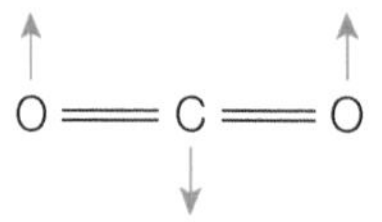 O=C=O	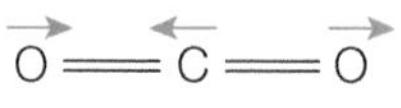 O=C=O
IR active or IR inactive			

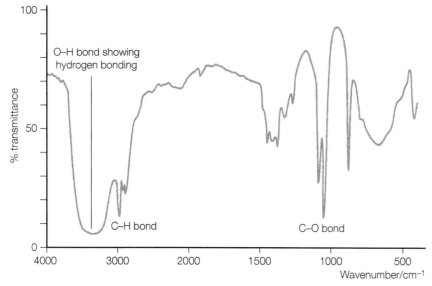

Figure 11.25 Infrared spectrum of ethanol

An IR spectrum is a plot that measures the percentage transmittance as a function of wavenumber. The location of each signal in an IR spectrum is reported in terms of a frequency-related unit termed the wavenumber. This is defined as the reciprocal of the wavelength (in cm) and it correlates with frequency and hence energy (Figure 11.26).

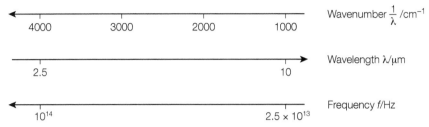

Figure 11.26 The relationship between wavenumber, wavelength and frequency

■ **QUICK CHECK QUESTIONS**

18 Outline how infrared waves differ from red light waves.

19 For light with a wavelength of 408 nm determine the wavelength, frequency and the wave number.

NATURE OF SCIENCE

The energy in any bond vibration depends on variables such as bond length, bond strength and the mass of the atoms at either end of the covalent bond. This means that each different bond will vibrate in a different way, involving different amounts of energy. The bonds can be regarded as behaving like springs and the IR stretching frequency increases in the order: single bonds (except those to hydrogen), double bonds, triple bonds and single bonds to hydrogen. Stronger bonds need more energy to make them vibrate, so they absorb a high frequency of IR radiation (higher wavenumber).

Table 11.4 shows the characteristic absorptions (in wavenumbers) for a selected range of functional groups. They are used for classification and identification.

Table 11.4 Characteristic ranges for infrared absorption due to stretching vibrations in organic molecules

Bond	Organic molecules	Wavenumber/cm⁻¹	Intensity
C—I	Iodoalkanes	490–620	Strong
C—Br	Bromoalkanes	500–600	Strong
C—Cl	Chloroalkanes	600–800	Strong
C—F	Fluoroalkanes	1000–1400	Strong
C—O	Alcohols, esters, ethers	1050–1410	Strong
C=C	Alkenes	1620–1680	Medium-weak; multiple bands strong
C=O	Aldehydes, ketones, carboxylic acids and esters	1700–1750	Strong
C≡C	Alkynes	2100–2260	Variable
O—H	Hydrogen bonding in carboxylic acids	2500–3000	Strong, very broad
C—H	Alkanes, alkenes, arenes	2850–3090	Strong
O—H	Hydrogen bonding in alcohols and phenols	3200–3600	Strong, broad
N—H	Primary amines	3300–3500	Medium, two bands

Each IR absorption has three characteristics: wave number, intensity (strong or weak) and shape: narrow or broad (caused by hydrogen bonding).

The intensity of the IR absorption is dependent on the dipole moment of the bond giving rise to the IR signal. $>C=O$ bonds produce strong IR absorptions, while non-polar $>C=C<$ bonds usually produce weak IR absorptions.

IR spectra are also used to identify unknown organic compounds by comparing their spectra with those of known compounds (in a database). The fingerprint region of the IR spectrum, in which each molecule gives a unique set of molecular absorption bands is most useful for this purpose.

IR spectroscopy is used in the chemical analysis of solids, liquids and gases, and is used to measure the concentration of carbon dioxide in the atmosphere and xenobiotics in the environment.

One method for analysing an IR spectrum is the following stepwise approaching which involves asking a series of questions:

1   Is there a strong carbonyl absorbance $(>C=O)$ between 1700 and 1750 cm^{-1}?
2   It could be an aldehyde, ketone, carboxylic acid or ester. Identifying the presence of the other bonds will confirm the identity of the functional group.
3   Is there a broad $-OH$ absorption band (3200–3600 cm^{-1}) and C$-$O band (1050–1410 cm^{-1})? If yes it is likely to be an alcohol, but if no go to 4.
4   If there is weak absorbance between 1620 and 1680 cm^{-1} it is due to the double bond $(>C=C<)$ of an alkene. If there are strong absorptions at 1450–1650 cm^{-1} it is aromatic. If there is a C$-$H stretch (2850–3090 cm^{-1}) and a band near 1450 cm^{-1} it is an alkane. If it is a simple IR spectrum with an absorption <1400 cm^{-1} it might be a halogenoalkane.

■ QUICK CHECK QUESTIONS

20 Explain why the stretching of the carbon–carbon double bond occurs between 1610 and 1680 wavenumbers, but the stretching of carbon–carbon triple bond occurs between 2100 and 2260 wavenumbers.

21 State the wavenumber (cm^{-1}) of the characteristic absorption bands of the functional groups in each of the three compounds: ethene, ethane-1,2-diol and ethanedioic acid.

22 The IR spectra of methyl propanoic acid ($CH_3CH(CH_3)COOH$) and ethyl methanoate ($HCOOC_2H_5$) contain absorptions in characteristic wavenumber ranges. Identify **two** wavenumber ranges common to the IR spectrum of both compounds and **one** wavenumber range found only in the IR spectrum of one compound.

23 The region below 1500 cm^{-1} is termed the fingerprint region. Explain the term fingerprint region.

24 The IR spectrum below is of a hydrocarbon with the molecular formula C_6H_{10}. Deduce the homologous series it belongs to and state which bonds are responsible for absorptions near 2100 cm^{-1} and near 3000 cm^{-1}.

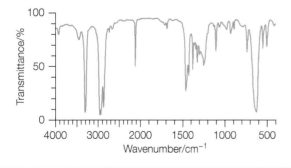

NMR spectroscopy

A hydrogen-1 atom consists of a single proton and a single electron. The proton can be viewed as a positively charged particle that spins about its axis. It generates a circulating electric current that produces a magnetic field. It behaves like a small bar magnet.

^{1}H NMR spectroscopy is concerned with the absorption and emission of energy (in the form of electromagnetic radiation) associated with the transition of the nuclei of hydrogen atoms between two energy levels, in an external magnetic field.

When the hydrogen atoms of a molecule (protons) are placed in an external magnetic field, they become aligned either in a low-energy orientation in the same direction as the magnetic field or in a higher energy orientation in opposition to the field (Figure 11.27). These are known as spin states and described by the spin quantum number.

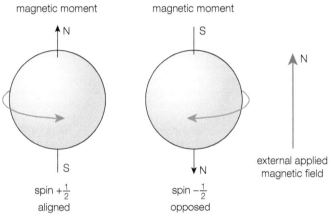

Figure 11.27 Spinning protons aligned with and opposed to an external **magnetic field** (the $\pm\frac{1}{2}$ represents the spin quantum number)

There is a small energy difference between the up and down states (Figure 11.28) and the absorbed radio waves rotate the spins of the protons.

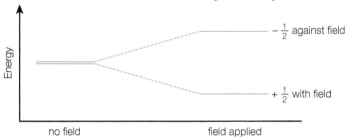

Figure 11.28 The spin states of protons in the absence and presence of a magnetic field

The upwards and downwards transitions between the energy levels represent the resonance state. There is a net absorption of energy because there is small excess of nuclei in the lower energy level (spin $-\frac{1}{2}$) (at equilibrium).

The sample (liquid, solid or solution) is placed in a constant external magnetic field (Figure 11.29) and subjected to radio waves of varying frequency (energy). When the frequency (energy) of the radiation exactly matches the energy differences between the two spins state for hydrogen atoms, energy is absorbed and resonance occurs.

The absorption of energy is plotted against the variation in frequency to give the NMR spectrum (Figure 11.30). The integration trace, or area under each NMR signal 7 (absorption), indicates the number of protons (hydrogen atoms) giving rise to the signal. These numbers provide the relative number, or ratio, of hydrogen atoms (protons) giving rise to each absorption (signal). In a ^{1}H NMR spectrum, each signal has three important characteristics: location, area and shape.

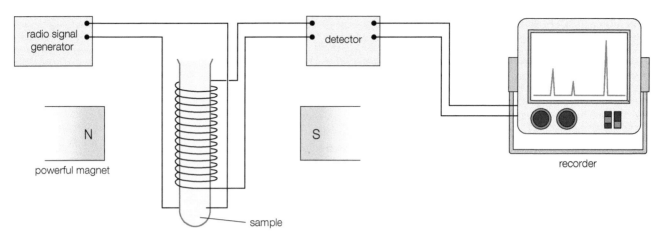

Figure 11.29 The basic features of an NMR spectrometer

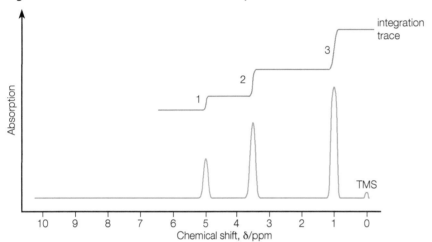

Figure 11.30 The low-resolution ¹H NMR spectrum of anhydrous ethanol, showing peaks and integration trace

Chemical compound nuclei, even of the same element, often give absorption peaks at slightly different frequencies owing to their different environments. These differences are known as chemical shifts and are measured relative to a standard, TMS (tetramethylsilane), that acts as an internal reference.

Chemical shifts (Table 11.5) have been recorded for protons in different chemical environments and are used to identify functional groups and other structural features that contain hydrogen atoms.

> **Expert tip**
>
> The left side of an NMR spectrum is described as downfield and the right side is described as upfield.

Table 11.5 Typical proton chemical shifts (δ) relative to TMS (tetramethylsilane)

Type of proton	Chemical shift (ppm)	Type of proton	Chemical shift (ppm)
$-CH_3$	0.9–1.0	$R-\overset{\overset{O}{\|}}{\underset{}{C}}-O-CH_2-$	3.7–4.8
$-CH_2-R$	1.3–1.4		
$-R_2CH$	1.5		
$RO-\overset{\overset{O}{\|}}{\underset{}{C}}-CH_2-$	2.0–2.5	$R-\overset{\overset{O}{\|}}{\underset{}{C}}-O-H$	9.0–13.0
$R-\overset{\overset{O}{\|}}{\underset{}{C}}-CH_2-$	2.2–2.7	$R-O-H$	1.0–6.0
		$-HC=CH_2$	4.5–6.0
benzene ring $-CH_3$	2.5–3.5	benzene ring $-OH$	4.0–12.0
$-C\equiv C-H$	1.8–3.1	benzene ring $-H$	6.9–9.0
$-CH_2-Hal$	3.5–4.4		
$R-O-CH_2-$	3.3–3.7	$R-\overset{\overset{O}{\|}}{\underset{}{C}}-H$	9.4–10.0

> ■ **QUICK CHECK QUESTIONS**
>
> 25 Predict the number of signals present in the low-resolution ^{1}H NMR spectrum of the following organic molecules:
>
> ethane, propan-2-ol, methanoic acid, 2,2-dimethylpropane, ethane-1,2-diol
>
> 26 Predict the number of signals present in the low-resolution ^{1}H NMR spectra of the following molecules:
>
> C_6H_6 (benzene), $CH_3CH_2OCH_3$, $CH_3CH_2OCH_2CH_3$, $CH_3CH_2CH_2CH_2COCH_3$, C_6H_5-$(CH_2)_4$-CH_3 (pentylbenzene)

The chemical shift increases as the electronegativity of the atom bonded to the hydrogen increases. For example, in a halogenoalkane, a carbon atom bonded to a fluorine atom is more electronegative than a carbon atom bonded to a chlorine atom. The more electronegative carbon atom draws the electron cloud away from an attached atom, reducing the screening and increasing the absorption frequency. This reduction of electron density around a hydrogen atom is known as deshielding.

The chemical shift (δ) is defined as the difference (in hertz) between the resonance frequency of the proton (hydrogen atom) being studied and that of TMS divided by the operating frequency of the NMR spectrometer, multiplied by 1 000 000.

> **Expert tip**
>
> NMR is often used to detect the hydrogen atoms in water molecules, and the analysis of water in the human body forms the basis of magnetic resonance imaging (MRI).

> ■ **QUICK CHECK QUESTIONS**
>
> 27 The ^{1}H NMR spectrum of a compound with a molecular formula of $C_4H_8O_2$ shows three major signals with chemical shifts and integration traces given below.
>
> | Chemical shift δ / ppm | Integration trace |
> | --- | --- |
> | 0.9 | 3 |
> | 2.0 | 2 |
> | 4.1 | 3 |
>
> a State the information that can be deduced from the number of peaks and the integration traces.
>
> b Draw and name the structural formula of **two** possible structural isomers of the compound.
>
> 28 Butanone, $CH_3C(O)CH_2CH_3$, is vaporized and analysed by mass spectrometry, ^{1}H NMR and IR spectroscopy.
>
> a State the information determined from the three analytical techniques and identify the type of radiation used in NMR.
>
> b State the formula of the unipositive ions for the following peaks in the mass spectrum: *m/z* 57 and *m/z* 43.

12.1 Electrons in atoms

Revised ☐

Essential idea: The quantized nature of energy transitions is related to the energy states of electrons in atoms and molecules.

Ionization energy

Revised ☐

- In an emission spectrum, the limit of convergence at higher frequency corresponds to the first ionization energy.
- Trends in first ionization energy across periods account for the existence of main energy levels and sub-levels in atoms.
- Successive ionization energy data for an element give information that shows relations to electron configurations.

■ Ionization energy

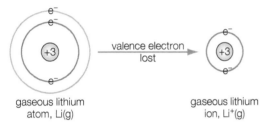

Figure 12.1 The concept of first ionization energy

Ionization energies may be measured experimentally inside a mass spectrometer, which vaporizes substances and then fires high-speed electrons at them to cause ionization.

■ Factors that influence ionization energy

■ The size of the nuclear charge

As the atomic number (number of protons) increases, the nuclear charge increases. The larger the positive charge, the greater the attractive electrostatic force between the nucleus and all the electrons. So, a larger amount of energy is needed to overcome these attractive forces if an electron is to be removed (during ionization). As the proton number increases across a row of the periodic table the ionization energy tends to increase.

■ Distance of outer electrons from the nucleus

The force of electrostatic attraction between positive and negative charges decreases rapidly as the distance between them increases. Hence, electrons in shells (main energy levels) further away from the nucleus are more weakly attracted to the nucleus than those closer to the nucleus. The further the outer electron shell is from the nucleus, the lower the ionization energy. Ionization energies tend to decrease down a group of the periodic table.

■ Shielding effect

Since all electrons are negatively charged, they repel each other. Electrons in full inner shells repel electrons in outer shells. The full inner shells of electrons prevent the full nuclear charge being experienced by the outer electrons. This is called shielding (Figure 12.2). The greater the shielding of outer electrons by the inner electron shells, the lower the electrostatic attractive forces between the

> **Key definition**
>
> **First ionization energy** (see Figure 12.1) of an atom – the minimum energy required to remove one mole of electrons from one mole of gaseous atoms and form one mole of unipositive ions (under standard conditions):
>
> $$M(g) \rightarrow M^+(g) + e^-$$

Expert tip

The gaseous symbols, (g), are essential in this equation because they are an important part of the definition.

nucleus and the outer electrons. The ionization energy is lower as the number of full electron shells between the outer electrons and the nucleus increases.

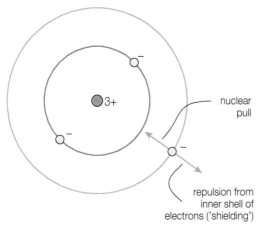

Figure 12.2 Electrostatic forces operating on the outer or valence electron in a lithium atom

■ **QUICK CHECK QUESTION**

1 Predict, with a reason, if the second ionization energy of sulfur is higher or lower than the first ionization energy of phosphorus. Assume both atoms are gaseous and in the ground state.

■ Periodic variation in first ionization energy

A graph of first ionization energies plotted against atomic numbers shows periodicity in periods 2 and 3. There is a clear periodic trend (Figure 12.3) in the first ionization energies of the atoms of the elements. The general trend is that first ionization energies increase from left to right across a period. This applies to periods 1, 2 and 3.

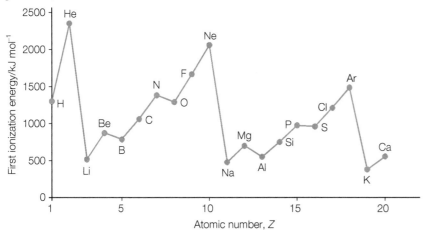

Figure 12.3 First ionization energies for periods 1, 2 and 3

The nuclear charge increases across a period as a proton is added to the next element. The electrons are added to the same outer shell and so the increase in electron–electron repulsion (shielding) rises slowly (Figure 12.4). As a result the atoms become smaller and outer electrons require increasing amounts of energy to release them.

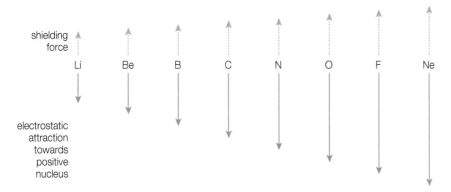

Figure 12.4 A diagram illustrating how the balance between shielding and nuclear charge changes across period 2

There is a rapid decrease in ionization energy between the last element in one period and the first element in the next period. For example, the first ionization energy for lithium is much smaller than the value for helium.

A helium atom has two electrons and these are in the first shell (main energy level) ($1s^2$). However, the lithium atom ($1s^2\ 2s^1$) has three electrons, but the third electron is in a 2s sub-level in the second shell (second main energy level), further away from the nucleus. So the force of electrostatic attraction between the positive nucleus and the outer electrons decreases because the distance between the nucleus and the outer electrons increases and the shielding by inner shells increases. These two factors outweigh the increase in nuclear charge.

There is a slight decrease in the first ionization energy between beryllium and boron. Although boron has one more proton than beryllium, there is a slight decrease in ionization energy on removal of the outer electron.

Beryllium has the electronic structure $1s^2\ 2s^2$ and the boron atom has the electronic structure $1s^2\ 2s^2\ 2p^1$ (Figure 12.5). The fifth electron in boron is in the 2p sub-level; which is slightly further away from the nucleus than the 2s sub-level.

There is less electrostatic attraction between the fifth electron and the nucleus because the distance between the nucleus and the outer electrons increases plus the shielding (electron–electron repulsion) by inner shells increases slightly and these two factors outweigh the increased nuclear charge.

There is a slight decrease in first ionization energy between nitrogen and oxygen. Oxygen has one more proton than nitrogen and the electron removed is in the same 2p sub-level.

However, the electron being removed from the nitrogen atom is from an orbital that contains an unpaired electron. However, the electron removed from the oxygen atom is from an orbital that contains a spin pair of electrons (Figure 12.6).

The additional repulsion between the spin pair of electrons in this orbital results in less energy being needed to remove an electron. Hence the ionization energy decreases from nitrogen to oxygen.

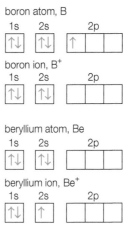

Figure 12.5 Orbital notations for boron and beryllium atoms and their unipositive ions

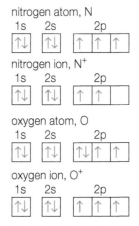

Figure 12.6 Orbital notation for nitrogen and oxygen atoms and their unipositive ions

---

Expert tip

These patterns in first ionization energy repeat themselves across period 3. However, the presence of the d-block elements in period 4 disrupts the pattern, as d-block elements have first ionization energies that are relatively similar and relatively high.

■ QUICK CHECK QUESTIONS

2   The orbital diagram below shows the electron configuration of a nitrogen atom.

   1s   2s $2p_x$ $2p_y$ $2p_z$

   | ↑↓ | ↑↓ | ↑ | ↑ | ↑ |

   a State the significance of *x*, *y* and *z* in describing the electrons in the 2p sub-level.

   b Describe the relative positions of the 2p orbitals.

   c Explain why the $2p_z$ electron for nitrogen is not placed in the $2p_x$ or $2p_y$ orbital.

3   Identify two mistakes in the orbital diagram below showing the electron configuration of a phosphorus atom and draw the correct orbital diagram.

   1s   2s $2p_x$ $2p_y$ $2p_z$ 3s $4p_x$ $4p_y$ $4p_z$

   | ↑↓ | ↑↓ | ↑↓ | ↑↓ | ↑↓ | ↑↓ | ↑↓ | ↑ | |

Figure 12.7 summarizes and explains the periodicity in first ionization energy from hydrogen to argon.

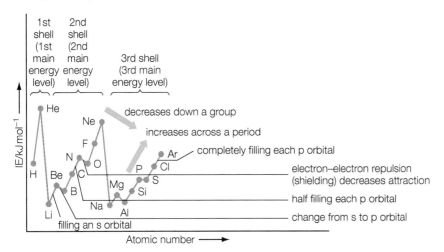

Figure 12.7 Explaining the periodicity in first ionization energy across the first three periods of the periodic table, the 2–3–3 pattern is due to the entry of electrons into the orbitals of s and p sub-levels

---

■ **QUICK CHECK QUESTION**

4   The first 20 elements show periodicity in the values of first ionization energy.

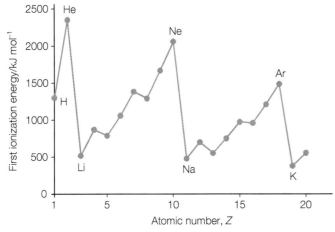

   a Write an equation (with state symbols) describing the first ionization energy of helium.

   b Predict, from the graph, an estimate for the value of the first ionization energy of rubidium.

   c Explain why the noble gases (group 18 elements) have the highest values of ionization in each period.

   d Explain why the alkali metals (group 1 elements) have the lowest values of ionization energy in each period.

   e Explain the general (overall) increase in ionization energy across the three periods.

   f Explain the decreases in ionization energy from beryllium to boron and nitrogen to oxygen.

■ Successive ionization energies

Successive ionization energies provide experimental data supporting the idea of electrons in main energy levels (shells) and sub-levels. Successive ionization energies can be measured by electron bombardment of gaseous atoms.

For example, the second and third ionization energies are described by the following equations:

$$M^+(g) \rightarrow M^{2+}(g) + e^-$$

$$M^{2+}(g) \rightarrow M^{3+}(g) + e^-$$

Successive ionization energies increase for all atoms because as more electrons are removed the remaining electrons experience an increasing effective nuclear charge and are held closer to the nucleus and hence more tightly.

A graph of the successive ionization energies for a potassium atom (Figure 12.8) provides experimental evidence of the number of electrons in each main energy level (shell). The large increases ('jumps') in ionization energy correspond to a change to a new inner shell, closer to the nucleus, with the electrons held more strongly. The graph shows that the electron arrangement of a potassium atom is 2, 8, 8, 1.

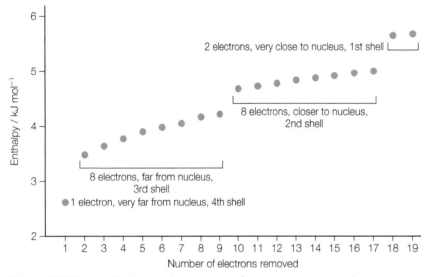

Figure 12.8 Successive ionization energies for a gaseous potassium atom

If the graph of the successive ionization energies is examined closely it can be seen that that the graph does increase regularly within a shell. There is evidence that the main energy levels (shells) are split into sub-levels. A characteristic 2,3,2 pattern can be seen.

Expert tip

A logarithmic scale is used because there is a wide range of values for ionization energies.

■ **QUICK CHECK QUESTIONS**

5 An element has the following first five successive ionization energies:

   Ionization energy (kJ mol⁻¹): 779, 2018, 2945, 11 779, 15 045

   Deduce, with a reason, the group to which the element belongs.

6 The table below shows the successive ionization energies for magnesium. Plot a graph of $\log_{10}$ of the successive ionization energies against the number of electrons removed. Annotate the graph with the main energy levels (shells) and explain why the graph always increases.

Number of electron removed	Ionization energy, kJ mol⁻¹
1	736
2	1 448
3	7 740
4	10 470
5	13 490
6	18 200
7	21 880
8	25 700
9	31 620
10	35 480
11	158 300
12	199 500

→

7 The successive ionization energies of titanium from the d-block are shown below. They can be determined by electron bombardment of gaseous titanium atoms.

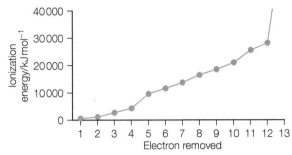

a State the full electron configuration of a titanium atom and identify the sub-level from which the electron is removed when the 3rd ionization energy is measured.

b Explain why there are relatively large differences between the 4th and 5th, and between the 10th and 11th ionization energies.

■ Bohr theory

Bohr's model of the hydrogen atom proposed that the electron behaved as a particle and moved in circular paths (orbits) around the stationary nucleus of an atom. The energies with each allowed orbit were of fixed value (quantized). Energy is only emitted or absorbed by an electron when it undergoes a transition from one orbit (allowed energy level) to another orbit. This energy is emitted or absorbed as a photon (electromagnetic wave) (Figure 12.9). The energy difference between two orbits or energy levels (ΔE) corresponds to the amount of energy absorbed or emitted during the electron transition.

$$\Delta E = E_2 - E_1 = h\nu$$

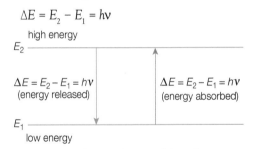

Figure 12.9 Electronic transitions between energy levels

NATURE OF SCIENCE

The Bohr model provided a theoretical explanation of line spectra. These are generated when a high voltage is applied to tubes that contain different gases under reduced pressure, the gases emit different colours of light. When the light is analysed with a prism or diffraction grating the characteristic emission spectra in the form of coloured lines is observed.

■ Determination of the ionization energy from an emission spectrum

The convergence limit for the Lyman series (Figure 12.10) in the hydrogen emission spectrum is where the electron responsible for the spectral lines have been excited into an energy level ($n \rightarrow \infty$) of sufficient energy that the electron no longer experiences the attractive force of the proton (nuclear charge).

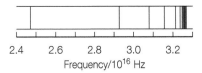

Figure 12.10 The Lyman series in the emission spectrum of atomic hydrogen

■ QUICK CHECK QUESTION

8 The energy of the electron in the second and third orbits (energy levels) of the hydrogen atom are 5.42×10^{-19} J and 2.41×10^{-19} J.

Calculate the wavelength of the radiation (photons) when an electron undergoes a transition from the third to the second orbit (energy level).

Hence, the first ionization energy for the atom of any element can be calculated from the convergence limit for the spectral lines at highest frequency (shortest wavelength).

It is also possible to estimate the value of the first ionization energy by plotting the frequency or energy (*y*-axis) versus change in frequency (Δv) between adjacent lines (*x*-axis). The continuum begins at the point extrapolated on the graph (Figure 12.11) where $\Delta v = 0$ or $\Delta E = 0$ and it is this value which is substituted into $E = hv$.

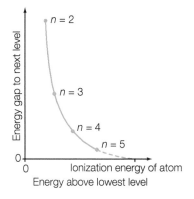

Figure 12.11 A graph showing how the value of ionization energy can be estimated by extrapolation

Expert tip

The ionization energy of gaseous hydrogen atoms can only be determined by studying the emission lines where the excited electrons undergo a transition back to their ground state ($n = 1$).

■ **QUICK CHECK QUESTION**

9 Calculate the ionization energy for hydrogen if the wavelength at the convergence limit is 91.2 nm.

Expert tip

Planck's equation, $E = hv$ and the value of Planck's constant in base SI units are given in the IB Chemistry *data booklet*. Planck's constant is the proportionality constant relating a photon's energy to its frequency.

■ Planck's equation

Planck's equation, $\Delta E = hv$, and the wave equation, $c = v\lambda$, can be used to solve a wide range of problems involving interconversion between energy, frequency and wavelength. Avogadro's constant can also be used to convert the energy of a single photon to an energy per mole of photons.

■ **QUICK CHECK QUESTIONS**

10 A sodium street lamp emits a characteristic yellow light of wavelength 588 nm. Calculate the energy (in kJ mol⁻¹) of the photons.

11 Neon gas is used in adverting signs and emits radiation at 616 nm. Calculate the frequency, the distance travelled in 90 s and the energy of a single photon (in J).

12 The bond enthalpy of a Cl–Cl bond is 243 kJ mol⁻¹. Calculate the maximum wavelength of light that would break one mole of these bonds to form individual chlorine atoms.

13 Calculate the wavelength in the electromagnetic spectrum that corresponds to an energy of 1609 kJ mol⁻¹. Identify the region of the electromagnetic spectrum this photon energy corresponds to.

■ **QUICK CHECK QUESTION**

14 The hydrogen atom is a quantum mechanical system with the electron energy levels shown in the diagram below. The energy levels are indicated by the principal quantum number, ***n*** (an integer).

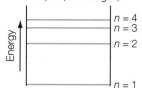

a Using the quantum numbers indicated, state which quantum numbers will involve the electronic transitions that requires the most and least energy.

b Arrange the following transitions in order of increasing wavelength of light absorbed:

I $n = 1$ to $n = 2$

II $n = 2$ to $n = 3$

III $n = 1$ to $n = 3$

NATURE OF SCIENCE

During the 1920s various experimental results suggested that electrons do not always behave in the way that particles should behave. It was shown that a beam of electrons could be diffracted by a crystal lattice, just as a beam of light could be diffracted by a grating. This showed that electrons have wave properties, similar to those of light. De Broglie suggested that the wavelength of the wave associated with a particle could be related to the particle's momentum. This new branch of physics is known as quantum mechanics. Quantized systems can only have a fixed and definite number of energy states – like rungs on a ladder.

13.1 First-row d-block elements

Revised ▢

Essential idea: The transition elements have characteristic properties; these properties are related to their all having incomplete d sub-levels.

Transition elements atoms

Revised ▢

■ Transition elements have variable oxidation states, form complex ions with ligands, have coloured compounds, and display catalytic and magnetic properties.
■ Zn is not considered to be a transition element as it does not form ions with incomplete d orbitals.
■ Transition elements show an oxidation state of +2 when the s electrons are removed.

▦ Electron configurations of the d-block elements

The d-block elements are located in groups 3 to 12 of the periodic table. In period 4 the d-block elements are the elements from scandium to zinc and they possess one or more 3d electrons in their configuration. The changes in the properties across a series of d-block elements are much less marked than the large changes across a p-block. This is because from one element to the next, as the atomic number of the nucleus increases by one, the extra electron goes into the inner d sub-level.

For transition metals in period 4, the outermost orbital that is occupied is always the 4s orbital. Note that in writing the electron configuration of a transition metal, the 3d electrons precede the 4s electrons because the configuration is written in order of increasing energy (see Table 13.1). For this reason, if a transition metal is ionized, the 4s electrons will be lost first. Note that the electron configurations of copper and chromium are exceptions to the general pattern, because of the stability conferred by a stable half-filled and filled d-sub-level respectively. However, when the d-block elements form cations it is the outer 4s electrons that are always lost first during ionization.

Table 13.1 Condensed electron configurations of the first-row d-block metals, where [Ar] represents the electron configuration of the noble gas argon

Element	Atomic number	Electron configuration		3d					4s
Sc	21	$[Ar]3d^14s^2$	[Ar]	↑					↑↓
Ti	22	$[Ar]3d^24s^2$	[Ar]	↑	↑				↑↓
V	23	$[Ar]3d^34s^2$	[Ar]	↑	↑	↑			↑↓
Cr	24	$[Ar]3d^54s^1$	[Ar]	↑	↑	↑	↑	↑↓	↑
Mn	25	$[Ar]3d^54s^2$	[Ar]	↑	↑	↑	↑	↑↓	↑↓
Fe	26	$[Ar]3d^64s^2$	[Ar]	↑↓	↑	↑	↑	↑↓	↑↓
Co	27	$[Ar]3d^74s^2$	[Ar]	↑↓	↑↓	↑	↑	↑↓	↑↓
Ni	28	$[Ar]3d^84s^2$	[Ar]	↑↓	↑↓	↑↓	↑	↑↓	↑↓
Cu	29	$[Ar]3d^{10}4s^1$	[Ar]	↑↓	↑↓	↑↓	↑↓	↑↓	↑
Zn	30	$[Ar]3d^{10}4s^2$	[Ar]	↑↓	↑↓	↑↓	↑↓	↑↓	↑↓

The chemistry of an atom is largely determined by its outer (valence) electrons because they are involved in bonding. The elements from scandium to zinc are similar in chemical and physical properties.

Transition metals

Transition metals are called d-block elements because they have a partially filled d sub-level – in one or more of their oxidation states. In the first row of the d-block, this definition includes all the metals except for zinc. Zinc is excluded because all its compounds are in the +2 oxidation state. Removal of two electrons gives an ion, Zn^{2+}, which has the electron configuration $[Ar]\ 3d^{10}$ where all the orbitals of the 3d sub-level are full.

Properties of transition metals

Transition metals share a number of common features:
- They are metals with useful mechanical properties and with high melting points.
- They form compounds in more than one oxidation state.
- Many transition metals form alloys, which are solutions of one metal in another. Their atoms can be readily incorporated into another's lattice because they have similar sizes.
- They form coloured compounds.
- They form a huge variety of complex ions with many different ligands.
- They form paramagnetic compounds (including complex ions) due to the presence of unpaired electrons.
- They act as catalysts, either as metals or as compounds.

Differences between transition and non-transition elements

Table 13.2 Differences between transition metals and non-transition metals

Transition elements	Non-transition elements (s and p-block elements)
All the transition elements are metals	Non-transition elements may be metals, non-metals or metalloids
The atomic radii generally show a small decrease across a series of transition elements	The atomic radii show a strong and regular decrease across a period
The ionization energies show an irregular but small increase across a period	The ionization energies show a regular and large increase across a period
They show a variety of stable oxidation states – they usually differ by 1	They usually show one stable oxidation state, or multiple oxidation states – usually differing by 2
Most of the aqueous solutions of the ions are coloured	Aqueous solutions of the ions are colourless
They form a large number of complexes	They usually form fewer complexes
Their elements and compounds often show catalytic activity	They usually lack catalytic activity
Their compounds are often paramagnetic	Their compounds are mostly diamagnetic

■ **QUICK CHECK QUESTION**

1 Give full electron configurations for the species Ti, Ti^{2+} and Ti^{4+}.

Common mistake

Not all first row d-block elements are transition metals. Zinc is a d-block metal but not a transition metal. The zinc(II) ion is its only oxidation state and has no unpaired 3d electrons.

■ **QUICK CHECK QUESTIONS**

2 Draw and label clearly the orbital diagram (arrows-in-boxes) to show the 3s, 3p, 3d and 4s orbitals of a copper atom. Explain why copper is classified as a d-block element and transition element.

3 By referring to their electronic configurations, explain why scandium is classified as a transition metal but zinc is not.

4 State the number of unpaired electrons in Mn^{2+}, Fe^{2+}, Ni and Cu^+.

5 Ammonium dichromate(VI), $(NH_4)_2Cr_2O_7$, reacts on heating to form chromium(III) oxide, nitrogen and water. Write an equation describing the reaction and show using oxidation states that it involves an oxidation of the cation by the anion.

NATURE OF SCIENCE

Chromium and copper are considered anomalous in the first row of the d-block since they have electron configurations ($3d^5 4s^1$ and $3d^9 4s^1$) that are not predicted by the Aufbau principle. Copper has two oxidation states: +1 (unstable in aqueous solution) and +2. The copper(I) ion is transitional (it has unpaired 3d electrons), but the copper(II) ion is not transitional (paired 3d electrons). Zinc is anomalous because it is not a transition element: it only forms one cation, Zn^{2+}, which is non-transitional ($3d^{10}$ configuration).

Variable oxidation states

All the d-block elements, except for scandium and scandium and zinc, can exist in more than one oxidation state (Figure 13.1). Transition elements show an oxidation state of +2 when the s-electrons are removed.

Sc	Ti	V	Cr	Mn	Fe	Co	Ni	Cu	Zn
								+1	
	+2	+2	+2	+2	+2	+2	+2	+2	+2
+3	+3	+3	+3	+3	+3	+3			
	+4	+4		+4					
		+5							
			+6	+6					
				+7					

Figure 13.1 Common oxidation states for the first row of the d-block

This is possible because the 3d and 4s electrons have similar energies and are available for bonding. It is seen that the maximum oxidation state rises to a peak on passing from scandium to manganese and then falls. The maximum oxidation number is never greater than the total number of 3d and 4s electrons, since further electrons would have to come from the argon core.

Transition metals form a variety of stable oxidation states, but metals in groups 1 and 2 (the s-block) can only form cations with oxidation states +1 and +2. This is because of the trends in the successive ionization energies. The variable oxidation states arise because the 3d electrons are comparable in energy to the 4s electrons and so can be removed during cation formation.

Expert tip

The first row transition metals show a range of positive oxidation states, but all of them exhibit the +2 oxidation state which corresponds to the removal of the 4s electrons (except for copper(II)).

Expert tip

The maximum oxidation state shown by some transition metals may be given by the sum of the number of 3d and 4s electrons.

Expert tip

The relative stability of oxidation states is very important for the feasibility of redox reactions and is discussed in terms of electrode potentials.

■ **QUICK CHECK QUESTIONS**

6   The successive ionization energies of titanium are shown below.

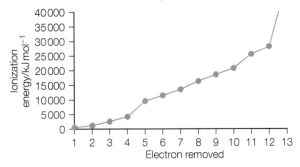

a Write an equation to describe the first and third ionization energies of titanium.

b Explain why there is relatively large difference between the fourth and fifth ionization energies.

7   Explain, by referring to relative values of successive ionization energies and electron configurations, why calcium compounds only exist in the +2 oxidation state but many transition metals exist in cations with range of stable oxidation states.

Paramagnetism

A substance which is attracted into a strong magnetic field is paramagnetic; if it is repelled it is diamagnetic. Transition metals and their ions are commonly paramagnetic, whereas many main group elements and their compounds are usually diamagnetic. Paramagnetism in transition metals and their compounds is due to unpaired electrons in their partially filled d orbitals. The paramagnetic character increases with number of unpaired electrons.

■ QUICK CHECK QUESTIONS

8 State and explain whether the complex ions $[Cu(H_2O)_5Cl]^+$ and $[Zn(H_2O)_6]^{2+}$ are diamagnetic or paramagnetic. They contain Cu^{2+} and Zn^{2+}, respectively.

9 Using iron as an example, describe the properties in which iron and its compounds are different from magnesium and its compounds.

Redox chemistry

Transition elements can undergo redox reactions (Topic 1 Stoichiometric relationships) because they have more than one stable oxidation state. During a redox reaction, one or more electrons pass from the reducing agent to the oxidizing agent. Much of the chemistry of these ions can be used to determine the concentration of a reducing or oxidizing agent in solution as redox titrations.

Solutions of potassium manganate(VII) are purple and contain manganese in its highest oxidation state of +7. The manganate(VII) ion (Figure 13.2) is a powerful oxidizing agent which is reduced to its lowest oxidation state of +2 (in acidic conditions). The +2 oxidation state is the most stable oxidation state of manganese in acidic aqueous solutions with its $3d^5$ electron arrangement.

Potassium manganate(VII) is often used in redox titrations to measure the concentration of reducing agents (though it is not a primary standard). There is no need for an indicator as these titrations are self-indicating because of the colour change from intense purple (MnO_4^-) to very pale pink (Mn^{2+}).

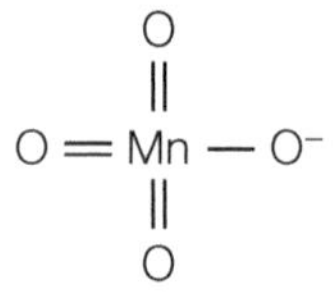

Figure 13.2 Structure of the manganate(VII) ion

The two most common oxidation states of chromium are +6 and +3. Two examples of chromium in the +6 oxidation state are the bright yellow sodium chromate(VI) (Na_2CrO_4) and the bright orange sodium dichromate(VI) ($Na_2Cr_2O_7$) (Figure 13.3).

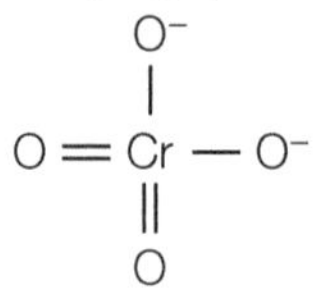

Figure 13.3 Structure of the chromate(VI) ion

These two different coloured forms of chromium in the +6 oxidation state exist in equilibrium in solution (Topic 7 Equilibrium). The equilibrium mixture is yellow in alkaline solutions and orange in acidic conditions.

Chromium(III) compounds are often green, for example, Cr_2O_3, although $[Cr(H_2O)_6]^{3+}$ is purple. Acidified dichromate(VI) is a powerful oxidizing agent and used in organic chemistry (Topic 10 Organic chemistry) for the oxidation of primary and secondary alcohols.

Ethanedioate ions being oxidized to carbon dioxide on reaction with acidified dichromate(VI) is another example of the redox chemistry of dichromate(VI)/chromium(III).

$$Cr_2O_7^{2-} + 3C_2O_4^{2-} + 14H^+ \rightarrow 2Cr^{3+} + 6CO_2 + 7H_2O$$

A redox titration which can be used to estimate the amount of iron(II) ions in solution is its reaction with dichromate(VI).

$$Cr_2O_7^{2-} + 6Fe^{2+} + 14H^+ \rightarrow 2Cr^{3+} + 6Fe^{3+} + 7H_2O$$

■ **QUICK CHECK QUESTIONS**

10 Outline the reasoning for the following in terms of condensed electronic configurations: $V^{3+}(aq)$ is can behave as a reducing agent (in acidified aqueous solution), whereas $Zn^{2+}(aq)$ does not behave as a reducing agent.

11 State the oxidation states of the transition metals in the following reaction and identify the oxidizing and reducing agents.

$$3MnO_4^- + 5FeC_2O_4 + 24H^+ \rightarrow 3Mn^{2+} + 5Fe^{3+} + 12H_2O + 10CO_2$$

■ Selected redox chemistry

■ Chromium

Chromium has the condensed electronic structure $3d^5 4s^1$ and its most important positive oxidation states are +2, +3 and +6 (Table 13.3). The dichromate(VI) ion in acidic solution is a powerful oxidizing agent:

$$Cr_2O_7^{2-} + 14H^+ + 6e^- \rightarrow 2Cr^{3+} + 7H_2O$$

Table 13.3 Summary of chromium chemistry

Oxide	Properties	Cation	Properties	Anion	Properties
CrO	Basic (reacts with acids)	Cr^{2+}	Readily oxidized		
Cr_2O_3	Amphoteric (reacts with bases and acids)	Cr^{3+}	Stable	Chromate(III), $[Cr(OH)_4]^-$	
CrO_3	Acidic (reacts with bases)			Chromate(VI), CrO_4^{2-}	Stable
				Dichromate(VI), $Cr_2O_7^{2-}$	Easily reduced

■ Manganese

Manganese has the outer electronic structure $3d^5 4s^2$ and it has compounds in all oxidation states from +2 to +7 (Table 13.4).

Table 13.4 Summary of manganese chemistry

Oxidation state	Ion	Comment
+2	Manganese(II), Mn^{2+}	Most stable oxidation state (3d sub-level half full); very pale pink in solution
+4	Manganese(IV) oxide, MnO_2	Polar covalent and stable; oxidizing agent
+6	Manganate(VI), MnO_4^{2-}	Unstable; only stable in alkaline solution
+7	Manganate(VII), MnO_4^-	Powerful oxidizing agent

■ Vanadium

Vanadium has the condensed electronic structure $3d^3\ 4s^2$ and its most important positive oxidation states are +2, +3, +4 and +5 (Table 13.5).

Table 13.5 Summary of vanadium chemistry

Oxidation state	d configuration	Comment
+2	d^3	V^{2+}(aq), powerful reducing agent; purple
+3	d^2	V^{3+}(aq), powerful reducing agent; green
+4	d^1	VO^{2+}, blue
+5	d^0	VO_2^{+}, yellow

13.2 Coloured complexes

Revised ☐

Essential idea: d orbitals have the same energy in an isolated atom, but split into two sub-levels in a complex ion. The electric field of ligands may cause the d orbitals in complex ions to split so that the energy of an electron transition between them corresponds to a photon of visible light.

Coloured complexes

Revised ☐

■ The d sub-level splits into two sets of orbitals of different energy in a complex ion.
■ Complexes of d-block elements are coloured, as light is absorbed when an electron is excited between the d orbitals.
■ The colour absorbed is complementary to the colour observed.

■ Formation of complex ions

Because of their small size and availability of empty orbitals, d-block ions can act as Lewis acids (electron pair acceptors) and attract species that are rich in electrons. Such species are known as ligands, which act as Lewis bases. Ligands are molecules or anions which contain a non-bonding or lone pair of electrons. These electron pairs can form coordinate covalent bonds with the metal ion to form complex ions.

The number of ligands bonded to the metal ion is known as the coordination number. Compounds with a coordination number of 6 are octahedral in shape, those with a coordination number of 4 are tetrahedral or square planar, whereas those with a coordination number of 2 are usually linear (Figure 13.4).

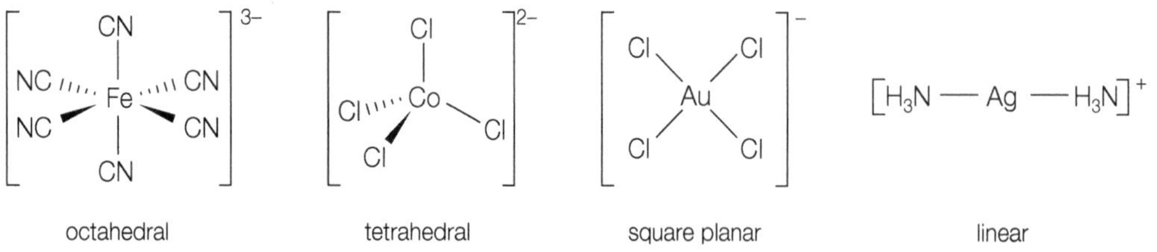

Figure 13.4 Common shapes of complex ions

Expert tip

Complex ions can be regarded as Lewis acid–base reactions. The ligands are Lewis bases (electron pair donors) and the transition metal ion (or atom) is a Lewis acid (electron pair acceptor).

Common mistake

An octahedron has six corners (for six ligands) and eight faces. A tetrahedron has four corners (for four ligands) and four faces.

> ■ **QUICK CHECK QUESTIONS**
>
> 12 Draw diagrams to show the shapes of the following complex ions:
> $[Cu(H_2O)_6]^{2+}$, $[Ag(CN)_2]^-$, $[CuCl_4]^{2-}$
>
> 13 State both the formula and the charge of the following complex ions formed from the components indicated:
>   a one titanium (III) ion and six water molecules
>   b one copper (II) ion and four iodide ions
>   c one cobalt (II) ion and four chloride ions
>   d one copper(II) ion, four ammonia molecules and two water molecules
>   e one iron(III) ion, one thiocyanate ion and five water molecules
>   f one copper(II) ion and two cyanide ions
>   g one iron atom and five carbon monoxide molecules.
>
> 14 Consider the transition metal complex $K_3[Cr(OH)_3Cl_3]$.
>   a Identify all the ligands in the complex and deduce the oxidation number of chromium.
>   b Write the full electron configuration of chromium in the complex and hence, determine the number of unpaired electrons in this ion.

■ Ligand exchange reactions

Reactions involving complex ions in solution often involve exchanging one ligand for another. These ligand exchange reactions are often reversible. The ammonia and water ligands are similar in size and both neutral. With these ligands, exchange reactions take place with no change in the coordination number of the metal ion and there is no change in the oxidation number of the transition metal ions.

$$[Cu(H_2O)_6]^{2+}(aq) + 4NH_3(aq) \rightleftharpoons [Cu(NH_3)_4(H_2O)_2]^{2+}(aq) + 4H_2O(l)$$

pale blue   deep royal blue

The chloride ion is larger than unchanged ligands, such as water, so fewer chloride ions can fit round a central metal ion. Ligand exchange involves a change in coordination number.

$$[Cu(H_2O)_6]^{2+}(aq) + 4Cl^-(aq) \rightleftharpoons [CuCl_4]^{2-}(aq) + 6H_2O(l)$$

pale blue   yellow

Expert tip

Ligand exchange reactions are not redox reactions. There is no change in oxidation number. In the example above, the copper ion is Cu^{2+} throughout. The total charge of the complex ion is obtained by summing the charge of the transition metal ion and any negatively charged ligands.

■ Naming complex ions

Anionic ligands have names ending in 'o': 'ide' → 'o', e.g. chloride → chloro, cyanide → cyano and hydroxide → hydroxo. Neutral ligands are named as the molecule with these notable exceptions: H_2O → aqua, NH_3 → ammine and carbon monoxide, CO → carbonyl.

The numbers of ligands in a complex are specified using Greek prefixes: di- for 2, tetra- for 4 and hexa- for 6.

The name of a cationic complex ion ends in the name of the central transition metal ion with the oxidation state shown as a Roman numeral in parentheses at the end of the metal's name, e.g. iron(III).

The name of an anionic complex ion ends in 'ate', sometimes the Latin name is used, e.g. chromium(II) → chromate(II), iron(II) → ferrate(II) and copper(I) → cuprate(I). Ligands are named before the central metal atom.

Expert tip

Transition metals react with carbon monoxide to give neutral complexes called carbonyls. The oxidation state of the transition metal remains at 0 and its electron configuration is increased to that of a noble gas (with 18 electrons) by the electron pair donated from each carbon monoxide ligand.

■ QUICK CHECK QUESTIONS

15 Write balanced equations to describe the following ligand exchange reactions:

hexaaquacobalt(II) ions with chloride ions

hexaaquairon(III) ions with thiocyanate ions

Explain why these are not redox reactions.

16 When aqueous copper(II) sulfate is added to concentrated hydrobromic acid (HBr), the following equilibrium involving ligand exchange is established:

$[Cu(H_2O)_6]^{2+} + Br^- \rightleftharpoons [Cu(H_2O)_5Br]^+ + H_2O$

a State the coordination number and charge of the copper ion in the complex ion $[Cu(H_2O)_5Br]^+$.

b Define the term ligand and describe the bond formed between the ligands and the copper ion.

c Name both complex ions.

Polydentate ligands

Ligands such as ammonia and cyanide ions are known as monodentate ligands as they use just one non-bonding (lone) pair to form a coordinate covalent bond to the transition metal ion. Some ligands contain more than one non-bonding (lone) pair and form two or more coordinate bonds to the metal ion. These polydentate ligands form very stable complexes (Figure 13.5). Chelates are complex ions are formed by bidentate and polydentate ligands, such as EDTA. Powerful chelating agents trap metal ions and effectively isolate them in solution.

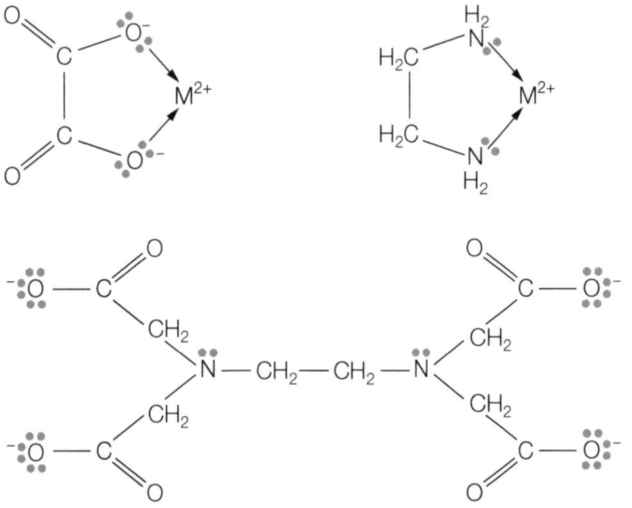

Figure 13.5 Polydentate ligands

Expert tip

Chelate complexes are often significantly more stable than complexes formed by monodentate ligands. This chelate effect can be explained by considering the entropy change for the ligand exchanges involved. When a bidentate ligand replaces monodentate ligands in a complex there is an increase in the number of molecules and ions. The entropy change for the reaction is positive and hence favourable.

■ QUICK CHECK QUESTION

17 Show that the number of molecules and ions in solution increases when hexaaminenickel(II) ions, $[Ni(NH_3)_6]^{2+}(aq)$ reacts with ethylenediamine (1,2-diaminoethane).

Three common examples are ethylenediamine (1,2-diaminoethane) $H_2NCH_2CH_2NH_2$, the oxalate (ethanedioate) ion $(COO^-)_2$ (both of which can use two non-bonding or lone pairs to form bidentate ligands), and EDTA (ethylenediaminetetraacetic acid), or its ion, $EDTA^{4-}$, which can act as a hexadentate ligand (Figure 13.6). EDTA only uses one of the two oxygens in each carboxylate group when binding to a metal ion.

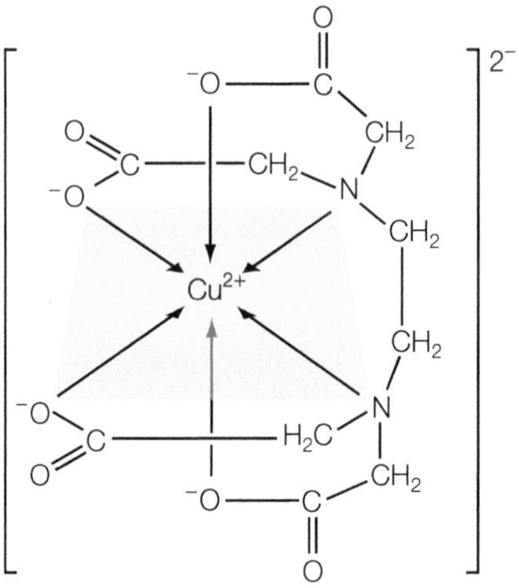

Figure 13.6 The $[Cu(EDTA)]^{2-}$ complex

■ **QUICK CHECK QUESTIONS**

18 State whether each ligand below is monodentate, bidentate or polydentate.

$$H_2N - \overset{\displaystyle H}{\underset{\displaystyle H}{C}} - COOH$$

glycine

2, 2′-bipyridine

$$H_3C - C \overset{\displaystyle H}{\underset{\displaystyle \ddot{:O}:}{C}} = C \underset{\displaystyle :\ddot{O}\bar{:}}{-} CH_3$$

19 Draw a structure of the glycinate anion ($NH_2CH_2CO_2^-$), which can act as a bidentate ligand (nitrogen-donor and oxygen-donor). Draw the two structural isomers of the square planar complex $Cu(NH_2CH_2CO_2)_2$.

■ Stereoisomerism

Some complex ions or complexes can exist as stereoisomers; they have the same formulae but different spatial arrangements of their atoms. There are two types: geometric and optical isomers.

■ *cis–trans* isomers

This type of isomerism only occurs in square planar or octahedral complexes. For example, in $[Pt(NH_3)Cl_2]$ which is square planar (Figure 13.7), there are two possible arrangements of the two different ligands. The fact that there are two isomers of $[Pt(NH_3)Cl_2]$ shows it is not tetrahedral.

$$\underset{\displaystyle cis}{Cl \overset{\displaystyle Cl}{\underset{\displaystyle Cl}{\diagdown} Pt \diagup} \overset{\displaystyle NH_3}{\underset{\displaystyle NH_3}{}}} \quad or \quad \underset{\displaystyle trans}{\overset{\displaystyle Cl}{\underset{\displaystyle H_3N}{} Pt \overset{\displaystyle NH_3}{\underset{\displaystyle Cl}{}}}}$$

Figure 13.7 *cis– trans* isomerism in diamminedichloroplatinum(II)

■ Optical isomerism

Octahedral complexes which have at least two bidentate ligands show optical isomerism. As a result of their three-dimensional shape, there are two possible mirror image isomers, which are non-superimposable and will rotate the plane of plane-polarized light in opposite directions. Figure 13.8 shows the two isomers from a complex with three ethylenediamine (en) ligands around an M^+ metal ion.

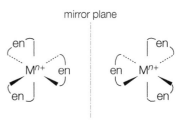

Figure 13.8 Optical isomerism in a complex ion with the bidentate ethylenediamine ligand

---

■ **QUICK CHECK QUESTIONS**

20 1,2-ethanediamine, $NH_2CH_2CH_2NH_2$, forms a complex ion with the cobalt(III) ion. The two molecular structures of the complex ions are shown below. State the type of isomerism shown and state one physical property that is different for the two isomers in aqueous solution.

21 The two geometric isomers of $[Pt (NH_3)_2Cl_2]$ are crystalline. One of the isomers is widely used as a drug in the treatment of cancer.

a Name the complex and draw both isomers of the complex.

b Deduce the polarity of each isomer, using a diagram for each isomer to support your answer.

c State a suitable physical method to establish the structures.

Aqua complexes

Transition metal ions form aqua complexes when dissolved in water. Examples include the hexaaqua ions, $[Fe(H_2O)_6]^{3+}$ and $Mn(H_2O_6)^{2+}$(aq). Solutions of the salts of iron(III) and chromium(III) are acidic because of hydrolysis. The hydrated iron(III) ion, for example, is an acid. Because of its polarizing power the result of a small cation with a large charge and hence a high charge density. The iron(III) ion draws electrons towards itself, so the water molecules can more easily give away protons than free water molecules (Figure 13.9).

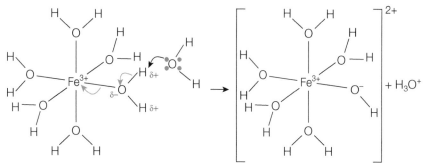

Figure 13.9 The polarization and deprotonation of a water molecule by an iron(III) ion

The hydrated ions in M^{2+} salts are far less acidic because of the smaller polarizing power of these metal cations.

> ■ **QUICK CHECK QUESTION**
>
> 22 Explain, with an equation, why an aqueous solution of copper(II) sulfate is slightly acidic.

Coloured ions

Coloured transition metal ions absorb radiation in the visible region of the electromagnetic spectrum with wavelengths between 400 nm and 700 nm. It is the d electrons in coloured compounds that absorb radiation as they undergo a transition from a low energy level in their ground state to a higher energy level (excited state).

According to quantum theory, there is a simple relationship between the size of the energy gap and the frequency of the radiation absorbed, $E = h\nu$ (Topic 2 Atomic structure).

Light absorbed by transition metal compounds and complex ions interact with the d electrons (Figure 13.10). Some of the wavelengths in the white light are absorbed leaving the complementary colour to be seen by the eye.

Figure 13.11 is a colour wheel diagram to help deduce the colour of a complex. When a particular colour is absorbed by a transition metal ion complex it will take on the colour of its complementary colour in the colour wheel. A complementary colour is the one directly opposite a particular colour.

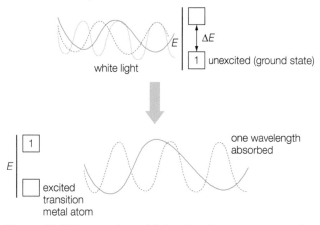

Figure 13.10 Interaction of light showing promotion of a d electron

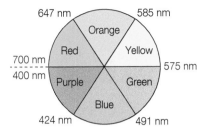

Figure 13.11 Colour wheel of complementary colours

Colour of transition metal complex ions

The colour of a transition metal complex ion depends on

■ the identity of the transition metal ion
■ the oxidation number of the metal
■ the identity of the ligands
■ the coordination number (geometry).

Colour in transition metal ions is usually due to electronic transition between d orbitals in the outer d sub-level; in a free (gaseous) atom all the five d orbitals have the same energy.

When a transition metal ion forms a complex ion the d orbitals are split by the electric field of the ligands into two groups with different energies. The size of the energy gap depends on the number and nature of the ligands in the complex. In an octahedral complex the orbitals of the d sub-level are split, three lower in energy and two higher in energy (Figure 13.12).

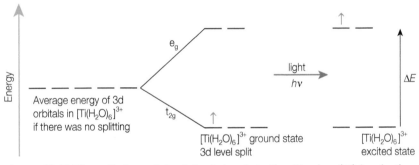

Figure 13.12 The splitting of the 3d sub-levels in the titanium(III) ion in the octahedral $[Ti(H_2O)_6]^{3+}(aq)$ complex

Expert tip

If all the d orbitals are full (Zn^{2+}) or empty (Sc^{3+}) there is no possibility of electronic transitions between them and the ions are colourless.

Splitting of the d orbitals

In the free transition metal ion (in the gas phase) the five 3d orbitals are degenerate (have identical energy). Note that three of the 3d orbitals ($3d_{xy}$, $3d_{yz}$ and $3d_{xz}$) have lobes that lie between the x, y and z-axes, but the other two 3d orbitals ($3d_{x^2-y^2}$ and $3d_{z^2}$) have lobes that lie along the axes (Figure 13.13).

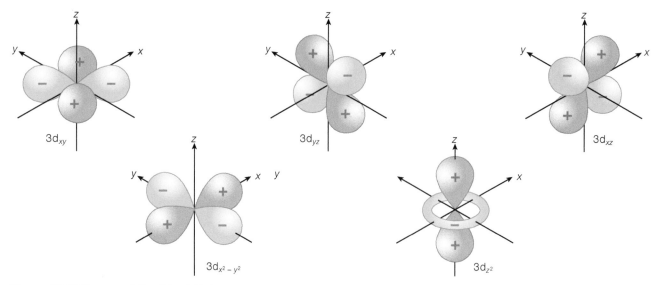

Figure 13.13 Shapes of the 3d orbitals

As the ligands approach the transition metal ion along the *x*-, *y*- and *z*-axes to form an octahedral complex the lone pairs on the ligands will repel the lobes of the $d_{x^2-y^2}$ and d_{z^2} orbitals causing the five d orbitals to split, three to lower energy and two to higher energy (Figure 13.14). Energy will be conserved during this process.

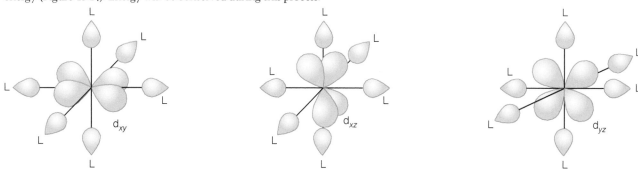

Figure 13.14 Ligands at the ends of the *x*-, *y*- and *z*-axes have weaker interactions with d_{xy}, d_{yz} and d_{xz} orbitals than with d_{z^2} and $d_{x^2-y^2}$ orbitals

▤ Spectrochemical series

Ligands can be arranged in order of their experimental ability to split the 3d sub-level in complex ions:

$$I^- < Br^- < S^{2-} < Cl^- < F^- < OH^- < H_2O < SCN^- < NH_3 < CN^- < CO$$

This order is known as the spectrochemical series. Halide ions are weak-field ligands and cause a small degree of splitting. Cyanide ions and carbon monoxide molecules are strong-field ligands and cause a large degree of splitting.

The energy of light waves (photons) absorbed increases (Figure 13.15) when ammonia is substituted for water as a ligand in the hexaaquacopper(II) ion during the ligand exchange reaction:

$$[Cu(H_2O)_6]^{2+}(aq) + 4NH_3(aq) \rightleftharpoons [Cu(NH_3)_4(H_2O)_2]^{2+}(aq) + 4H_2O(l)$$

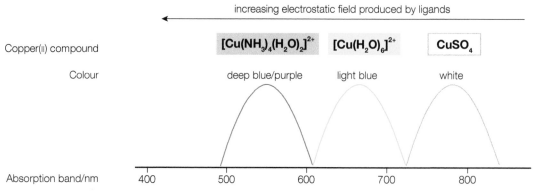

Figure 13.15 Absorption spectra and colours of selected copper(II) compounds showing the effect of a stronger ligand field

NATURE OF SCIENCE

A number of different theories have been proposed to explain the bonding and properties of d-block metals in complex ions and complexes. Many of these models are based on valence bond (VB) and molecular orbital (MO) theories. The simplest is crystal field theory, a simple electrostatic model, based upon ligands behaving as point charges. However, it cannot explain the order of all the ligands in the spectrochemical series or account for the formation of carbonyl complexes.

Spectrophotometry

If visible light of increasing frequency is passed through a sample of a coloured complex ion, some of the light is absorbed. The amount of light absorbed is proportional to the concentration of the absorbing species. Some complexes have only pale colours and do not absorb light strongly. In these cases a suitable ligand is added to intensify the colour. Absorption of visible light is used in spectrometry to determine the concentration of coloured ions.

The general method is as follows:
1   Add an appropriate ligand to intensify the colour.

2   Make up solutions of known concentration.

3   Measure absorbance or transmission.

4   Plot graph of results (calibration line).

5   Measure absorbance of unknown and compare.

Catalysis

■ Heterogeneous catalysts

A heterogeneous catalyst is in a different physical state from the reactants. Generally, a heterogeneous catalyst is a solid while the reactants are gases. Heterogeneous catalysts are used in large scale continuous processes (Table 13.6), such as the Haber and Contact processes (Topic 7 Equilibrium). Heterogeneous catalysts work by adsorbing reactant molecules onto the surface and weakening the bonds.

Table 13.6 Some transition metal based industrial catalysts

Process	Reaction catalysed	Products	Catalyst
Haber	$N_2(g) + 3H_2(g) \rightleftharpoons 2NH_3(g)$	Ammonia	Iron, Fe
Contact	$2SO_2(g) + O_2 \rightleftharpoons 2SO_3(g)$	Sulfuric acid	Vanadium(V) oxide, V_2O_5
Hydrogenation of unsaturated oils to harden them	$RCH=CHR' \rightarrow RCH_2CH_2R'$	Semi-solid unsaturated fat	Nickel, Ni
Hydrogenation	Alkene to alkane	Alkane	Nickel, Ni
Ziegler–Natta polymerization of alkenes	$nCH_2=CHR \rightarrow -[CH_2-CHR]_n-$	Stereoregular polymer	Complex of $TiCl_3$ and $Al(C_2H_5)_3$

■ Homogeneous catalysts

A homogeneous catalyst is in the same physical state as the reactants. Typically, the reactants and the catalyst are dissolved in aqueous solution. A homogeneous catalyst works by forming an intermediate with the reactants, which then breaks down to form the products. Transition metal ions can be homogeneous catalysts because they can gain and lose electrons as they change from one stable oxidation state to another.

The direct oxidation of iodide ions by persulfate ions, for example, is very slow, because of repulsion between anions:

$$S_2O_8^{2-}(aq) + 2I^-(aq) \rightarrow 2SO_4^{2-}(aq) + I_2(aq)$$

The reaction is catalysed by iron(III) ions in the solution. The iron(III) ions are reduced to iron(II) ions as they oxidize iodide ions to iodine molecules. The persulfate ions oxidize the iron(II) ions back to iron(III) ions ready to oxidize more of the iodide ions:

$$2Fe^{2+}(aq) + S_2O_8^{2-}(aq) \rightarrow 2SO_4^{2-}(aq) + 2Fe^{3+}(aq)$$

and

$$2Fe^{3+}(aq) + 2I^-(aq) \rightarrow 2Fe^{2+}(aq) + I_2(aq)$$

These two reactions are rapid because they involve attraction between oppositely charged ions.

■ **QUICK CHECK QUESTION**

29 In aqueous solution, $S_2O_8^{2-}$ ions are reduced to SO_4^{2-} ions by I^- ions.

  a Write an ionic equation to describe this reaction.

  b Suggest why the reaction has a high activation energy with a low rate.

  c Iron(II) salts can catalyse this reaction. Write two equations to explain the role of the iron(II) ion as a catalyst in this reaction.

■ Magnetic properties

Iron, nickel and cobalt show ferromagnetism. This is a permanent and very strong type of magnetism. In this type of magnetism the unpaired electrons align parallel to each other in domains (small areas of the crystal) regardless of whether an external magnetic or electric field is present.

Many complex ions of transition metals contain unpaired electrons. The unpaired electrons spin on their axis and generate a small magnetic field. They line up in an applied magnetic field to make the transition metal complex weakly magnetic when a strong magnetic field is applied, that is, they reinforce the external magnetic field. This type of weak magnetism is known as paramagnetism (Figure 13.16).

The greater the number of unpaired electrons there are in the complex ion the greater the magnetic moment of the complex. When all the electrons in a transition metal complex ion are paired up the complex is said to be diamagnetic. In spin pairs the opposing spins means no magnetic field is generated.

Complex ions containing first row transition metal ions (3d complexes) are high spin with weak field ligands and low spin with strong field ligands. High-spin complex ions are paramagnetic and low-spin complexes are diamagnetic. Note that high and low-spin states occur only for 3d metal complexes with between 4 and 7d electrons.

Cobalt(II) complex ions ([Ar] 3d⁷) can be paramagnetic or diamagnetic (Figure 13.17). The five 3d orbitals are split by the ligands according to the spectrochemical series. If the ligands are low in the spectrochemical series, for example, water molecules, they will split the 3d orbitals, leading to a small energy gap, so the electrons can occupy all the 3d orbitals and there will be four unpaired 3d electrons. This is known as weak ligand crystal field splitting.

Ligands high in the spectrochemical series, such as cyanide ions, will cause a larger degree of d–d splitting, leading to a larger energy gap. Only the lower 3d orbitals will be occupied and the complex will be diamagnetic as there are no unpaired 3d electrons.

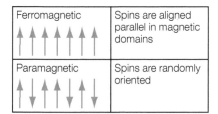

| Ferromagnetic | Spins are aligned parallel in magnetic domains |
| Paramagnetic | Spins are randomly oriented |

Figure 13.16 Ferromagnetism and paramagnetism

Expert tip

Compounds containing scandium(III), titanium(IV), copper(I) or zinc ions have no unpaired electrons and therefore their compounds are diamagnetic.

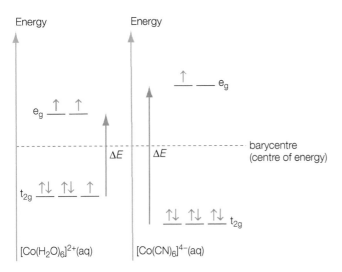

Figure 13.17 d orbital energy diagrams for high- and low-spin Co^{2+} complexes, d^7

Greater splitting within the 3d sub-level means photons with higher energies are absorbed to promote d–d electronic transitions. Table 13.7 summarizes how a large and small energy gap will affect the colour of the complex ion.

Table 13.7 Effect of energy gap on the colour of a complex ion

Property	Large energy gap	Smaller
Energy of light absorbed	Higher energy	Lower energy
Frequency	Higher	Lower
Wavelength	Smaller	Longer
Colour of light absorbed	Blue end of visible spectrum	Red end of visible spectrum
Colour observed	Yellow or red	Cyan or blue

■ QUICK CHECK QUESTIONS

30 The CrF_6^{4-} ion is known to have four unpaired electrons. State and explain whether the fluoride ligand produces a strong or a weak field.

31 The table shows the properties of three iron(II) complex ions where X, Y and Z are three different ligands. All three cationic complexes have a coordination number of 6.

Complex ions	Wavelength of maximum adsorption/nm	Paramagnetic properties
$[FeX_6]^{2+}$	675	Yes
$[FeY_3]^{2+}$	455	No
$[FeZ_6]^{2+}$	415	No

a State three factors that will affect the colours of transition metal complex ions.

b State and explain which ligand(s), X, Y or Z, is(are) a bidentate ligand.

c All these ligands are electron pair donors. Arrange the ligands in increasing order (from lowest to highest) of the energy difference they produce between the two sets (e_g and t_{2g}) of 3d orbitals in an octahedral complex. Explain your answer.

d In all three complex ions, the electronic configuration of the Fe^{2+} ion is [Ar] $3d^6$. By considering the paramagnetic properties, suggest possible electronic arrangements of the 3d-electrons of Fe^{2+} ions in $[FeX_6]^{2+}$ and $[FeZ_6]^{2+}$.

Chemical bonding and structure

14.1 Further aspects of covalent bonding and structure

Essential idea: Larger structures and more in-depth explanations of bonding systems often require more sophisticated concepts and theories of bonding.

Further aspects of covalent bonding and structure

- Sigma and pi bonds.
- Formal charge.
- Delocalization and resonance structures.

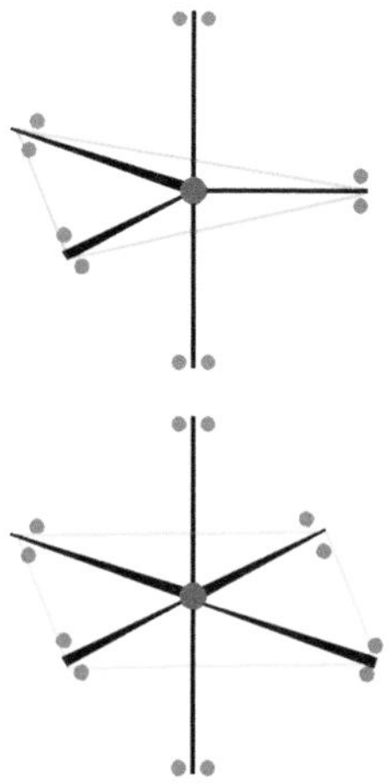

■ VSEPR theory

VSEPR theory can be used to predict the shapes of molecules and polyatomic ions having five or six electron domains around the central atom (Figure 14.1). The two basic geometries are those of a triangular bipyramid and an octahedron.

Table 14.1 summarizes how the numbers of bonding and lone pairs of electrons determine the geometries of molecules with five and six electron domain.

Table 14.1 Summary of molecular shapes for species with five and six centres of negative charge

Total number of electron pairs	Number of electron domains		Molecular shape	Examples
	Bonding pairs	Lone pairs		
5	2	3	Linear	ICl_2^-, XeF_2 and I_3^-
5	3	2	T-shaped	ClF_3 and BrF_3
5	4	1	See-saw (distorted tetrahedral)	SF_4
5	5	0	Trigonal bipyramidal	PCl_5
6	6	0	Octahedral	SF_6 and PF_6^-
6	5	1	Square pyramidal	BrF_5 and ClF_5
6	4	2	Square planar	XeF_4 and ICl_4

Figure 14.1 The basic shapes for molecules with five and six electron pairs

Expert tip

In a trigonal bipyramid molecule the axial and equatorial positions are not equivalent: lone pairs always go in the equatorial position (in the middle). This rule only holds for five electron domains.

■ Exceptions to the octet rule

Sometimes the octet rule cannot be used to predict the formula of covalent molecules because there are covalent molecules that do not obey the octet rule. There are three main types of exceptions to the octet rule:

- Molecules in which a central atom has less than an octet (incomplete octet), for example, $BeCl_2(g)$ and $BCl_3(l)$.
- Molecules in which a central atom has more than eight electrons in its valence shell. This occurs in atoms in period 3 (and later periods) due to the presence of low energy empty d orbitals. Examples include $PCl_5(g)$ and $SF_6(g)$.
- Molecules with an odd number of electrons (radicals). Examples of radicals, include $NO(g)$, $NO_2(g)$ and $ClO_2(g)$ (Figure 14.2).

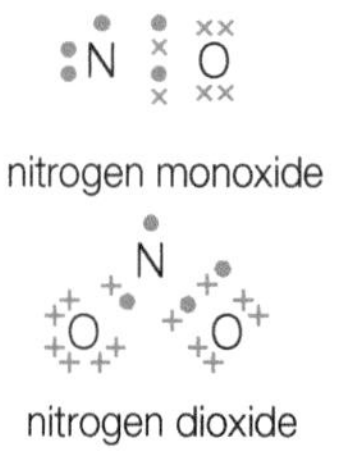

nitrogen monoxide

nitrogen dioxide

Figure 14.2 Lewis structures (electron dot diagrams) of nitrogen monoxide and nitrogen dioxide molecules

■ QUICK CHECK QUESTION

1 Draw the Lewis structures for xenon oxytetrafluoride $XeOF_4(g)$ and the tetrafluoroborate ion, $BrF_4^-(g)$. Deduce the molecular shapes and bond angle(s).

■ **QUICK CHECK QUESTIONS**

2 Phosphorus forms anions containing halogens. PF_4^{n-} has a see-saw structure. State the total number of electron domains and bond pairs and lone pairs around the central phosphorus atom. State the value of n in this anion.

3 Deduce the shape of, and the bond angle in, the species $XeCl_2$ and BrF_2^+.

4 Iodine is relatively insoluble, but this can be improved by complexing the iodine with iodide to form soluble triiodide ions: $I_2 + I^- \rightleftharpoons I_3^-$. Draw the Lewis (electron dot) structure of triiodide, I_3^-, and state its shape.

Table 14.2 shows examples of molecules without the noble gas configuration.

Table 14.2 Examples of molecules with incomplete and expanded octets

Compound	'Dot-and-cross' diagram	Comments
$BeCl_2(g)$	$\overset{\times\times}{\underset{\times\times}{\times}Cl\times} \; \overset{\bullet}{\underset{\bullet}{Be}} \; \overset{\times\times}{\underset{\times\times}{\times}Cl\times}$	After bonding, the central beryllium atom has four valence electrons. After bonding, the central boron atom has only six valence electrons.
$BCl_3(l)$	boron trichloride	Beryllium and boron are period 2 elements with a second shell as the valence shell ($n = 2$) which can hold a maximum of $2n^2 = 8$ electrons.
		(Beryllium and boron form polar covalent compounds due to the high ionization energies involved in forming Be^{2+} and B^{3+}, respectively.)
		They often have incomplete valence shells (fewer than 8 electrons) in their compounds. These electron-deficient compounds are very reactive.
$PF_5(g)$	phosphorus pentafluoride PF_5	After bonding, the central phosphorus atom has 10 valence electrons. After bonding, the central sulfur atom has 12 valence electrons.
		Phosphorus and sulfur are Period 3 elements with a third shell (main energy level) the valence shell ($n = 3$) which can hold a maximum of $2n^2 = 18$ electrons.
$SF_6(g)$	sulfur hexafluoride SF_6	There are low-energy 3d orbitals to hold the extra electrons required for phosphorus and sulfur to expand their octets and show their maximum oxidation state.

Common mistake

There are a number of examples of molecules and ions that break the octet rule and expand the octet. If the central atom (in a molecule or ion) is from period 3, 4, etc. it may have up to 18 electrons in its outer (valence) shell.

Expert tip

It is important to identify the central atom(s) when answering questions about molecular shapes (VSEPR theory). The central atom may have an expanded octet if it is from period 3 onwards.

■ Delocalization

Resonance is a model that uses two or more Lewis structures to represent or describe an ion or molecule where the structure cannot be adequately described by a single Lewis (electron dot structure). They can also be described by an equivalent model involving delocalization of pi (π) electrons.

In molecular orbital theory, resonance-stabilized structures are described in terms of 'delocalized' pi orbitals where the π electron clouds extend over three or more atoms. Delocalized pi orbitals are formed by the overlap and merging of parallel p orbitals. Delocalization of pi electrons can occur whenever alternating single and double carbon–carbon bonds occur between carbon atoms.

Figures 14.3, 14.4 and 14.5 show the delocalization of pi electrons in the benzene molecule, carboxylate ion and carbonate ion.

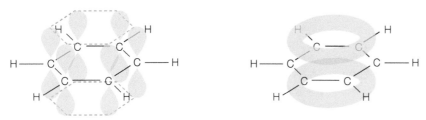

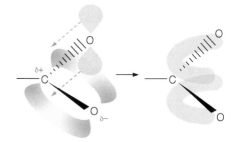

Figure 14.3 Overlapping of p_z orbitals to form a delocalized cyclic pi orbital in the benzene molecule

Figure 14.4 Delocalization in a carboxylate ion

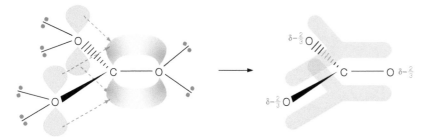

Figure 14.5 Delocalization in a carbonate ion

Formal charge

It is often possible to write two (or more) different Lewis structures (electron dot diagrams) for a molecule or ion different in the arrangement of valence electrons.

The formal charge on an atom in a Lewis structure is the charge it would have if the bonding electrons were shared equally. The formal charge of an atom is the number of valence electrons in the free atom minus the number of electrons assigned to that atom (Figure 14.6).

All lone pairs (non-bonding) electrons are assigned to the atoms in which they are found. Half of the bonding electrons are assigned to each atom in the bond. Formal charge is defined as shown in Figure 14.6.

$$\textbf{formal charge} = \begin{bmatrix}\text{number}\\\text{of valence}\\\text{electrons in}\\\text{free atom}\end{bmatrix} - \underbrace{\frac{1}{2}\begin{bmatrix}\text{number}\\\text{of}\\\text{bonding}\\\text{electrons}\end{bmatrix} - \begin{bmatrix}\text{number}\\\text{of lone}\\\text{pair}\\\text{electrons}\end{bmatrix}}_{\text{number of electrons assigned to atom}}$$

Figure 14.6 The calculation of the formal charge on an atom

In general, when several Lewis structures are possible, the most stable Lewis structure is the one with no formal charges or the structure in which:

- the atoms bear the smallest formal charges
- the negative charges appear on the more electronegative atoms.

The sum of the formal charges of the atoms in a Lewis structure must equal zero for a molecule, or must equal the ionic charge for a polyatomic ion.

NATURE OF SCIENCE

The concept of formal charge is a simple book-keeping method to determine if an atom within a molecule or ion is neutral or has a positive or negative charge. It provides integer charges only but is a good starting point for determining the electron distribution and hence predicting chemical and physical properties. It is a simple but powerful bonding model. If two chemical bonding models are otherwise equivalent and produce the same results, it is considered best to pick the model that is the simplest. This is an application of Occam's razor.

QUICK CHECK QUESTION

5 a Use the VSEPR theory to deduce and explain the shape of the nitrate(III) ion, NO_2^-.

b Draw two equivalent Lewis (electron dot) structures and state the approximate value of the bond angle. Draw a delocalized representation showing for the ion showing the partial charges.

Expert tip

The relationship for calculating formal charge is one which needs to be learnt, together with the criteria for deducing which are the most appropriate Lewis structures for a molecule or ion.

Worked example

Draw two possible Lewis structures of the azide ion, N_3^-, and using formal charges, state and explain which is the likely Lewis structure of the azide ion.

	I	II
	$[\bar{N}\equiv N-\bar{N}]^-$	$[\bar{N}=N=\bar{N}]^-$
FC:	0 +1 −2	−1 +1 −1

Lewis structure II is more likely since it has a smaller difference in formal charges.

■ **QUICK CHECK QUESTIONS**

6 Two possible Lewis structures of the chlorate(V) ion, ClO_3^-, are shown below.

a Using VSEPR theory, state the electron domain geometry, molecular shape and the O–Cl–O bond angle in structure I.

b Applying the concepts of formal charge and the octet rule, state and explain which is likely to be the Lewis structure of the ClO_3^- ion.

7 Mercury and silver salts containing the fulminate ion, CNO^-, are used in explosive detonators.

a Draw out the three possible Lewis (electron dot) structures of the fulminate ion and label clearly the formal charges on each of the three atoms.

b State and explain which is the most likely Lewis (electron dot) structure of the fulminate ion.

8 Nitrogen monoxide, •NO, is produced in internal combustion engines at high temperatures. Draw a Lewis (electron dot) structure of nitrogen monoxide and calculate the formal charge on the nitrogen and oxygen atoms.

Covalent bonds

Revised ☐

In order to form a covalent bond, the two atoms must come close enough for their atomic orbitals with unpaired electrons to overlap and merge. Too large an overlap will result in a strong repulsion between the bonding nuclei. The most stable or equilibrium situation is when there is partial overlapping of the two atomic orbitals (Figure 14.7).

The overlapping and merging of atomic orbitals result in the formation of a new molecular orbital surrounding both nuclei. The pair of electrons in the covalent bond is shared between two atoms in the volume where the orbitals overlap.

The covalent bond in a hydrogen molecule is formed by the overlapping and merging of two 1s orbitals. The two electrons are shared between the two hydrogen atoms with high electron density along the inter-nuclear axis. ·

The molecular orbital (MO) and valence bond (VB) theories are both based upon the combination of atomic orbitals and the build-up of electron density between nuclei.

■ Sigma (σ) and pi (π) bonds

There are two types of molecular orbitals: sigma bonds and pi bonds, depending on the type of atomic orbital overlapping.

■ Sigma (σ) bonds

A sigma bond is formed by the direct head-on/end-to-end overlap of atomic orbitals, resulting in electron density concentrated between the nuclei of the bonding atoms.

■ *s–s overlapping*: this involves the overlap of two half-filled s orbitals along the inter-nuclear axis as shown in Figure 14.8.

Figure 14.8 Forming a sigma bond by s–s overlapping

■ *s–p overlapping*: this involves the overlap of the half-filled s orbital of one atom and the half-filled p orbital of another atom (Figure 14.9).

Figure 14.9 Forming a sigma bond by s–p overlapping

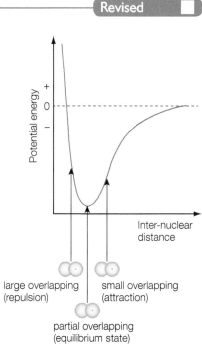

Figure 14.7 Overlap of atomic orbitals

Expert tip

In general, the greater the atomic orbital overlap the stronger is the bond formed between two atoms. This explains why multiple bonding is most common in period 2 with relatively small atoms.

- *p–p overlapping*: this involves the overlap of the half-filled p orbital of one atom and the half-filled p orbital of another atom (Figure 14.10).

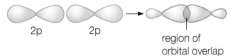

2p   2p   region of orbital overlap

Figure 14.10 Forming a sigma bond by p–p overlapping

■ Pi (π) bonding

A pi bond is formed by the sideways overlap of p atomic orbitals, resulting in electron density above and below the plane of the nuclei of the bonding atoms (Figure 14.11).

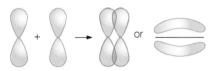

p orbital   p orbital   p–p overlapping

Figure 14.11 Formation of a pi bond

Table 14.3 A summary of the differences between σ and π bonds

Sigma (σ) bond	Pi (π) bond
This bond is formed by the axial overlap of atomic orbitals.	This bond is formed by the sideways overlap of atomic orbitals.
This bond can be formed by the axial overlap of s with s or of s with a hybridized orbital.	This bond involves the sideways overlap of parallel p orbitals only.
The bond is stronger because overlapping can take place to a larger extent.	The bond is weaker because the overlapping occurs to a smaller extent.
The electron cloud formed by axial overlap is symmetrical about the inter-nuclear axis and consists of a single electron cloud.	The electron cloud is discontinuous and consists of two charged electron clouds above and below the plane of the atoms.
There can be a free rotation of atoms around the σ bond.	Free rotation of atoms around the π bond is not possible because it would involve the breaking of the π bond.
The σ bond may be present between the two atoms either alone or along with a π bond.	The π bond is only present between two atoms with a σ bond, i.e. it is always superimposed on a σ bond.
The shape of the molecule or polyatomic ion is determined by the σ framework around the central atom.	The π bonds do not contribute to the shape of the molecule.

> ### ■ QUICK CHECK QUESTION
>
> 9   Compare the formation of a sigma and a pi bond between two carbon atoms in a molecule.

14.2 Hybridization

Essential idea: Hybridization results from the mixing of atomic orbitals to form the same number of new equivalent hybrid orbitals that can have the same mean energy as the contributing atomic orbitals.

Hybridization

- A hybrid orbital results from the mixing of different types of atomic orbitals on the same atom

■ Hybridization

In all molecules (except the hydrogen molecule), some of the atomic orbitals involved in covalent bonding combine (after electron promotion) to form a new set of atomic orbitals. The process of combining two or more atomic orbitals that have similar energies is called hybridization. (It is part of the valence bond (VB) theory) and used to explain the geometrical shapes of common molecules.)

> **Expert tip**
>
> The hybrid orbitals are more effective in forming stable bonds than pure atomic orbitals (especially in smaller atoms). Hybridized orbitals are always equivalent in energy and shape.

The hybridization of s and p orbitals produces three types of hybrid orbitals: sp, sp^2 and sp^3 hybridized orbitals. The number of hybrid orbitals formed by hybridization equals the total number of atomic orbitals that are hybridized.

sp hybrid orbitals are formed by the hybridization of one s and one p orbital; sp^2 hybrid orbitals are formed by the hybridization of one s and two p orbitals and sp^3 hybrid orbitals are formed by the hybridization of one s and three p orbitals.

■ sp hybrid orbitals

The hybridization of one s atomic orbital with one p atomic orbital gives rise to two sp hybrid orbitals. To minimize electron repulsion, the two sp hybrid orbitals are linear and able to form covalent bonds at 180° to each other (Figure 14.12).

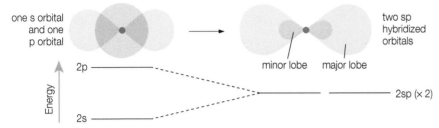

Figure 14.12 The formation of sp hybridized orbitals from an s orbital and a p orbital

This type of hybridization is used to explain the linear shape of the beryllium chloride molecule ($BeCl_2(g)$) and chemical bonding in the nitrogen molecule (N_2) and the ethyne molecule (C_2H_2) (Figures 4.13 and 4.14).

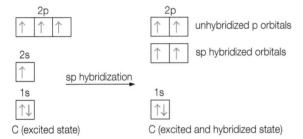

Figure 14.13 Excited carbon atom and sp hybridized orbitals

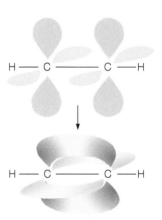

Figure 4.14 Pi bond formation in ethyne, C_2H_2

■ sp^2 hybrid orbitals

The hybridization of one s atomic orbital with two p atomic orbitals gives rise to three sp^2 hybrid orbitals (Figure 4.15). The three hybrid orbitals are co-planar (lie in a plane) at an angle of 120° to each other. The sp^2 hybrid orbitals are oriented towards the corners of an equilateral triangle.

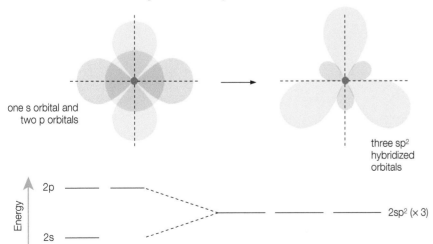

Figure 4.15 The formation of sp^2 hybridized orbitals from an s orbital and two p orbitals

This type of hybridization is used to explain the trigonal planar shape of the boron trifluoride molecule, $BF_3(g)$ and the bonding in ethene (C_2H_4) (Figures 14.16 and 14.17).

Figure 14.16 Excited carbon atom and an sp² hybridized carbon atom

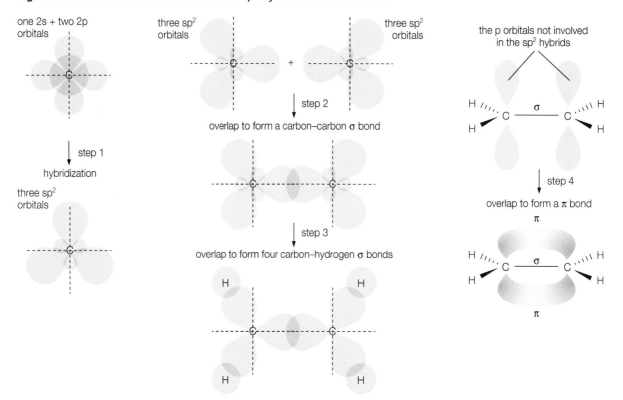

Figure 14.17 The formation of the ethene molecule from sp² hybridized orbitals

---

■ **QUICK CHECK QUESTIONS**

11 State the number of sigma bonds and pi bonds in $H_2CC(CH_3)CHCHCl$.

12 a Draw a Lewis structure for the ethene molecule, C_2H_4.

  b State the hybridization of the two carbon atoms in ethene and the total number of sigma and pi bonds in an ethene molecule.

  c With reference to the ethene molecule, explain how a sigma and a pi bond are formed in the carbon–carbon and carbon–hydrogen bonds.

13 1,3-butadiene, C_4H_6 is a conjugated hydrocarbon with conjugated double bonds. The structure of 1,3-butadiene is shown below.

```
     H       H
     |       |
     C = C   C = C
H ─ C   C ─ C ─ H
         |   |
         H   H
```

  a Explain what is meant by the terms conjugation and delocalization.

  b State the type of hybridization present in the three carbon atoms of 1,3-butadiene.

  c Predict and explain the bond length for the carbon–carbon bond in a molecule of 1,3-butadiene.

  d Describe how the hybridized orbitals are formed from the ground state of a carbon atom.

Although the hybridization model has proven to be a powerful theory, especially
in organic chemistry, it does have its limitations. For example, it predicts that
H_2O and H_2S molecules will have electron domains based on bond angles slightly
smaller than the tetrahedral angle of 109.5°, due to the greater repulsion by the
lone pair. This is a good model for the water molecule (104.5°) but the bond angle
in the hydrogen sulfide molecule is only 92°, suggesting that the p orbitals (which
are 90° apart) are a better model of the electron distribution around the sulfur
atoms than sp^3 hybridized orbitals.

sp³ hybridization

The hybridization of one atomic s orbital with three p atomic orbitals produces
four sp^3 hybrid orbitals. Each sp^3 hybrid orbital contains one electron and is
directed to the corner of a tetrahedron (Figure 14.18).

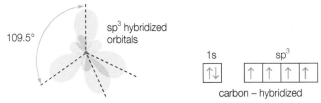

Figure 14.18 The formation of sp³ hybridized orbitals from an s orbital and
three p orbitals

This type of hybridization is used to explain the bonding in methane
(Figures 14.19), ammonia and water molecules.

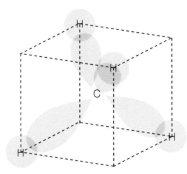

Figure 14.19 sp³ hybrid orbitals in a
methane molecule; showing the overlap
to form sigma bonds with four H atoms

Table 14.4 Summary of hybridization in carbon

Type of hybridized orbital	sp³	sp²	sp
Atomic orbitals used	s, p, p, p	s, p, p	s, p
Number of hybridized orbitals	4	3	2
Number of atoms bonded to the carbon atom	4	3	2
Number of σ bonds	4	3	2
Number of unhybridized p orbitals	0	1	2
Number of π bonds	0	1	2
Bonding arrangements	Tetrahedral; four single bonds only	Trigonal planar; two single bonds and a double bond	Linear; one single bond and a triple bond or two double bonds
Example	CH_4, C_2H_6 and CCl_4	$H_2C=CH_2$ and H_2CO	$H−C\equiv C−H$, $H_2C=C=CH_2$

Relationship between type of hybridization, Lewis structure and molecular shapes

VSEPR theory can be used to identify the hybridization state of the central atom
in a molecule or polyatomic ion (Table 14.5).

Table 14.5 The relationship between the Lewis structure and the hybridization of the central atom

Hybridization state of central atom	Number of electron domains	Number of covalent bonds	Number of lone pairs	Shape	Examples
sp	2	2	0	Linear	BeF_2, CO_2
sp²	3	3	0	Trigonal planar	BF_3, graphite, fullerenes, SO_3 and CO_3^{2-}
sp²	3	2	1	V-shaped or bent	SO_2 and NO_2^-
sp³	4	4	0	Tetrahedral	CH_4, diamond, ClO_4^- and SO_4^{2-}
sp³	4	3	1	Pyramidal	NH_3, HF_3, PCl_3 and H_3O^+
sp³	4	2	2	V-shaped or bent	H_2O, H_2S and NH_2^-

■ QUICK CHECK QUESTIONS

14 a Nitrogen oxychloride has the formula NOCl. Given that nitrogen is the central atom, draw Lewis (electron dot) structure for nitrosyl chloride.

   b State the shape of the molecule and predict the bond angle.

   c State the meaning of the term *hybridization* and identify the type of hybridization shown by the nitrogen and chlorine atoms in nitrogen oxychloride.

15 State the type of hybridization of the carbon and nitrogen atoms in ethanamide, CH_3CONH_2.

16 One resonance structure of tetrasulfur tetranide, S_4N_4, is shown below.

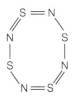

   a State the number of sigma (σ) bonds, lone pairs and pi (π) bonds present in S_4N_4.

   b Suggest the bond angle of N–S–N.

   c State and explain the hybridization of the nitrogen atoms in S_4N_4.

   d Draw an equivalent resonance structure for the S_4N_4 molecule.

17 Using the table given below, state the number of carbon atoms that are sp, sp² and sp³ hybridized in the following molecules.

Molecule	sp hybridized	sp² hybridized	sp³ hybridized
Butanal $CH_3(CH_2)_2CHO$			
Pentane, C_5H_{12}			
Ethylbenzene, C_6H_5-C_2H_5			

18 Draw a Lewis structure, state the molecular geometry and bond angles, hybridization models and types of bonds in triazene, H_2N–N=NH.

19 a State the number of sigma and pi bonds in a molecule of propene (structure shown below).

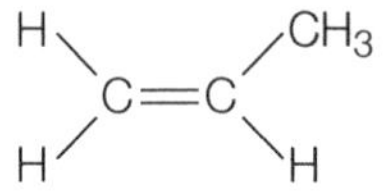

   $$H_2C=CH-CH_3$$

   b Sketch two diagrams showing the formation of the pi bond.

   c Explain why the pi bond is described as localized.

Valence bond (VB) theory and molecular orbital (MO) theory

▦ VB theory

Valence bond theory treats every bond as the sharing of electron density between two atoms as a result of constructive interference of their atomic orbitals. The shapes of many molecules, polyatomic ions and multiple bonds are explained by sp, sp^2 and sp^3 hybridization models.

▦ MO theory

In molecular orbital theory the bonding molecular orbitals arise from constructive interference between two atomic orbitals; anti-bonding molecular orbitals result from destructive interference. Each molecular orbital is associated with the entire molecule.

Molecules are stable when the bond order is greater than 0. Bond order is half of the difference between the numbers of electrons present in the bonding and anti-bonding orbitals.

If all the molecular orbitals in a molecule are doubly occupied, the substance is diamagnetic and repelled by a strong magnetic field. However, if one or more molecular orbitals are singly occupied the substance is paramagnetic and attracted by a strong magnetic field, for example, O_2.

> **Expert tip**
>
> Two molecular orbitals are the most important to consider: the highest occupied molecular orbital (HOMO) and the lowest unoccupied molecular orbital (LUMO). These are involved in the absorption and emission of radiation (Option C Energy).

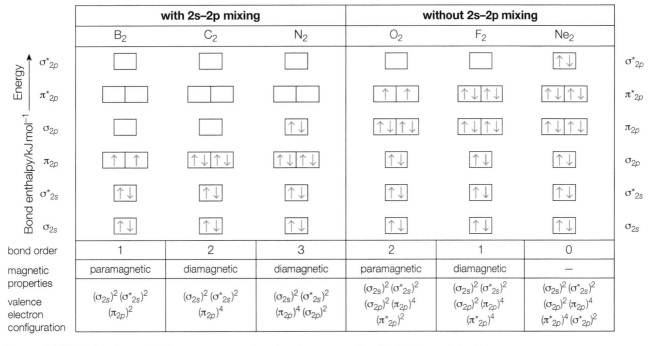

Figure 14.20 Molecular orbital occupancy and molecular properties for B_2 through to Ne_2

▦ Ozone and ultraviolet radiation

The surface of the Earth is protected from harmful ultraviolet radiation by the ozone layer in the stratosphere.

Oxygen and ozone molecules absorb ultraviolet radiation (of different wavelengths) and undergo dissociation:

$$O_2(g) \rightarrow 2O{\bullet}(g)$$

$$O_3(g) \rightarrow O{\bullet}(g) + O_2(g)$$

▩ Comparison of photodissociation in ozone and diatomic oxygen

Both the formation of oxygen-free radicals – from diatomic oxygen – and the decomposition of ozone into diatomic oxygen and a free radical require light of a specific wavelength (frequency):

Photodissociation of diatomic oxygen, O_2		**Photodissociation of ozone, O_3**	
Equation	$O_2 + UV \rightarrow O\bullet + O\bullet$	Equation	$O_3 + UV \rightarrow O_2 + O\bullet$
UV wavelength, λ, nm	(shorter than) 242	UV wavelength, λ, nm	300
Frequency, v, s^{-1} (higher energy radiation)	$v = \dfrac{c}{\lambda}$ $= \dfrac{3.00 \times 10^8\,m\,s^{-1}}{242 \times 10^{-9}\,m}$ $= 1.24 \times 10^{15}\,s^{-1}$	Frequency, v, s^{-1}	$v = \dfrac{c}{\lambda}$ $= \dfrac{3.00 \times 10^8\,m\,s^{-1}}{300 \times 10^{-9}\,m}$ $= 1 \times 10^{15}\,s^{-1}$
Bond enthalpy/kJ mol^{-1}	498	Bond enthalpy/kJ mol^{-1}	399
Covalent bond types	One σ bond and one π bond (bond order = 2)	Covalent bond types	One σ and 2 π electrons delocalized over three oxygen atoms – bonds intermediate between a double and single bond (bond order = 1.5)

The wavelength of the light necessary to cause the dissociation of diatomic oxygen is lower than the wavelength of the light needed for the dissociation of ozone, O_3. This means more energy (higher frequency) is needed to break the bond in the oxygen molecule than in the oxygen bond in the ozone molecule. This makes sense as the double bond in the diatomic oxygen molecule is stronger (and shorter) than the combined single and delocalized bond (or resonance of the two π bonding electrons) in the ozone molecule.

Worked example

Calculation of wavelength of one photon to dissociate O_2 when the bond enthalpy is 498 kJ mol^{-1}.

Energy of one photon is

$$\frac{498\,000\,J\,mol^{-1}}{6.02 \times 10^{23}\,mol^{-1}} = 8.27 \times 10^{-19}\,J$$

$$E = hv \text{ and } v = \frac{c}{\lambda}$$

$$\therefore E = \frac{hc}{\lambda}$$

$$\lambda = \frac{hc}{E}$$

$$= \frac{6.63 \times 10^{-34}\,J\,s \times 3.00 \times 10^8\,m\,s^{-1}}{8.27 \times 10^{-19}\,J}$$

$$= 2.41 \times 10^{-7}\,m = 241\,nm$$

Worked example

Calculation of bond enthalpy in kJ mol^{-1} for ozone when a wavelength of 300 nm is needed to dissociate the bond.

$$E = \frac{hc}{\lambda}$$

$$= \frac{6.63 \times 10^{-34}\,J\,s \times 3.00 \times 10^8\,m\,s^{-1}}{300 \times 10^{-9}\,m}$$

$$= 6.63 \times 10^{-19}\,J \text{ (one photon/one bond)}$$

for 1 mole of bonds:

$$6.03 \times 10^{-19}\,J \times 6.02 \times 10^{23}\,mol^{-1}$$

$$= 399\,126\,J\,mol^{-1} = 399\,kJ\,mol^{-1}$$

▩ Catalysis of ozone depletion

Chlorofluorocarbons (CFCs) such as dichlorodifluoromethane, CCl_2F_2, have their carbon–chlorine bond broken homolytically by the absorption of ultraviolet radiation in the stratosphere:

$CCl_2F_2(g) \rightarrow \bullet CClF_2(g) + \bullet Cl(g)$

Ozone molecules are rapidly removed by reactions with free radicals (reactive species with unpaired electrons), such as chlorine atoms or molecules of nitrogen oxides (NO_x).

These free radicals can take part in a rapid chain reaction, which converts ozone molecules to oxygen molecules, and regenerates the free radical (represented by $\bullet X$).

$\bullet X(g) + O_3(g) \rightarrow \bullet XO(g) + O_2(g)$ (X = $Cl\bullet$, $NO\bullet$ or $\bullet NO_2$)

$XO\bullet(g) + O\bullet(g) \rightarrow O_2(g) + X\bullet(g)$

The overall reaction (cancelling the radical, $X\bullet$, and the intermediate, $XO\bullet$) is:

$O_3(g) + O\bullet(g) \rightarrow 2O_2(g)$

The free radical ($Cl\bullet$, $NO\bullet$ or $\bullet NO_2$) catalyses the decomposition of ozone and is regenerated in the second step and therefore acts as a homogeneous catalyst.

▪ QUICK CHECK QUESTIONS

20 Nitrogen monoxide participates in ozone layer depletion. In this process, ozone (trioxygen) is converted into oxygen (dioxygen).

  a Write a balanced equation to describe this overall process.

  b Using chemical equations, describe the mechanism of ozone depletion when nitrogen monoxide acts as a homogeneous catalyst.

21 The ozone–oxygen cycle is the process by which ozone is continually regenerated in Earth's stratosphere, converting ultraviolet radiation (UV) into heat.

  Step 1 $O_2 \rightarrow O\bullet + O\bullet$

  Step 2 $O\bullet + O_2 \rightarrow O_3$

  Step 3 $O_3 \rightarrow O_2 + O\bullet$

  a Use Table 11 of the IB Chemistry *data booklet* to calculate the minimum wavelength of radiation needed to break the O=O in Step 1.

  b With reference to the bonding in oxygen and ozone molecules, explain whether step 3 absorbs UV radiation of a higher or lower frequency than step 1.

22 Chlorofluorocarbons, CFCs, were responsible for most of the destruction of the ozone layer. Using relevant chemical equations, explain how chlorine radicals destroy the ozone layer. State the role of chlorine atoms in this process.

Topic **15** Energetics/thermochemistry

15.1 Energy cycles

Essential idea: The concept of the energy change in a single step reaction being equivalent to the summation of smaller steps can be applied to changes involving ionic compounds.

Energy cycles

- Representative equations (e.g. $M^+(g) \rightarrow M^+(aq)$) can be used for enthalpy/energy of hydration, ionization, atomization, electron affinity, lattice, covalent bond and solution.
- Enthalpy of solution, hydration enthalpy and lattice enthalpy are related in an energy cycle.

■ Born–Haber cycles

Born–Haber cycles are energy cycles describing the formation of ionic compounds. They are graphical representations of Hess's law and are normally drawn relative to the arbitrary zero enthalpy content of a pure element in its most stable form under standard conditions. Positive enthalpy changes involve going up the diagram and negative changes involve going down the diagram (Figure 15.1).

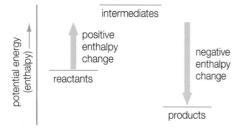

Figure 15.1 Energy changes on an energy level diagram

The Born–Haber cycle for a binary ionic compound is constructed by:
1 forming the cations from the solid metal (atoms) in the correct amounts for the formula of the ionic solid
2 then forming the anions, remembering that these ions involve electron affinities not ionization energies
3 then bringing the gaseous ions together to form a lattice.

The stepwise reaction pathway has to be equal to the direct route of enthalpy of formation of the solid from its elements, so the lattice enthalpy can be determined.

Direct pathway = indirect pathway

Enthalpy of formation = formation of ions + lattice energy

> **Expert tip**
>
> Hess's law can be used to find any missing value from a Born Haber cycle, not just a lattice enthalpy.

Note that the large exothermic value for the lattice enthalpy of the ionic compound (Figure 15.2) helps to compensate for the endothermic steps that lead to the formation of gaseous ions.

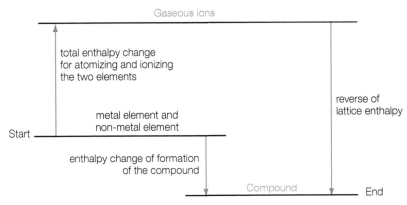

Figure 15.2 The main features of a generalized Born–Haber cycle

The lattice enthalpy is the enthalpy change when one mole of an ionic crystal is separated into its gaseous ions (at infinite distance).

■ **QUICK CHECK QUESTIONS**

1 Explain why lattice enthalpies are always positive.
2 Write equations describing the lattice enthalpies of magnesium oxide, aluminium fluoride, potassium sulfide and sodium peroxide.

Figure 15.3 shows the Born–Haber cycle for sodium chloride. The unknown value is usually the lattice enthalpy, but the cycle can be used to calculate any individual enthalpy change.

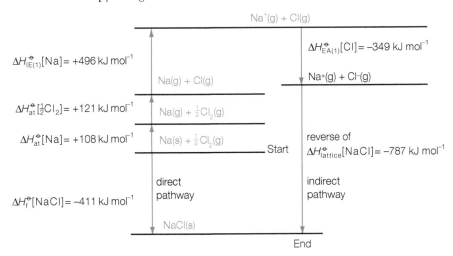

Figure 15.3 The Born–Haber cycle for sodium chloride

Expert tip

Conversion of a lattice to gaseous ions requires an input of energy into the system and is thus endothermic, so by the normal conventions of thermodynamics lattice enthalpies are given a positive sign. Hess's law can be used to define the reverse of lattice enthalpy: the conversion of gaseous ions to a lattice. This would be exothermic – same size, but opposite (negative) sign to the lattice enthalpy.

The enthalpy of atomization is the enthalpy change when one mole of gaseous atoms is formed from the element (under standard conditions). For diatomic molecules this is equal to half the bond enthalpy.

Common mistake

The definition of enthalpy change of atomization is for the formation of 1 mole of gaseous atoms. It may or may not be formed from one mole of atoms or molecules.

■ QUICK CHECK QUESTIONS

3   a Write equations with state symbols to describe the enthalpies of atomization for nitrogen, barium, iodine and scandium.

    b Explain why the value is zero for neon.

4   The lattice enthalpy of an ionic solid cannot be measured directly by experiment.

    a Define lattice enthalpy.

    b Name all the enthalpy changes that need to be measured to calculate the lattice enthalpy of an ionic compound MX [$M^+ X^-$]. Write an equation with state symbols to represent all the enthalpy changes that occur.

    Assume M and X are solids under standard conditions.

5   Using the following data, together with data from the IB Chemistry *data booklet*, draw a Born–Haber cycle and determine the experimental lattice enthalpy of magnesium oxide.

    Enthalpy change of atomization of magnesium: $+148\,kJ\,mol^{-1}$; enthalpy change of formation of magnesium oxide: $-602\,kJ\,mol^{-1}$ and the second ionization energy of magnesium: $+1451\,kJ\,mol^{-1}$.

6   a Draw a labelled Born–Haber cycle for the formation of barium bromide from barium metal and bromine gas.

    b Calculate the enthalpy of formation for barium bromide, $BaBr_2$.

    Lattice enthalpy of barium bromide = $1950\,kJ\,mol^{-1}$; atomization energy of barium = $175\,kJ\,mol^{-1}$; 1st ionization energy of barium = $503\,kJ\,mol^{-1}$; 2nd ionization energy of barium = $965\,kJ\,mol^{-1}$; bromine bond enthalpy = $193\,kJ\,mol^{-1}$ and electron affinity of bromine = $-325\,kJ\,mol^{-1}$.

7   a Use the following data to calculate the enthalpy of formation of gold(II) chloride, $AuCl_2(s)$.

    Enthalpy of atomization (Au): $368\,kJ\,mol^{-1}$; calculated lattice enthalpy (AuCl$_2$): $-2180\,kJ\,mol^{-1}$; electron affinity (Cl): $-349\,kJ\,mol^{-1}$; bond energy (Cl$_2$): $242\,kJ\,mol^{-1}$ and ionization energies (Au): 890, 1980, $2900\,kJ\,mol^{-1}$.

    b Determine whether a gold chloride with the formula $AuCl_2$ is predicted by calculation to be an energetically stable solid.

Expert tip

The bond enthalpy of a diatomic molecule is twice the enthalpy change of atomization. For example, the bond energy for the oxygen molecule [$O_2(g) \rightarrow 2O(g)$] is twice the enthalpy change of atomization for oxygen $\frac{1}{2}O_2(g) \rightarrow O(g)$].

The lattice enthalpy is a measurement of the strength of the ionic bonding within an ionic lattice. Lattice enthalpy is directly proportional to the product of the charges on the ions and inversely proportional to the distance between the ions. Hence lattice enthalpy increases when the ions have a higher charge and smaller ionic radii (Figure 15.4).

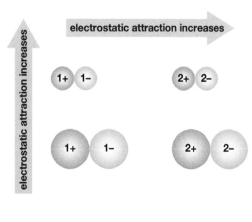

Figure 15.4 The effect of ion size and ion charge on electrostatic attraction and hence lattice enthalpy

8   Explain why the lattice enthalpies of the sodium halides decrease from
    sodium fluoride, NaF, to sodium iodide, NaI.

Compound	Experimental value/kJ mol^{-1}
NaF	+930
NaCl	+790
NaBr	+754
NaI	+705

9   State the factors that will affect the lattice enthalpy of an ionic compound.
    Hence predict how the value of lattice enthalpy of potassium chloride, KCl,
    will be compared to that of calcium chloride, $CaCl_2$.

■ Solubility of ionic solids in liquids

The enthalpy of solution is the energy change (absorbed or released) when 1 mole
of an ionic solid is dissolved in sufficient water to form a very dilute solution. The
enthalpy of solution of an ionic substance in water depends on the lattice enthalpy
of the compound and the enthalpy change of hydration of its ions (Figure 15.5).
Enthalpies of solution can be positive (endothermic) or negative (exothermic).
Substances with large positive values of enthalpy of solution are likely to be
relatively insoluble.

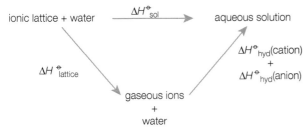

Figure 15.5 A generalized energy cycle to show the dissolving (dissolution)
of an ionic solid

10 At 298 K, the enthalpy of solution of calcium chloride is −123 kJ mol^{-1} and
    the lattice enthalpy of this salt is +225 kJ mol^{-1}. The enthalpy of hydration of
    the calcium ion is −1650 kJ mol^{-1}. Calculate the enthalpy of hydration of the
    chloride ion.

The enthalpy of hydration is the heat released when one mole of gaseous ions are
hydrated (under standard conditions). Ion–dipole forces are formed between the
oxygen atoms of the water molecule and the cations and between the hydrogen
atoms of the water molecules and the anions. The enthalpy of hydration of an ion
is directly proportional to the charge of the ions and inversely proportional to the
ionic radius (Figure 15.6).

11 The enthalpy of hydration for the chloride ion is −359 kJ mol^{-1} and that for
    the bromide ion is −328 kJ mol^{-1}. Explain why the chloride ion has the more
    negative value.

Generally, an ionic compound is soluble if the enthalpy of solution is negative, an
exothermic process. This implies that the sum of the hydration energies is greater
than its lattice enthalpy, so that the positive enthalpy needed to break up the
lattice is more than recovered from the negative enthalpy of hydration of the ions.
However, solubility is determined by the Gibbs free energy change, ΔG, which also
involves an entropy change, ΔS.

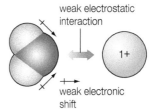

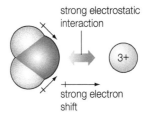

Figure 15.6 The effect
of charge density of a
cation on the hydration
enthalpy

Water is a polar solvent with partial
negative charges on the oxygen atom
and partial positive charges on the
hydrogen atoms. Water molecules
orientate themselves so that their
partial charges surround cations and
anions, forming a hydration shell
(Figure 15.7). The enthalpy change of
hydration is a measure of the amount
of energy released during hydration.

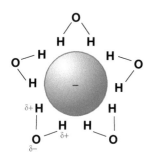

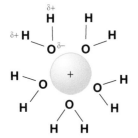

Figure 15.7 Hydrated anion and cation showing hydration shells

15.2 Entropy and spontaneity

Revised ☐

Essential idea: A reaction is spontaneous if the overall transformation leads to an increase in total entropy (system plus surroundings). The direction of spontaneous change always increases the total entropy of the universe at the expense of energy available to do useful work. This is known as the second law of thermodynamics.

Entropy and spontaneity

Revised ☐

- Entropy (S) refers to the distribution of available energy among the particles. The more ways the energy can be distributed the higher the entropy.
- Gibbs free energy (G) relates the energy that can be obtained from a chemical reaction to the change in enthalpy (ΔH), change in entropy (ΔS), and absolute temperature (T).
- Entropy of gas > liquid > solid under same conditions.

■ Spontaneity

A spontaneous process (Figure 15.8) is one which is able to proceed on its own, though it may be immeasurably slow unless enough activation energy is supplied. Most reactions which are spontaneous are exothermic. However, some spontaneous reactions are endothermic and the direction of some spontaneous reactions may be reversed by a change in temperature.

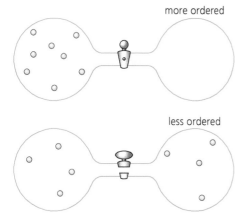

Figure 15.8 Gases spread out spontaneously by diffusion to maximize entropy

In addition to the enthalpy change (ΔH) another thermodynamic function known as entropy (ΔS) is involved in determining the direction of spontaneous change.

> ■ **QUICK CHECK QUESTION**
>
> **13** Distinguish between spontaneous and non-spontaneous reactions.

■ Entropy

Entropy (S) is a state function that can be visualized as a measure of disorder or randomness in a system. The greater the disorder in an isolated system, the higher is its entropy. An isolated system is a thermodynamic system that cannot exchange either energy or matter outside the boundaries of the system.

Entropy depends on the number of arrangements of atoms, ions or molecules among the energy levels. The greater the number of arrangements open to a system, the greater is the entropy.

For a given substance entropy values increase from solids (most order), to liquids to gases (most disorder) (Figure 15.9).

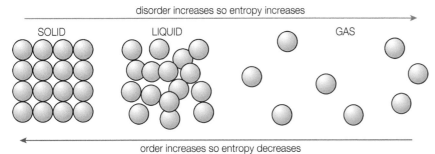

Figure 15.9 Effect of changes in state on entropy

The entropy of a substance is affected by changing temperature (Figure 15.10). Increasing temperature increases the translational, rotational and vibrational motion of the molecules. This leads to a greater disorder and an increase in the entropy. Similarly, the entropy decreases as the temperature is lowered.

> ■ **QUICK CHECK QUESTION**
>
> **14** The molar entropy values for steam, water and ice are +189, +70 and +48 J K^{-1} mol^{-1}. Explain what the data suggests about the structures of these substances.

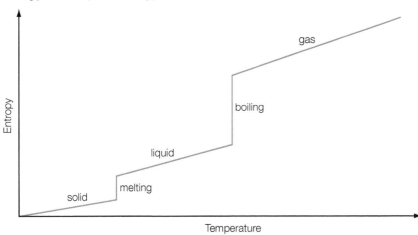

Figure 15.10 Entropy changes with temperature

The sign of the entropy change can usually be determined by examining the stoichiometry and state symbols of the chemical equation. If the products are more disordered than the reactants, then the entropy change (of the system), ΔS, is positive. If the products are less disordered than the reactants, then the entropy change (of the system), ΔS, is negative. Some of the more common examples are summarized in Table 15.1.

Worked example
$SO_3(g) \rightarrow 2SO_2(g) + O_2(g)$
$\Delta S > 0$ because 1 mol of gas produces 3 mol of gas.

Worked example
$H_2O(l) \rightarrow H_2O(s)$
$\Delta S < 0$ because liquid (disordered) turns into solid (more ordered).

Table 15.1 Qualitative entropy changes (of the system) for some common reactions and physical changes

Chemical reaction or physical change	Entropy change	Example
Melting	Increase	$H_2O(s) \rightarrow H_2O(l)$
Freezing	Decrease	$H_2O(l) \rightarrow H_2O(s)$
Boiling	Large increase	$H_2O(l) \rightarrow H_2O(g)$
Condensing	Large decrease	$H_2O(g) \rightarrow H_2O(l)$
Sublimation	Very large increase	$I_2(s) \rightarrow I_2(g)$
Vapour deposition	Very large decrease	$I_2(g) \rightarrow I_2(s)$
Dissolving a solute to form a solution	Generally an increase (except with highly charged ions)	$NaCl(s) + (aq) \rightarrow NaCl(aq)$
Precipitation	Large decrease	$Pb^{2+}(aq) + 2Cl^-(aq) \rightarrow PbCl_2(s)$
Crystallization from a solution	Decrease	$NaCl(aq) \rightarrow NaCl(s)$
Chemical reaction: solid or liquid forming a gas	Large increase	$CaCO_3(s) \rightarrow CaO(s) + CO_2(g)$
Chemical reaction: gases forming a solid or liquid	Large decrease	$2H_2S(g) + SO_2(g) \rightarrow 3S(s) + 2H_2O(l)$
Increase in number of moles of gas	Large increase	$2NH_3(g) \rightarrow N_2(g) + 3H_2(g)$

■ **QUICK CHECK QUESTION**

15 In the following reactions or processes, state whether there is a decrease or increase in the entropy of the system as the reaction proceeds.

a $Na_2S_2O_3(aq) \rightarrow Na_2S_2O_3(s) + (aq)$

b $SOCl_2(l) + H_2O(l) \rightarrow SO_2(g) + 2HCl(g)$

c $N_2(g) + 3H_2(g) \rightarrow 2NH_3(g)$

d $Pb^{2+}(aq) + 2I^-(aq) \rightarrow PbI_2(s)$

e $6SOCl_2(l) + FeCl_3.6H_2O(s) \rightarrow 12HCl(g) + 6SO_2(g) + FeCl_3(s)$

f $C_6H_6(l) \rightarrow C_6H_6(g)$

g $MgO(s) + CO_2(g) \rightarrow MgCO_3(s)$

h $Cu(s) (100\,°C) \rightarrow Cu(s) (25\,°C)$

i $2NH_3(g) + CO_2(g) \rightarrow H_2O(l) + NH_2CONH_2(aq)$

▨ Entropy and probability

Entropy is a measure of the randomness or disorder in a system, but this is a reflection of the number of ways that the energy in a system can be arranged. The larger the number of ways the energy can be distributed, the larger the entropy. In statistical terms, the larger the number of ways the energy can be distributed for a particular value of the total energy, the larger is the probability of finding the system in that state, and the larger is its entropy.

Energy is in the form of 'packets' known as quanta and so it is possible to count the different possible arrangements of energy quanta. Imagine two molecules with four quanta of energy between them. Table 15.2 shows there are five ways in which the energy (four quanta) can be divided between the two molecules.

Table 15.2 Distribution of four quanta between two molecules

Quanta possessed by first molecule	Quanta possessed by second molecule
0	4
1	3
2	2
3	1
4	0

However, if the two molecules absorb six quanta of energy then there are now seven ways of distributing the six quanta. The number of ways of arranging the energy and hence the entropy has increased (Table 15.3).

Table 15.3 Distribution of six quanta between two molecules

Quanta possessed by first molecule	Quanta possessed by second molecule
0	6
1	5
2	4
3	3
4	2
5	1
6	0

■ **QUICK CHECK QUESTION**

16 a Deduce how many ways there are of arranging four quanta of energy, A, B, C and D, in two boxes, 1 and 2.

  b Deduce the probability of finding all the quanta in box 2.

  c Deduce the percentage of the arrangements that are the sum of the 3:1 and 2:2 arrangements.

The quanta of energy could be electronic, vibrational, rotational and/or translational energy (Figure 15.11). The size of the quanta, that is, the difference between the energy levels increases in the order:

translational < rotational < vibrational < electronic

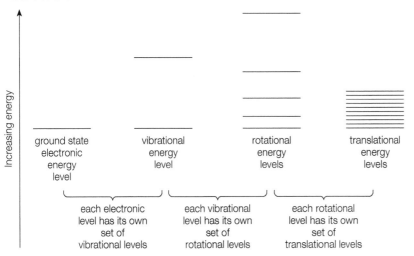

Figure 15.11 Each electronic energy level has several vibrational, rotational and translational energy levels associated with it

■ **QUICK CHECK QUESTION**

17 Describe the melting of ice in a glass of water in terms of changes in the structure of the system and changes in the energy distribution.

■ Absolute entropy values

When considering enthalpies, there is no explicit zero point – chemists have to define one. In contrast, entropy has a zero value, the entropy of a perfect crystal of a substance at absolute zero is the lowest possible and is taken as zero.

Thus, the standard entropy of a substance is the entropy change per mole that accompanies the heating of the substance from $0\,K$ to $298\,K$. Unlike enthalpy (H), absolute values of entropy (S) can be measured experimentally and many such values have been tabulated. Unlike $\Delta H_f^\ominus$ values which can be positive or negative, $S^\ominus$ values are always positive.

Because entropy is a state function, the standard entropy change for a reaction can be determined by calculating the difference between the total entropy of the products and the reactants.

In symbols this can be expressed as:

$$\Delta S^\ominus = \sum S^\ominus_{[products]} - \sum S^\ominus_{[reactants]}$$

Note that while the enthalpy of formation of an element is zero by definition, elements do have $S^\ominus$ values which must be included in the calculations.

Common mistake

When calculating an entropy change from entropy values, do not forget to multiply each entropy value by the coefficient in the balanced equation.

Worked example
Calculate the entropy change for the reaction.

$4Fe(s) + 3\,O_2(g) \rightarrow 2Fe_2O_3(s)$

$S^\ominus\,[Fe(s)] = 27.15\,J\,K^{-1}\,mol^{-1}$; $S^\ominus\,[O_2(g)] = 205.0\,J\,K^{-1}\,mol^{-1}$ and

$S^\ominus\,[Fe_2O_3(s)] = 89.96\,J\,K^{-1}\,mol^{-1}$.

$\Delta S^\ominus = 2S^\ominus\,[Fe_2O_3] - 4S^\ominus\,[Fe] - 3S^\ominus\,[O_2]$

$\quad = 2 \times 89.96\,J\,K^{-1}\,mol^{-1} - 4 \times 27.15\,J\,K^{-1}\,mol^{-1} - 3 \times 205.0\,J\,K^{-1}\,mol^{-1}$

$\quad = -543.7\,J\,K^{-1}\,mol^{-1}$

▣ Entropy and the second law of thermodynamics

The first law of thermodynamics simply says that energy is a constant – it can be changed from one form to another but the total amount does not change. This means that it is not possible to predict the direction of change in a system simply by knowing the sign of the enthalpy change – we have to know how both enthalpy and entropy change.

To predict whether a change will take place or not, the entropy changes in the system and surroundings (Figure 15.12) must be considered and the second law says that in any spontaneous change the *total* entropy change of the system and surroundings must be positive.

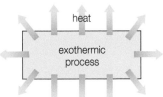

energy released to surroundings, the
entropy of which therefore increases

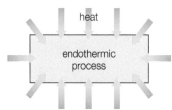

energy absorbed from surroundings, the
entropy of which therefore decreases

Figure 15.12 Changes in the entropy of the surroundings during exothermic and endothermic reactions

The total entropy change for a process is given by the relationship: $\Delta S_{total} = \Delta S_{system} + \Delta S_{surroundings}$, and the second law of thermodynamics states that for a spontaneous chemical change or physical process, $\Delta S_{total} > 0$.

In the case of an isolated system (one which can exchange neither matter nor energy with its surroundings) the situation is simple because for a spontaneous change: $\Delta S_{total} = \Delta S_{system} + \Delta S_{surroundings} = \Delta S_{system} > 0$.

For other cases the situation is more complex because we cannot neglect the surroundings. However we can simplify matters by considering a closed system at constant temperature and pressure. Since a closed system can only change the entropy of the surroundings by transfer of energy, it follows that:

$$\Delta S_{surroundings} = \frac{-\Delta H_{system}}{T}$$

This may seem restrictive but a closed system at constant temperature and pressure is how most chemical reactions occur in the laboratory.

Worked example

The enthalpy change of melting ice is $+6.0\,kJ\,mol^{-1}$. Calculate the entropy change of the system on melting 1 mole of ice at its melting point of 273 K.

$$\Delta S_{system} = \frac{\Delta H}{T} = +\frac{6\,000}{273} = +22\,J\,K^{-1}\,mol^{-1}$$

▣ Free energy

The direction of spontaneous change is determined by:

$$\Delta S_{total} = \Delta S_{system} + \Delta S_{surroundings} > 0$$

Now in a closed system at constant temperature and pressure we have:

$$\Delta S_{surroundings} = -\frac{\Delta H_{system}}{T}$$

So the equation becomes:

$$\Delta S_{total} = \Delta S_{system} - \Delta S_{surroundings} = -\frac{\Delta H_{system}}{T}$$

■ **QUICK CHECK QUESTIONS**

18 Deduce whether the thermal decomposition of calcium carbonate is feasible at 27°C and 927°C.

$$CaCO_3(s) \rightarrow CaO(s) + CO_2(g) \qquad \Delta H = +178\,kJ\,mol^{-1}; \Delta S_{system} = 160.4\,JK^{-1}\,mol^{-1}$$

Calculate the entropy change in the surroundings and the total entropy change for the system at both temperatures.

19 An apparatus consists of two bulbs of the same volume connected by a tap. Initially the tap is closed with one bulb containing neon gas and the other argon gas. Both bulbs are at the same temperature and pressure.

 a Outline what happens when the tap is opened and describe the equilibrium composition of the system.

 b State the signs of ΔH, ΔS and ΔG for the process.

 c State whether this is consistent with the second law of thermodynamics.

Or if we just assume it is all for the system: $\Delta S_{total} = \Delta S - \frac{\Delta H}{T} > 0$, which can be rearranged to give $\Delta H - T\Delta S < 0$. This is such a useful function that we give it a new name – the Gibbs free energy change, ΔG.

The Gibbs equation incorporates enthalpy and entropy which both determine if a reaction is spontaneous and is written in the form: $\Delta G = \Delta H - T\Delta S$, where T is the absolute temperature (K). ΔG and ΔH will both have values of $kJ\,mol^{-1}$ and ΔS will have units of $J\,mol^{-1}K^{-1}$. The condition for spontaneous change in any closed system at constant temperature and pressure is therefore that ΔG must be negative.

Worked example		

Geometric isomer	cis-but-2-ene	trans-but-2-ene
$\Delta H_f^{\ominus}/kJ\,mol^{-1}$	−5.7	−10.1
$S^{\ominus}/J\,K^{-1}\,mol^{-1}$	301	296

cis-but-2-ene   trans-but-2-ene

Calculate the Gibbs free energy change, $\Delta G^{\ominus}$, for the transition of cis-but-2-ene to trans-but-2-ene and that of trans-but-2-ene to cis-but-2-ene. These are known as isomerization reactions. Hence identify the more thermodynamically stable geometric isomer.

cis-but-2-ene $\rightarrow$ trans-but-2-ene, $\Delta G^{\ominus}_{c-t}$

trans-but-2-ene $\rightarrow$ cis-but-2-ene, $\Delta G^{\ominus}_{t-c}$

Using

$$\Delta H^{\ominus}_{c-t} = \Sigma\,\Delta H^{\ominus}_{f\,products} - \Sigma\Delta H^{\ominus}_{f\,reactants}$$

$$\Delta H^{\ominus}_{c-t} = [(-10.1\,kJ\,mol^{-1}) - (-5.7\,kJ\,mol^{-1})]$$

$$= -4.4\,kJ\,mol^{-1}$$

Using

$$\Delta S^{\ominus}_{c-t} = \Sigma S^{\ominus}_{products} - \Sigma S^{\ominus}_{reactants}$$

$$\Delta S^{\ominus}_{c-t} = [296 - 301]\,J\,K^{-1}\,mol^{-1}$$

$$= -5\,J\,K^{-1}\,mol^{-1}$$

Using the Gibbs equation:

$$\Delta G^{\ominus} = \Delta H^{\ominus} - T\Delta S^{\ominus}$$

$$\Delta G^{\ominus}_{c-t} = \Delta H^{\ominus}_{c-t} - T\Delta S^{\ominus}_{c-t} = (-4.4) - 298(-0.005)$$

$$= -2.91\,kJ\,mol^{-1}$$

$$\Delta G^{\ominus}_{t-c} = -\Delta G^{\ominus}_{c-t} = -(-2.91\,kJ\,mol^{-1})$$

$$= +2.91\,kJ\,mol^{-1}$$

and since $\Delta G^{\ominus}_{c-t}$ is negative, then trans-but-2-ene is more thermodynamically stable than the cis isomer. (This is largely because the repulsion between the two methyl groups is reduced.)

■ **QUICK CHECK QUESTION**

20 Explain why ΔG must be negative or zero for a spontaneous process or chemical reaction to occur.

Expert tip

ΔH and ΔS do not vary with temperature (at least over small ranges of temperature unless there is a change in the physical state of one of the reactants or products), but the value of ΔG varies significantly with temperature.

NATURE OF SCIENCE

There are two related definitions of entropy: the thermodynamic definition and the statistical mechanics definition. The thermodynamic definition describes how to measure the entropy of an isolated system in thermodynamic equilibrium. It makes no reference to the atomic and molecular nature of matter. The statistical definition was developed later by analysing the statistical behaviour of the particles in the system. The thermodynamic definition of entropy provides the experimental definition of entropy, while the statistical definition of entropy extends the concept providing a deeper understanding of its nature and connecting with quantum mechanics.

ΔG is not only a measure of spontaneity for a reaction (Figure 5.13), but it is also a measure of useful work (Figure 5.14). The Gibbs free energy for a reaction is equal to the maximum possible work (from a reaction at equilibrium).

ΔG
$\Delta G^{\ominus}$ positive –
    spontaneous change impossible

0 ——————————

$\Delta G^{\ominus}$ negative –
    spontaneous change possible

Figure 15.13 A diagram showing the possibility of change related to the value of ΔG

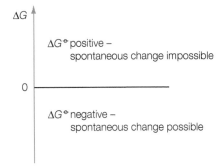

total energy change
at constant temperature
ΔH

| ΔG | $T\Delta S$ |

free energy available   un-free energy:
for useful work   not available for work

Figure 15.14 A diagram showing how an enthalpy change can be split into two parts: $\Delta H = \Delta G + T\Delta S$

Thermodynamics is not concerned with how fast reactions take place. Free energy values do not indicate how fast a reaction will occur.

Expert tip

When using the Gibbs equation the temperature must be expressed in kelvin and if the entropy change is expressed in $J\,mol^{-1}\,K^{-1}$, it should be divided by 1000 to convert to $kJ\,mol^{-1}\,K^{-1}$.

The sign of the Gibbs free energy change, ΔG, depends on the values of the enthalpy change, the entropy change and the temperature. It is the signs of ΔH and ΔS that determine whether a reaction is spontaneous or not. The various combinations are shown in Table 15.4.

Table 15.4 The four types of thermodynamic reactions

Enthalpy change, ΔH	Entropy change, ΔS	Gibbs free energy change, ΔG	Spontaneity
Positive (endothermic)	Positive (products more disordered than reactants)	Depends on the temperature	Spontaneous at high temperatures, when $T\Delta S > \Delta H$
Positive (endothermic)	Negative (products less disordered than reactants)	Always positive	Never spontaneous
Negative (exothermic)	Positive (products more disordered than reactants)	Always negative	Always spontaneous
Negative (products less disordered than reactants)	Negative (products less disordered than reactants)	Depends on the temperature	Spontaneous at low temperatures, when $T\Delta S < \Delta H$

The contribution of ΔS works in the opposite way to that of ΔH because of the minus sign in the ($-T\Delta S$) term of the Gibbs equation: $\Delta G = \Delta H - T\Delta S$. The absolute temperature, T, determines the relative importance of these two terms. As the temperature increases, the $T\Delta S$ term becomes more important. The contributions to ΔG are summarized in Figure 15.15.

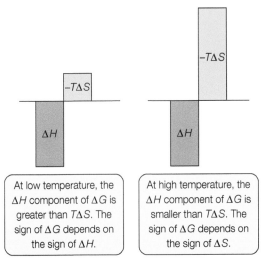

At low temperature, the ΔH component of ΔG is greater than $T\Delta S$. The sign of ΔG depends on the sign of ΔH.

At high temperature, the ΔH component of ΔG is smaller than $T\Delta S$. The sign of ΔG depends on the sign of ΔS.

Figure 15.15 How the temperature can affect the sign of ΔG for a reaction in which ΔH and ΔS are both negative

These relationships between ΔG and temperature for endothermic and exothermic reactions can also be shown graphically (Figure 15.16).

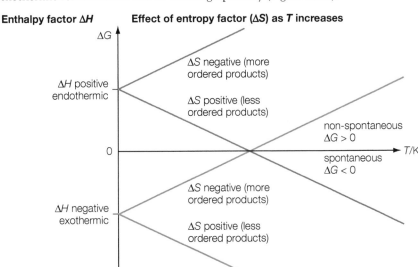

Figure 15.16 A graphical summary of the conditions for spontaneity (assuming that the values of the entropy and enthalpy changes remain constant with a change in temperature)

> ■ **QUICK CHECK QUESTION**
>
> **21** A certain reaction is non-spontaneous under standard conditions, but becomes spontaneous at higher temperatures. State what conclusion may be drawn under standard conditions.

If ΔG is negative (<0), the reaction or process is spontaneous and if ΔG is positive (>0), the reaction or process is non-spontaneous. If ΔG is zero the reaction is at equilibrium and there is no net reaction in either direction.

■ QUICK CHECK QUESTIONS

22 At all temperatures below 100 °C, steam at atmospheric pressure condenses spontaneously to form water. Explain this observation in terms of $\Delta G^{\ominus}$ and calculate the enthalpy of vaporization of water at 100 °C. This is the energy required to convert one mole of water at 100 °C to one mole of steam at 100 °C. $S^{\ominus}$ [H$_2$O(g)] = 189 J K^{-1} mol^{-1} and $S^{\ominus}$ [H$_2$O (l)] = 70 J K^{-1} mol^{-1}.

23 Explain why the change of 1 mole of diamond to graphite ($\Delta G^{\ominus}$ = −2 kJ mol^{-1}) is feasible at all temperatures but the rate is not measureable under standard conditions. $S^{\ominus}$ [C(s) graphite] = 6 J K^{-1} mol^{-1} and $S^{\ominus}$ [C(s) diamond] = 3 J K^{-1} mol^{-1}.

24 The reaction between 1 mole of calcium oxide and carbon dioxide to form calcium carbonate ($\Delta H^{\ominus}$ = −178 kJ mol^{-1}) stops being spontaneous above a certain temperature, T. Determine the value of temperature, T. $S^{\ominus}$ [CO$_2$(g)] = 214 J K^{-1} mol^{-1} and $S^{\ominus}$ [CaO(s)] = 40 J K^{-1} mol^{-1} and $S^{\ominus}$ [CaCO$_3$(s)] = 90 J K^{-1} mol^{-1}.

25 The diagram below shows how the Gibbs free energy change for a particular gas phase reaction varies with temperature.

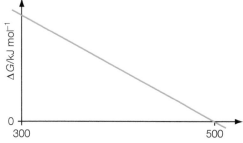

a Explain, with the aid of a thermodynamic equation, why this line obeys the mathematical equation for a straight line, $y = mx + c$.

b Explain why the magnitude of ΔG decreases as T increases in this reaction.

c State the spontaneity of this reaction at temperatures lower than 500 K.

■ Gibbs free energy change of formation

The Gibbs free energy change of formation, $\Delta G_f^{\ominus}$, is the Gibbs free energy change when a mole of a compound is formed from its elements in their standard states. It can be calculated from the known values of $\Delta H_f^{\ominus}$ of the compound and of $S^{\ominus}$ for the compound and its elements discussed above. Alternatively chemists can choose to define the Gibbs free energy change of formation, $\Delta G_f^{\ominus}$, for an element as zero (c.f. $\Delta H_f^{\ominus}$) and tabulate values of $\Delta G_f^{\ominus}$ on that basis.

Compounds which have positive values of $\Delta G_f^{\ominus}$ are thermodynamically unstable compared to their elements and should decompose spontaneously (but very often do not because the process is too slow). Compounds with negative values of $\Delta G_f^{\ominus}$ are thermodynamically stable and cannot decompose to the elements under standard conditions (Figure 15.17).

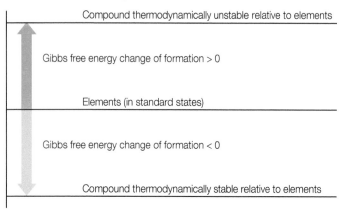

Figure 15.17 A diagram illustrating compounds with negative and positive Gibbs free energy changes of formation

An important use of free energies of formation lies in the fact that, because ΔG is a state function, the standard free energy change of any reaction can be determined by calculation from the free energies of formation of all the substances in the chemical equation, just as we used Hess's law to calculate enthalpy changes, so that, $\Delta G^\ominus = \Sigma \Delta G^\ominus_{f[products]} - \Sigma \Delta G^\ominus_{f[reactants]}$.

Worked example

Calculate the Gibbs free energy change, $\Delta G^\ominus$, for the following reaction under standard conditions:

$C_8H_{18}(l) + 12.5\ O_2(g) \rightarrow 8CO_2(g) + 9H_2O(g)$

$\Delta G^\ominus = 8\Delta G^\ominus_f(CO_2(g)) + 9\Delta G^\ominus_f(H_2O(g)) - \Delta G^\ominus_f(C_8H_{18}(l)) - 12.5\ \Delta G^\ominus_f(O_2(g))$

$\qquad = 8(-394.4\,kJ\,mol^{-1}) + 9(-228.57\,kJ\,mol^{-1}) - (1.77\,kJ\,mol^{-1}) - 25/2\ (kJ\,mol^{-1})$

$\qquad = -5214.1\,kJ\,mol^{-1}$

■ QUICK CHECK QUESTIONS

26 Calculate the Gibbs free energy change for the following reaction under standard conditions:

$Cu(s) + H_2O(g) \rightarrow CuO(s) + H_2(g)$ $\Delta G^\ominus_f(CuO(s)) = -129.7\,kJ\,mol^{-1}$

$\Delta G^\ominus_f(H_2O(g)) = -228.6\,kJ\,mol^{-1}$.

Comment on your answer.

27 Explain why the Gibbs free energy of formation for $Br_2(l)$ is not zero.

■ Gibbs free energy change and the position of equilibrium

For a chemical reaction or physical process to occur, the sign of the Gibbs free energy change, ΔG must be negative under the conditions chosen. Consider a simple reaction of the form: $A \rightarrow B$.

We could calculate $\Delta G^\ominus$ for this reaction knowing $\Delta G^\ominus_f$ for the reactant A and the product B, and the analysis above suggests that the reaction will be spontaneous if $\Delta G^\ominus$ is negative. However, many reactions are equilibria – the reaction stops part way and this needs a thermodynamic explanation.

We have seen above that the standard Gibbs free energy change of any reaction can be determined by calculation from the Gibbs free energies of formation of all the substances involved. $\Delta G^\ominus_f$ corresponds to the formation of a mole of material in its pure form in the standard state, whereas in a reaction, the products and starting materials are mixed, and their free energies depend on the composition of the mixture (Figure 15.18)

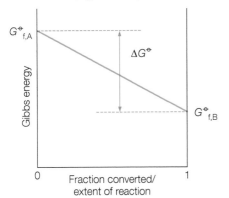

Figure 15.18 Change in Gibbs free energy as A is converted to B

The blue line in the diagram corresponds to an imaginary experiment under standard conditions in which we start with 1 mol of A in a beaker, and gradually throw it away while slowly adding B to another beaker. In the real reaction if we start with pure A and allow some of it to react to form B, we form a mixture of A and B. Since spontaneous mixing must by definition have a negative ΔG, there is an extra decrease in ΔG.

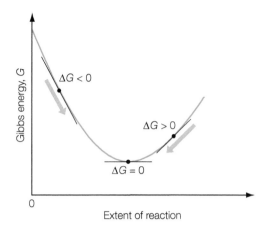

Figure 15.19 Representation of the free energy changes during a reaction whose standard free energy change is negative; equilibrium corresponds to zero gradient

Now imagine that we start with pure B. Since the standard free energy change for conversion of B to A is positive, we might imagine that it will not happen. However, conversion of some of B to A produces a mixture and the Gibbs free energy is lowered by the mixing.

When ΔG attains a minimum value the mixture has a constant composition (Figure 5.19). The composition at this point of minimum ΔG corresponds to equilibrium because ΔG is positive for a change in either direction. For a system at equilibrium, any change would lead to an increase in G and therefore neither the forward or backward reaction is spontaneous.

A detailed analysis of the effect of mixing on free energy changes leads to the equation which relates the standard Gibbs free energy change to the equilibrium constant: $\Delta G^{\ominus} = -RT\ln K$; rearranging $K = e^{\left(\frac{\Delta G}{RT}\right)}$.

The importance of this equation lies in the fact that a simple table of ΔG_f values for compounds allows chemists to predict equilibrium constants for any reactions between those compounds.

Worked example

At 25 °C the equilibrium constant for this reaction, $CO(g) + 2H_2(g) \rightleftharpoons CH_3OH(g)$, has the value 2.1×10^4. Calculate $\Delta G^{\ominus}$ for this reaction at this temperature.

$$\Delta G^{\ominus} = -RT\ln K = (-8.31\,J\,mol^{-1}\,K^{-1})(298\,K)(\ln 2.1 \times 10^4)\left(\frac{1\,kJ}{1000\,J}\right)$$

$$= -25\,kJ\,mol^{-1}$$

Worked example

Calculate the equilibrium constant for the reaction,

$$Fe_3O_4(s) + 2C(s) \rightleftharpoons 3Fe(s) + 2CO_2(g)$$

given $\Delta G_f^{\ominus}(Fe_3O_4(s)) = -1162.0\,kJ\,mol^{-1}$ and $\Delta G_f^{\ominus}(CO_2(g)) = -394.4\,kJ\,mol^{-1}$.

$$\Delta G^{\ominus} = -RT\ln K; \ln K = -\frac{\Delta G^{\ominus}}{RT} \rightarrow K = e^{-\frac{\Delta G^{\ominus}}{RT}}$$

$$\Delta G^{\ominus} = 2(-394.4\,kJ\,mol^{-1}) - (-1162.0\,kJ\,mol^{-1}) = 373.2\,kJ\,mol^{-1}$$

$$K = e^{-\frac{372.2 \times 1000}{8.314 \times 298}}$$

$$= e^{-150} = 7.2 \times 10^{-66}$$

■ **QUICK CHECK QUESTIONS**

28 Calculate $\Delta G^{\ominus}$ for the following acid–base reaction at 298 K. $K_b = 1.8 \times 10^{-5}$.

$$NH_3(aq) + H_2O(l) \rightleftharpoons NH_4^+(aq) + OH^-(aq)$$

29 The Gibbs free energy change for a reaction is $-36.2 \, \text{kJ mol}^{-1}$. Calculate the value of the equilibrium constant at 298 K.

30 If $K_c \ll 1$ for a reaction state, what you know about the sign and magnitude of ΔG.

31 If $\Delta G \ll 0$, deduce what you know about the magnitudes of K and Q for a reaction.

32 The diagram below shows how the free energy changes during a reaction: $A + B \rightarrow C$; on the left are pure reactants, and on the right is the pure product.

State the significance of the minimum in the plot and state what x represents.

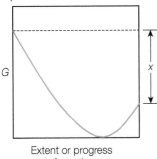

Extent or progress
of reaction

Topic **16** Chemical kinetics

16.1 Rate expression and reaction mechanism

Essential idea: Rate expressions can only be determined empirically and these limit possible reaction mechanisms. In particular cases, such as a linear chain of elementary reactions, no equilibria and only one significant barrier, the rate equation is equivalent to the slowest step of the reaction.

Rate expression and reaction mechanism

- Reactions may occur by more than one step and the slowest step determines the rate of reaction (rate-determining step/RDS).
- The molecularity of an elementary step is the number of reactant particles taking part in that step.
- The order of a reaction can be either integer or fractional in nature. The order of a reaction can describe, with respect to a reactant, the number of particles taking part in the rate-determining step.
- Rate equations can only be determined experimentally.
- The value of the rate constant (k) is affected by temperature and its units are determined from the *overall* order of the reaction.
- Catalysts alter a reaction mechanism, introducing a step with lower activation energy.

▓ Rate expressions

The rate of any equation may be expressed in the form rate = $k[A]^x[B]^y$... etc., where the terms in square brackets stand for the molar concentrations of reactants, k is the rate constant and x and y are the **orders of reaction with respect to the individual reactants**.

Rate expressions describe the relationship between the rate of a reaction and the concentrations of its reactants (Table 16.1).

> ### Key definitions
>
> **Order of reaction with respect to a reactant** – the *power* to which the concentration of that reactant is raised in the rate law and it describes the effect of a concentration change of that reactant on the reaction rate at a given temperature.
>
> **Overall order of reaction** – the sum of all the powers to which the concentrations of those reactants (including catalyst) that appear in the rate law have been raised.

Table 16.1 The relationship between order, concentration and rate of reaction

Order	Rate expression	Nature of dependence
Zero	Rate = $k[A]^0$	Rate is independent of the concentration of A; we could make any change to the concentration of A and the rate would remain the same
First	Rate = $k[A]^1$	Rate changes as the concentration of A changes; for example, if the concentration of A is doubled, the rate doubles
Second	Rate = $k[A]^2$	Rate changes as the square of the change in concentration of A; if the concentration of A is tripled (threefold increase), the rate increases by nine times (ninefold, as $3^2 = 9$)

The rate constant, k, is a numerical value included in the rate expression. It is characteristic of a particular reaction: each reaction has its own unique rate constant in terms of a value and associated units and it is constant at a particular temperature. The rate constant does *not* depend on the extent of the reaction or vary with the concentrations of the reactants.

The units of the rate constant depend on the overall order of the reaction. This is the sum of the individual orders for the reactants. The appropriate units (Table 16.2) can determined by rearranging the rate equation and substituting in the units of concentration ($mol\,dm^{-3}$) and rate ($mol\,dm^{-3}\,s^{-1}$).

> **Expert tip**
>
> The rate equation for any particular reaction is dependent upon its mechanism and can only be established experimentally. It cannot be predicted from the stoichiometry.

Table 16.2 The units of the rate constant for reactions of differing orders

Zero order	First order	Second order	Third order	nth order
Rate = $k[A]^0$	Rate = $k[A]$	Rate = $k[A]^2$	Rate = $k[A]^3$	Rate = $k[A]^n$
Units of rate = $mol\,dm^{-3}\,s^{-1}$	Units of rate $\frac{\text{(concentration)}}{} = s^{-1}$	Units of rate $\frac{\text{(concentration)}^2}{} = mol^{-1}\,dm^3\,s^{-1}$	Units of rate $\frac{\text{(concentration)}^3}{} = mol^{-2}\,dm^6\,s^{-1}$	Units of rate $\frac{\text{(concentration)}^n}{} = (mol\,dm^{-3})^{1-n}\,s^{-1}$

Rate expressions are used (together with the rate constant) to predict the rate of a reaction from a mixture of reactants of known concentrations.

There are two variables in a rate expression: rate and the concentrations of reactants and two constants, the rate constant and the order of the reaction. The knowledge of any three values in the rate equation allows the calculation of the fourth value.

A rate expression will help to propose a mechanism for the reaction: this is a description of the intermediates and the simple reactions (known as elementary steps) by which many reactions occur.

The differences between the rate of a reaction and a rate constant are summarized below in Table 16.3. The concept and definition of rate and rate constant are different. The units of rate are $mol\,dm^{-3}\,s^{-1}$ and the units of a rate constant depend on the overall order of the reaction.

Table 16.3 Differences between the rate and rate constant of a reaction

Rate of reaction	Rate constant, k
It is the speed at which the reactants are converted into products at a specific time during the reaction.	It is a constant of proportionality in the rate expression.
It depends upon the concentration of reactant species at a specific time.	It refers to the rate of reaction when the concentration of every reacting species is unity (one).
It generally decreases with time.	It is constant and does not vary during the reaction.

> **Expert tip**
>
> There are two variables in a rate expression: rate and the concentrations of reactants and two constants, the rate constant and the order of the reaction. The knowledge of any three values in the rate equation allows the calculation of the fourth value.

■ **QUICK CHECK QUESTIONS**

1 The disproportionation reaction between bromide ions and bromate(V) ions in acid solution

$5Br^-(aq) + BrO_3^-(aq) + 6H^+(aq) \rightarrow 3Br_2(aq) + 3H_2O(l)$

obeys the following rate expression:

rate = $k\,[BrO_3^-]^2\,[Br^-]\,[H^+]^2$

If the concentration of the bromate(V) ion is reduced by one half and the concentration of the bromide ion is doubled at constant pH, what would happen to the reaction rate?

A It would decrease by half.

B It would increase by a factor $\frac{5}{4}$.

C It would double.

D It would decrease by a factor 4.

2 The reaction between A and B obeys the rate expression given below:

rate = $k[A]^2[B]^2$

Which of the following are correct statements about this reaction?

I The reaction is second order with respect to both A and B.

II The overall order of reaction is 4.

III Increasing the concentration of A by 3× has the same effect on the rate of reaction as a threefold increase in the concentration of B.

A I only

B I and II

C I, II and III

D I and III

■ Graphical representations of reactions of different orders

Graphs of the concentrations of a reactant (Figure 16.1a, b and c) (or products) against time elapsed since the start of the reaction have different shapes for zero, first and second-order reactions.

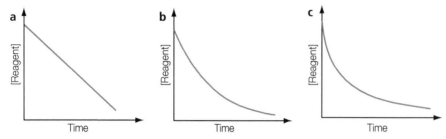

Figure 16.1 Graphs of reactant concentration against time for **a**, a zero-order reaction; **b**, a first-order reaction; and **c**, a second-order reaction

First-order reactions have a half-life which is independent of the concentration of reactant (Figure 16.2). The half-life is the time for the concentration of a reactant to halve. First-order reactions have a constant half-life which can be calculated from the following expression:

$$t_{\frac{1}{2}} = \frac{0.693}{k}$$

where $t_{\frac{1}{2}}$ = half-life (in seconds) and k is the rate constant.

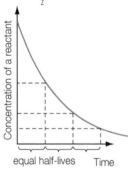

Figure 16.2 Concentration–time curve for a first-order reaction, showing how the half-life remains constant as concentration decreases

The greater the value of the rate constant, k, the more rapid the exponential decrease in the concentration of the reactant (Figure 16.3). Hence, a first-order rate constant is a measure of the rate: the greater the value of the rate constant, the faster the reaction.

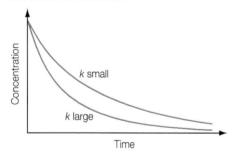

Figure 16.3 First-order concentration–time graphs with different rate constant values

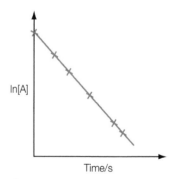

Figure 16.4 Determining the rate constant for a first-order reaction by plotting the natural log of concentration against time

Another way of establishing that a reaction is first order with respect to the reactant is to plot a graph of the natural logarithm of the concentration against the time. For a first-order reaction the graph (Figure 16.4) will take the form of a straight line with a slope (gradient) with a value of $-k$.

Graphs of the rates of a reaction against concentration have different shapes for zero, first and second-order reactions (Figure 16.5).

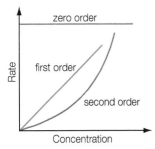

Figure 16.5 Rate–concentration graphs for zero-order, first-order and second-order reactions

Determining orders and rate constants

Use the following approaches for continuous experiments where there is one experiment and many readings are recorded during the reaction. The approach below may also be useful for practical investigations.

1 Plot the reactant concentration against time – half-life method:
   - Is it an approximately linear relationship (line of best fit)? If so, then the order (with respect to that reactant) is zero.
   - If it is not, the order is first or second, measure and tabulate half-lives.
   - Are they constant? If so, then the reaction is first order and the rate constant is given by the following relationship; $k = \dfrac{\ln 2}{t_{\frac{1}{2}}}$, where $t_{\frac{1}{2}}$ is the half-life.
   - If not, determine the initial concentration multiplied by the half-life for several values of the half-life.

Or

2 Plot the reactant concentration against time – alternative plots:
   - Is it an approximately linear relationship (line of best fit)? If so, then the order (with respect to that reactant) is zero.
   - If not, plot ln [reactant] against time.
   - Is it an approximately linear relationship (line of best fit)? If so it is first order (with respect to that reactant) and the gradient is equivalent to the rate constant, k.
   - If not, plot the reciprocal of the concentration of reactant against time. Is it a straight line? If so it is second order and the gradient is equivalent to the rate constant, k.

A summary of the key aspects of the different graphical methods of determining orders of reaction is given in Table 16.4.

Table 16.4 Summary of graphical methods of finding orders and half-lives

Overall reaction rate	Zero order	First order	Simple second order*
	A → products	A → products	A → products
			A + B → products
Rate expression	Rate = $k[A]_a$**	Rate = $k[A]$	Rate = $k[A]^2$
			Rate = $k[A][B]$
Data to plot for a straight-line graph	[A] versus t	ln[A] versus t	$\dfrac{1}{[A]}$ versus t
Slope or gradient equals	$-k$	$-k$	$+k$
Changes in the half-life as the reactant is consumed	$\dfrac{[A]_0}{2k}$	$\dfrac{\ln 2}{k}$	$\dfrac{1}{k[A]_0}$
Units of k	mol dm^{-3} s^{-1}	s^{-1}	dm^3 mol^{-1} s^{-1}

*A simple second-order reaction is a reaction which is second order with respect to one reactant; that is, rate = $k[A]^2$.

**$[A]_0$ is the initial concentration (at time $t = 0$).

■ QUICK CHECK QUESTIONS

3 The results table below shows data on the determination of the rate of the reaction

$H_2O_2(aq) + 2H^+(aq)\ 2I^-(aq) \rightarrow 2H_2O(l) + I_2(aq)$

The time taken to produce a fixed amount of iodine was determined for reaction mixtures in which only the $[I^-(aq)]$ was varied.

$[I^-(aq)]$ / $moldm^{-3}$	Time/s
0.100	24
0.080	30
0.060	39
0.040	58
0.020	120

Plot a graph of 1 / time (proportional to rate) against $[I^-(aq)]$ and hence determine the order of reaction with respect to $[I^-(aq)]$.

4 Hydrogen peroxide decomposes to release oxygen:

$2H_2O_2(aq) \rightarrow 2H_2O(l) + O_2(g)$

The following results were obtained from a study of the rate of reaction using different concentrations of hydrogen peroxide.

$[H_2O_2(aq)]/moldm^{-3}$	Reaction rate/$moldm^{-3}s^{-1}$
0.04	2.9×10^{-5}
0.08	5.9×10^{-5}
0.12	8.9×10^{-5}
0.16	1.21×10^{-4}
0.20	1.51×10^{-4}
0.24	1.77×10^{-4}

a Use a graphical method to find the order of reaction with respect to hydrogen peroxide.

b From the graph, determine a value for the rate constant at this temperature.

■ Initial rates method

The order of reaction may also be determined by the initial rates method, in which the initial rate is measured for several experiments using different concentrations of reactants. The initial rates can be found from the gradient of the concentration–time graph (Figure 16.6).

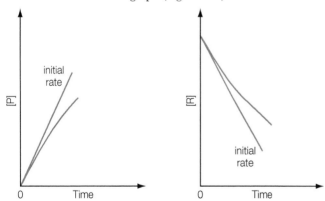

Figure 16.6 The initial rate is found from the gradient at the start of the reaction, when $t = 0$

In the initial rates approach the concentration of a reactant is changed while the others are fixed, so that a systematic set of results is obtained allowing the individual effects of each concentration to be deduced.

Worked example

Chemical reaction: A + B → C + D; rate expression = k [A]^x [B]^y [C]^z

Experiment	[A] mol dm^{-3}	[B] mol dm^{-3}	[C] mol dm^{-3}	Initial rate of reaction/ mol dm^{-3} s^{-1}
1	x	y	z	r
2	$2x$	y	z	$2r$
3	x	$2y$	z	$2r$
4	x	$2y$	$2z$	$8r$

ORDER WITH RESPECT TO REACTANT A

Use data from experiments 1 and 2.

[B] and [C] remained constant; [A] doubled and the rate of reaction doubled and therefore, the reaction is first order with respect A.

ORDER WITH RESPECT TO REACTANT B

Use data from experiments 1 and 3.

[A] and [C] remained constant; [B] doubled and the rate of reaction doubled and therefore the reaction is first order with respect to B.

ORDER WITH RESPECT TO REACTANT C

Use data from experiment 3 and 4.

[A] and [B] remained constant, [C] doubled and the rate of reaction increased by four times. Therefore the reaction is second order with respect to C.

Hence rate expression = k[A] [B] [C]^2.

Expert tip

When establishing individual orders from initial rate data, identify experiments where the concentration of one reactant is changed while the rest are kept constant.

■ QUICK CHECK QUESTIONS

5 For the reaction 2A + B → C + D the following results were obtained for kinetic runs at the same temperature:

Experiment	Initial [A]/ mol dm^{-3}	Initial [B]/ mol dm^{-3}	Initial rate/ mol dm^{-3} s^{-1}
1	0.150	0.25	1.4×10^{-5}
2	0.150	0.50	5.6×10^{-5}
3	0.075	0.50	2.8×10^{-5}
4	0.075	0.25	7.0×10^{-6}

a Find the order with respect to A, the order with respect to B and the overall order of the reaction and the rate expression for the reaction.

b Find the rate constant.

c Find the initial rate of a reaction, when [A] = 0.120 mol dm^{-3} and [B] is 0.220 mol dm^{-3}.

6 The following data were collected at a certain temperature for the rate of disappearance of NO in the reaction. Determine the rate expression and the value of the rate constant.

$2NO(g) + O_2(g) \rightarrow 2NO_2(g)$

Experiment	[NO] in mol dm^{-3}	[O$_2$] in mol dm^{-3}	initial rate in mol dm^{-3} s^{-1}
1	0.0126	0.0125	1.41×10^{-4}
2	0.0252	0.0250	1.13×10^{-3}
3	0.0252	0.0125	5.64×10^{-4}

■ Reaction mechanisms

Many reactions proceed via a series of elementary steps, which collectively are termed the mechanism of the reaction. In many cases one step is the rate-determining step (the slowest step) which controls the overall rate of the reaction.

Expert tip

Be careful here; although molecularity and order can be the same in a number of cases, they are determined differently. Order is an experimentally determined parameter, whereas molecularity is a term that applies to an elementary step in a theoretically suggested reaction mechanism.

The mechanism is a description of how a reaction takes place showing step by step the bonds that break and new bonds that form together with any intermediates. Some mechanisms involve heterolytic bond breaking and formation of ionic intermediates. Other mechanisms involve homolytic bond breaking with free radical intermediates.

A proposed mechanism must meet three criteria:

1 Its steps must add up (sum) to the overall stoichiometric equation.

2 Its steps must be unimolecular (Figure 16.7), involving one particle that undergoes decomposition (A → B + C) or rearrangement (A → B), or bimolecular, when two species collide and react (Figure 16.8).

3 The arrangement of fast and slow steps must lead to a rate expression that agrees with the experimental one. Only reactants in the rate-determining step will appear in the rate expression. The order with respect to a reactant will be zero if it is not in the rate-determining step.

The number of particles involved in the slow rate-determining step is known as the **molecularity of the reaction**. The value of molecularity will be one for a unimolecular step and two for a bimolecular step.

Important supporting information about proposed reaction mechanisms is derived from studying intermediates, using radioactive isotopes (to determine which bonds are broken in a reaction), or using chiral (optically active) reagents. The last method was useful in establishing the S_N1 and S_N2 mechanisms in organic chemistry (Figure 16.9).

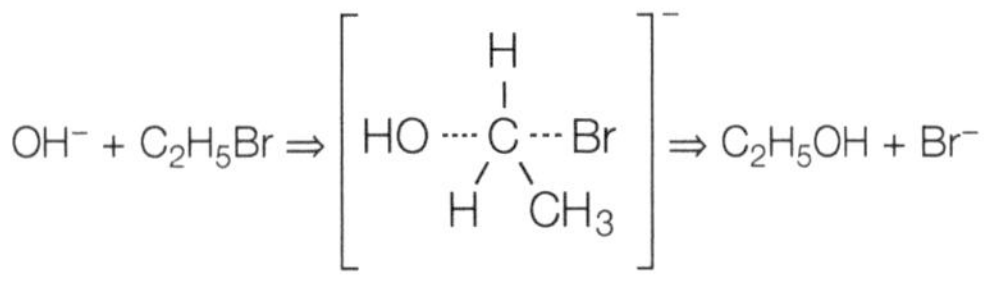

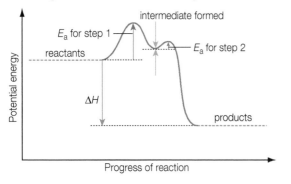

$$OH^- + C_2H_5Br \Rightarrow \left[\begin{array}{c} H \\ | \\ HO \cdots C \cdots Br \\ / \ \backslash \\ H \ \ CH_3 \end{array} \right]^- \Rightarrow C_2H_5OH + Br^-$$

transition state

Figure 16.9 S_N2 mechanism

■ Enthalpy level diagrams

Reactions can be described by enthalpy level diagrams which show the enthalpies (potential energies) of the substances involved on a vertical scale and extent or progress of reaction along the horizontal scale. Figure 16.10 shows an enthalpy level diagram for a reaction involving two transition states and an intermediate.

Figure 16.10 Enthalpy level diagram for an exothermic reaction

The activation energy, E_a, represents the combined minimum kinetic energy that reactants need for a successful collision and hence reaction. The activation energy enables bonds to break.

The *transition state* is a hypothetical species formed during a reaction that is extremely unstable and cannot be isolated. The transition state (or activated complex) represents a point of maximum potential energy in a reaction pathway diagram.

An intermediate species formed during a reaction is usually unstable but can be isolated and studied. An intermediate represents a point of minimum potential energy in a reaction pathway diagram

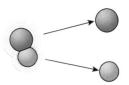

Figure 16.7 Unimolecular step (decomposition)

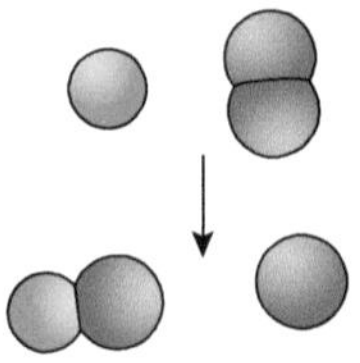

Figure 16.8 Bimolecular step

Key definition

Molecularity of a reaction – the number of particles involved in the rate-determining step of a reaction. In the case of a single-step reaction, molecularity equals the overall order.

Expert tip

One difference between gaseous and solution reactions is that many reactions in solution involve simple ions, or activated complexes (transition states) which are ionic. Hence the rate the may be affected by the nature of the solvent.

Expert tip

The sign and size of the enthalpy change, ΔH, has no effect on the rate of reaction.

■ Transition state theory

In a bimolecular step the particles will react to form an activated complex or transition state (Figure 16.11). This species contains bonds that are in the process of being broken and formed during the reaction. An activated complex cannot be isolated and studied. An activated complex will break down to reform the products or the original reactants. An activated complex is a high-energy and unstable species located at the peak of an energy or enthalpy level diagram.

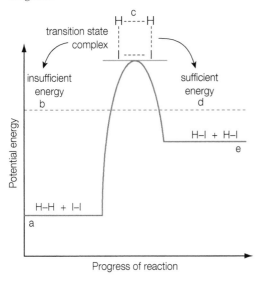

Figure 16.11 Enthalpy or potential energy level diagram for the formation of hydrogen iodide

Expert tip

Activation energy is the difference in potential energy (enthalpy) between the reactants and the transition state.

Steps with high activation energy include reactions between two molecules, between ions of the same charge and if a bond is broken to release free radicals. Steps with low activation energy include reactions between two free radicals, between ions of opposite charge and acid-base reactions.

■ Complex reactions

Complex reactions consist of several elementary steps, each with its own molecularity. Complex reactions may have fractional orders for some reactants. Two models, or approximations, are used to describe the mechanism of a complex reaction.

Consider a reaction in which reactant R forms a product P through a linear series of intermediates, $I_1, I_2, I_3, \ldots, I_n$:

$$R \xrightarrow{k_0} I_1 \xrightarrow{k_1} I_2 \xrightarrow{k_2} I_3 \xrightarrow{k_3} \cdots \xrightarrow{k_{n-1}} I_n \xrightarrow{k_n} P$$

The entire reaction sequence may be described as a single profile containing the individual steps showing the low energy intermediates (Figure 16.12).

To derive a rate expression it is assumed that a slow rate-determining step exists. It is also assumed that all the steps before the rate-determining step are in equilibrium:

$$R \underset{}{\overset{K_0}{\rightleftharpoons}} I_1 \underset{}{\overset{K_1}{\rightleftharpoons}} I_2 \underset{}{\overset{K_2}{\rightleftharpoons}} \cdots \underset{}{\overset{K_{n-1}}{\rightleftharpoons}} I_n \underset{}{\overset{K_n}{\rightleftharpoons}} P$$

In a reaction where data suggest a slow rate-determining step does not exist, then the steady state approximation is assumed meaning that the concentration of intermediates is constant during the reaction. The rate of formation of an intermediate is equal to the rate of its decomposition. Enzyme kinetics assumes the steady state approximation.

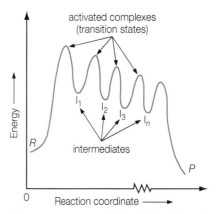

Figure 16.12 Energy (enthalpy) level diagram for a complex reaction with a sequence of intermediates

■ Chain reactions

A few photochemical reactions are complex reactions known as chain reactions. These are exothermic reactions that have a high activation energy. Light or ultraviolet radiation is absorbed and breaks a covalent bond in one of the reactant molecules, removing an atom (free radical). This free atom (radical) is reactive enough to react with another reactant molecule and produce another free atom (radical). The process then repeats building up into a chain reaction and the radiation has provided the energy to overcome the activation energy barrier.

Examples of chain reactions include the chlorination of alkanes and the depletion of the ozone layer by chlorine atoms (free radicals) released by the photolysis of chlorofluorocarbons (CFCs) (Figure 16.13).

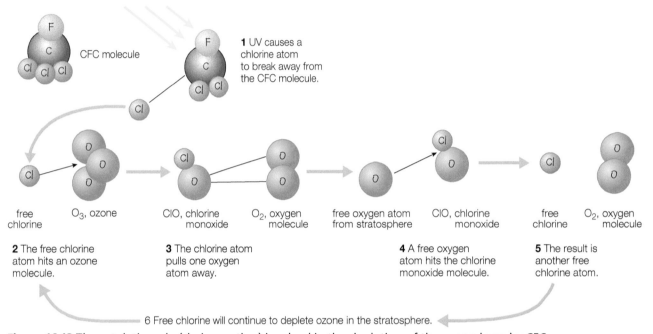

Figure 16.13 The catalytic cycle (chain reaction) involved in the depletion of the ozone layer by CFCs

---

■ QUICK CHECK QUESTIONS

7 The mechanism for the reaction of nitrogen dioxide with carbon monoxide to form nitric oxide and carbon dioxide is thought to be

$NO_2(g) + NO_2(g) \rightarrow NO_3(g) + NO(g)$ slow

$NO_3(g) + CO(g) \rightarrow CO_2(g) + NO_2(g)$ fast

a Write the rate expression expected for this mechanism.

b Draw a potential energy level profile to illustrate the multi-step reaction.

8 The mechanism for the decomposition of hydrogen peroxide is thought to consist of three elementary steps:

$H_2O_2 \rightarrow 2OH$

$H_2O_2 + OH \rightarrow H_2O + HO_2$

$HO_2 + OH \rightarrow H_2O + O_2$

The rate expression is found experimentally to be:

rate = $k\,[H_2O_2]$.

Which step is the rate-determining step?

9 Propose the elementary steps involved in the mechanisms of each of the following three reactions.

	Reaction	Rate expression
a	$A + B_2 + C \rightarrow AB + BC$	rate = $k[A][C]$
b	$2A + B \rightarrow A_2B$	rate = $k[A]^2[B]$
c	$2H_2 + 2NO \rightarrow 2H_2O + N_2$	rate = $k[NO]^2[H_2]$

■ Catalysts

A catalyst increases the rate of reaction by providing a new alternative mechanism (pathway) in which the particles require less kinetic energy to react and form the product (Figure 16.14).

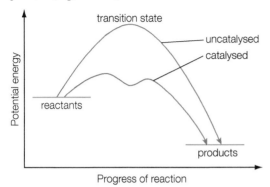

Figure 16.14 General enthalpy level diagram for uncatalysed and catalysed reactions of an exothermic reaction

There are two very different types of catalysis or catalysts: homogeneous and heterogeneous.

Heterogeneous catalysts are in a different phase to the reactants (often the catalyst is solid and the reactants gases) while homogeneous catalysts are in the same phase (often in solution) and achieve their catalytic action via the temporary formation of low energy intermediates (Figure 16.15).

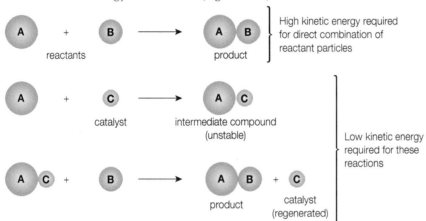

Figure 16.15 The principle of homogeneous catalysis

Heterogeneous catalysts are used in large-scale continuous industrial processes. In the Haber process for ammonia manufacture the nitrogen and hydrogen gases flow through a reactor containing a finely divided iron catalyst.

Heterogeneous catalysts work by adsorbing (Figure 16.16) gaseous reactant molecules at active sites on the surface of the solid. Many transition metals can act as a catalyst for the addition of hydrogen to an alkenyl functional group in unsaturated compounds, such as ethene, by adsorbing hydrogen molecules which probably split up into single atoms held on the surface of the metal crystals. The hydrogen atoms react with the alkene to form the addition compound and then desorb from the surface (Figure 16.17).

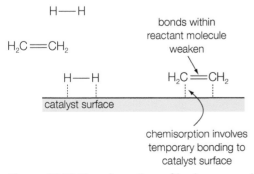

Figure 16.16 Chemisorption of hydrogen molecules onto active site of catalytic metal surface

product molecule desorbed
(i.e. leaves the catalyst surface)

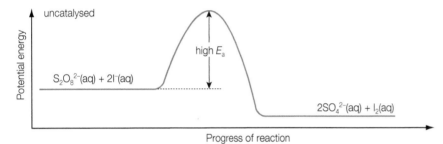

Figure 16.17 Desorption of ethane molecules from the active site of the catalytic metal surface

Transition metal ions can be effective homogeneous catalysts because they can gain and lose electrons as they change from one stable oxidation state to another. The oxidation of iodide ions by peroxydisulfate(VI) ions, for example:

$$S_2O_8^{2-}(aq) + 2I^-(aq) \rightarrow 2SO_4^{2-}(aq) + I_2(aq)$$

The uncatalysed (direct) reaction is relatively slow due to the reaction between ions of the same charge. The reaction is catalysed by iron(III) ions in the solution. A likely mechanism is that iron(III) ions are reduced to iron(II) ions as they oxidize iodide ions to iodine molecules. The peroxydisulfate(VI) ions oxidize the iron(II) ions back to iron (III) ions ready to oxidize more of the iodide ions (Figure 16.18).

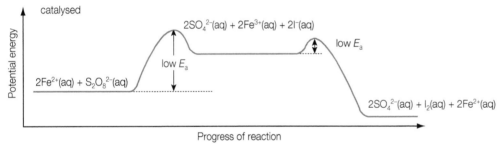

Figure 16.18 Enthalpy (energy level) diagrams for the uncatalysed and catalysed reactions between peroxydisulfate(VI) and iodide ions

Enzymes are complex protein molecules that have an active site that allows a specific molecule to be bonded by intermolecular forces onto the surface of the enzyme. The activation energy of the reaction is lowered. Enzymes are usually very specific and catalyse only one particular reaction. For example, the enzyme catalase catalyses the decomposition (disproportionation) of water.

$$2H_2O_2(aq) \rightarrow 2H_2O\ (l) + O_2(g)$$

Reaction	Activation energy/kJ mol^{-1}
Uncatalysed	75
With platinum	49
With catalase	23

16.2 Activation energy

Essential idea: The activation energy of a reaction can be determined from the effect of temperature on reaction rate.

Activation energy

- The Arrhenius equation uses the temperature dependence of the rate constant to determine the activation energy.
- A graph of $\ln k$ against $\frac{1}{T}$ is a linear plot with gradient $-\frac{E_a}{R}$ and intercept $\ln A$.
- The frequency factor (or pre-exponential factor) (A) takes into account the frequency of collisions with proper orientations.

▓ The effect of temperature

The rate constants of reactions generally increase exponentially with temperature (Figure 16.19). The dependence of the rate constant on temperature is given by the Arrhenius equation (derived from collision theory)

$$k = A e^{-\frac{E_a}{RT}}$$

where k is the rate constant, R is the gas constant and T the absolute temperature. A is a constant known as the Arrhenius constant and is related to the steric requirement for a reaction.

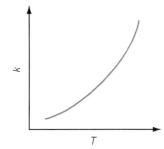

Figure 16.19 Plot of rate constant, k, against absolute temperature, T

> **Common mistake**
>
> In the Arrhenius equation, the unit of temperature is kelvin and the unit of activation energy is joules per mole ($J\,mol^{-1}$).

The equation can be expressed in a logarithmic form (by taking logarithms to the base e of both sides)

$$\ln k = \ln A - \frac{E_a}{RT} = \text{constant} - \frac{E_a}{RT}$$

This equation gives a straight line if $\ln k$ is plotted against reciprocal of the absolute temperature, $\frac{1}{T}$ (Figure 16.20). The gradient of the line is $-\frac{E_a}{R}$, which gives a value for the activation energy. Extrapolating the graph back to the $\ln k$ axis will give an intercept with a value equal to $\ln A$.

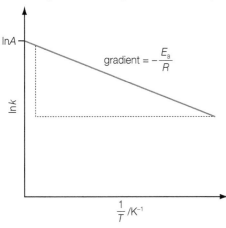

Figure 16.20 An Arrhenius plot of ln k against T^{-1}

If Arrhenius plots are drawn on the same axes for two reactions with different activation energies (Figure 16.21) the reaction with the higher activation energy has a steeper gradient.

This indicates that the rate constant and the initial rate will change with temperature much more quickly than for the reaction with the lower activation energy. This is because the value of the activation energy is given by the expression $-R \times$ gradient.

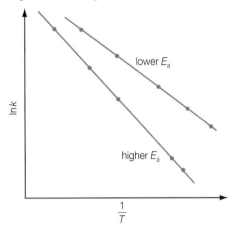

Figure 16.21 Arrhenius plots for two reactions with different activation energies

The activation energy an also be calculated (non-graphically) from two values of the rate constant, k_1 and k_2, at only two different temperatures, T_1 and T_2, by using the following formula:

$$\ln k_2 - \ln k_1 = -\frac{E_a}{R}\left(\frac{1}{T_2} - \frac{1}{T_1}\right)$$

■ **QUICK CHECK QUESTIONS**

10 A first order reaction has a rate constant of $2.2 \times 10^{-3}\,s^{-1}$ at 298 K, and $9.3 \times 10^{-3}\,s^{-1}$ at 325 K. Use this information to calculate the activation energy E_a for the reaction.

11 An Arrhenius plot ($\ln k$ versus $\frac{1}{T}$) of data for a given reaction gives a straight line of gradient $-5000\,K$. Calculate the activation energy E_a for the reaction.

12 Wilkinson's catalyst is a complex containing the transition metal rhodium, bonded to one chloride and three triphenylphosphine ligands, $[RhCl(PPh_3)_3]$, where the phenyl group, $-C_6H_5$, is written as Ph. In solution, this compound is a homogeneous catalyst for the hydrogenation of alkenes.

   a State three characteristic features of the function of a catalyst.

   b Outline what is meant by the term homogeneous.

   c Deduce the oxidation state and the co-ordination number of rhodium in Wilkinson's catalyst.

   d State what feature of the triphenylphosphine molecule enables it to act as a ligand.

   In the mechanism of the reaction of cyclohexene with hydrogen, catalysed by Wilkinson's catalyst, the intermediate $[RhCl(H)_2(PPh_3)_3]$ is formed.

   e Write an overall equation for the reaction of cyclohexene with hydrogen.

   f By stating a reagent and an observation, give a chemical test which would show that this hydrogenation reaction has gone to completion.

   g State which feature of rhodium chemistry allows Wilkinson's catalyst to function.

17.1 The equilibrium law

Essential idea: The position of equilibrium can be quantified by the equilibrium law. The equilibrium constant for a particular reaction only depends on the temperature.

The equilibrium law

- Le Châtelier's principle for changes in concentration can be explained by the equilibrium law.
- The position of equilibrium corresponds to a maximum value of entropy and a minimum in the value of the Gibbs free energy.
- The Gibbs free energy change of a reaction and the equilibrium constant can both be used to measure the position of an equilibrium reaction and are related by the equation, $\Delta G = -RT \ln K$.

▦ The equilibrium law and Le Châtelier's principle

The equilibrium law based on equilibrium concentrations allows us to provide mathematical evidence for Le Châtelier's principle. To remind you of the ground covered in Topic 7 Equilibrium, the reaction quotient, Q, which is the ratio of non-equilibrium concentrations at any given time during the reaction, helps us predict the 'direction' a reaction mixture will take:

- if $Q < K_c$ then the reaction proceeds towards the products so that the concentrations of products increase and the concentrations of reactants decreasec as then the Q value increases to the K_c value; when the value of the value of the equilibrium law is equal to K_c then the equilibrium law is obeyed.
- if $Q > K_c$ then the reaction proceeds towards the reactants so that the concentrations of products decrease and the concentrations of reactants increase as then the Q value decreases to the K_c value, which means the equilibrium law is obeyed
- if $Q = K_c$ then the reaction is at equilibrium and no net reaction occurs; the reactant and product concentrations remain constant.

The above considerations emphasize that an equilibrium system is dynamic with the forward and reverse reactions occurring at the same time, but when the conditions are changed the system responds in accordance with Le Châtelier's principle. This can be explained and understood using the equilibrium law expression or the relative rates of the reverse and forward reactions. The synthesis of ammonia from its elements at constant temperature is used as an example.

$$N_2(g) + 3H_2(g) \rightleftharpoons 2NH_3(g)$$

▦ The effect of changing concentration of a reactant

▦ Equilibrium law argument

For example, consider introducing a reactant, nitrogen gas, at constant volume. Using the equilibrium law argument; adding more nitrogen at constant volume increases its concentration, since there are more gas molecules per unit volume.

$$K_c \neq \frac{[NH_3(g)]^2}{[N_2(g)][H_2(g)]^3}$$

So at the moment of adding the nitrogen gas the equilibrium law is no longer obeyed (the bottom line is numerically too large). The chemical system changes until the equilibrium law is again obeyed by decreasing the concentrations of nitrogen and hydrogen gases and increasing the concentration of ammonia gas

(Figure 17.1a). All concentrations are changed: $[NH_3(g)]$ is increased significantly; $[N_2(g)]$ is increased compared with its initial value and $[H_2(g)]$ is decreased. The value of the equilibrium, K_c, is unchanged and equilibrium position moves to the right and both rates will be higher.

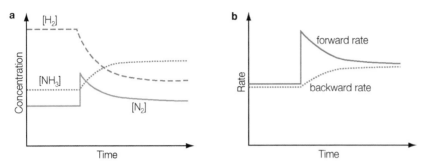

Figure 17.1 a The effect on the component concentrations of adding nitrogen (at constant volume and temperature) into an equilibrium mixture of nitrogen, hydrogen and ammonia; **b** the effect of this addition on the rates of the forward and backward reactions.

■ Rates argument

Adding more nitrogen gas increases its concentration and an increase in concentration causes an increase in the rate of the forward reaction (Figure 17.1b). This uses up nitrogen and hydrogen so the forward rate decreases, and it makes more ammonia, so the reverse rate increases. The decreasing forward rate and the increasing reverse rate finally become equal when a new equilibrium is re-established (at constant temperature) (Figure 17.1b).

■ The effect of changing concentration of a product

■ Equilibrium law argument

Removing the product, ammonia, at constant volume will decrease its concentration because there will fewer gas molecules per unit volume.

$$K_c \neq \frac{[NH_3(g)]^2}{[N_2(g)][H_2(g)]^3}$$

Therefore, the equilibrium law is no longer satisfied; the top line is numerically too small. The chemical system will change until the equilibrium law is again obeyed by decreasing the concentrations of nitrogen and hydrogen and increasing the concentration of ammonia (Figure 17.2a). The result is a change in all concentrations – they are all decreased, but ammonia is decreased more than nitrogen and hydrogen. The equilibrium constant, K_c, is unchanged. The equilibrium position shifts to the right and both the forward and reverse rates decrease.

> **Expert tip**
>
> The equilibrium system can partially but not completely remove the effect of changes in concentration.

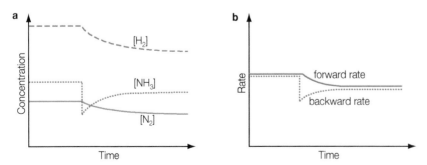

Figure 17.2 a The effect on the component concentrations of removing ammonia (at constant volume and temperature) from an equilibrium mixture of nitrogen, hydrogen and ammonia; **b** the effect of this removal on the rates of the forward and backward reactions

Rates argument

Removing ammonia decreases the reverse rate. The forward rate continues, making ammonia more quickly than it is decomposing. The forward rate slows, while the reverse rate increases until they are equal again (Figure 17.2b).

Table 17.1 summarizes the effects of altering the concentration of a component of an equilibrium mixture on the equilibrium positon for a general reversible reaction:

$$A + B \rightleftharpoons C + D$$

where

$$K_c = \frac{[C][D]}{[A][B]}$$

Table 17.1 The changes involved in component concentrations when a reactant is supplemented or removed

If A is added to reaction	If A is removed from reaction
The forward reaction works to remove excess A	The backward (reverse) reaction works to replace A
A and B react together to produce more C and D	C and D react together to produce more A and B
At equilibrium, there will be more C and D but less B compared with the *original* equilibrium	At equilibrium, there will be more B but less C and D compared with the *original* equilibrium
Position of equilibrium shifts from *left* to *right*	position of equilibrium shifts from *right* to *left*
K_c remains the same	K_c remains the same

The effect of increasing the pressure on the system

For a gas phase reaction, changing the pressure effectively alters the concentrations of the components of the mixture, and similar arguments can be applied to explain the changes predicted by Le Châtelier's principle. The same, industrially significant, equilibrium system will be used as an example.

Equilibrium law argument

Using the equilibrium law argument: doubling the pressure doubles all the concentrations of the three gases (assuming ideal behaviour by the gases), but these appear in the equilibrium law expression raised to different powers.

$$K_c \neq \frac{[NH_3(g)]^2}{[N_2(g)][H_2(g)]^3}$$

The top line of this expression has two concentration terms multiplied, while the bottom line has four, hence the numerical value of the top line will be less significantly affected than the bottom line. The chemical system reacts, reducing the concentration of the two reactants and making more ammonia (product). This results in an increase in all concentrations, but more for ammonia than the two reactants, nitrogen and hydrogen (Figure 17.3a).

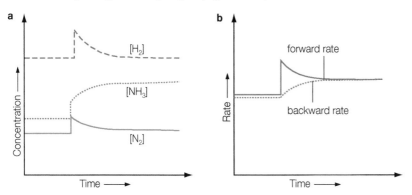

Figure 17.3 a The effect of increasing pressure (at constant volume and temperature) on concentrations in an equilibrium mixture of nitrogen, hydrogen and ammonia; **b** the effect on rates on forward and backward reactions

The value of the equilibrium constant, K_c, is unchanged and the equilibrium position moves to the right and both rates of the forward and reverse reactions will be greater.

■ Rates argument

The forward reaction depends on collisions between nitrogen and hydrogen molecules; these will happen more often if the gases are compressed into a smaller volume (greater concentration). The reverse reaction, the decomposition of ammonia molecules, involves the collision of fewer molecules and so will not, at first, be changed by the increase in pressure. However, as more ammonia is formed, the reverse rate increases until it equals the forward rate and equilibrium is reached again.

■ Calculating equilibrium concentrations and values of K_c

The fact that such a highly significant relationship as that of the equilibrium constant can be established for *any* reversible reaction occurring under conditions where equilibrium can be reached is very useful in quantitative chemistry. The calculations involved at IB level are confined to those relating to homogeneous equilibrium systems. Such calculations require knowledge of the relevant chemical equation and the ability to write equilibrium expressions.

> **Expert tip**
>
> The effect of introducing a noble, unreactive gas into a gaseous equilibrium mixture is sometimes discussed; adding a noble gas at constant volume into a fixed container. If the volume of the container remains the same, then none of the concentrations will change and so there will be no change in the system.

■ QUICK CHECK QUESTIONS

1 A mixture of nitrogen and hydrogen was sealed into a steel vessel and maintained at 1000 K until equilibrium was reached.

$N_2(g) + 3H_2(g) \rightleftharpoons 2NH_3(g)$

The contents were then analysed and the results are in the table below.

Component of mixture	Equilibrium concentration/mol dm^{-3}
$N_2(g)$	0.142
$H_2(g)$	1.84
$NH_3(g)$	1.36

a Write an expression for K_c for this reaction.

b Calculate the value for K_c at 1000 K.

2 Nitrogen(II) oxide, NO, is a pollutant released into the atmosphere from car exhausts. It is also formed when nitrosyl chloride, NOCl, dissociates:

$2NOCl(g) \rightleftharpoons 2NO(g) + Cl_2(g)$

To study this reaction, different amounts of the three gases were placed in a closed container and allowed to come to equilibrium at 503 K and at 738 K.

The equilibrium concentrations of the three gases at each temperature are given below.

Temperature/K	Concentration/mol dm^{-3}		
	NOCl	NO	Cl_2
503	2.33×10^{-3}	1.46×10^{-3}	1.15×10^{-2}
738	3.68×10^{-4}	7.63×10^{-3}	2.14×10^{-4}

a Write the expression for the equilibrium constant, K_c, for this reaction.

b Calculate the value of K_c at each of the two temperatures given.

c Is the forward reaction endothermic or exothermic?

3 Hydrogen can be generated in a reaction between carbon monoxide and water at 500 °C. At equilibrium under these conditions the following concentrations were found in the mixture.

$[CO] = 0.150 \, mol \, dm^{-3}$

$[H_2O] = 0.0145 \, mol \, dm^{-3}$

$[H_2] = 0.200 \, mol \, dm^{-3}$

$[CO_2] = 0.0200 \, mol \, dm^{-3}$

a Write the balanced equation and the equilibrium expression for the reaction.

b Calculate the equilibrium constant for the reaction at 500 °C.

Answering more difficult questions depends on your being able to use the equation for the reaction to work out the concentrations of the various substances in the equilibrium mixture. In order to determine the value of an equilibrium constant, known amounts of reactants are allowed to reach equilibrium. The amount of one of the substances in the equilibrium mixture is determined. The others can be found from the stoichiometric equation using an 'ICE table'.

> **Expert tip**
>
> The values in an ICE table can be in terms of amounts (mol) or molar concentration, but values used in the equilibrium expression for the equilibrium constant, K_c, must be in terms of molar concentrations.

■ Using ICE tables to solve equilibrium problem for equilibrium constant, K_c, or equilibrium amounts

I = *initial* concentration: note that the initial concentration of the reactants are usually given, while the initial concentrations of products are assumed to be zero useless otherwise specified.

C = *change* in concentration: where actual amounts of product are not known, assign the change as the variable x, using the stoichiometry of the reaction to assign change for all species.

E = *equilibrium* concentration.

The essential stages of the ICE method are as follows:
1 Tabulate the known initial and equilibrium concentrations of all species involved in the equilibrium.
2 For those species for which both the initial and equilibrium concentrations are known, calculate the change in concentration that occurs as the system reaches equilibrium.
3 Use the stoichiometry of the equation to calculate the changes in concentration for all the other species.
4 From the initial concentrations and the changes in concentration, calculate the equilibrium concentrations. Then calculate K_c.

Worked example

A mixture of 5.00×10^{-3} mol of H_2 and 2.00×10^{-3} mol of I_2 were placed in a $5.00\,dm^{-3}$ container at $448\,°C$ and allowed to come to equilibrium. Analysis of the equilibrium shows that the concentration of HI is $1.87 \times 10^{-3}\,mol\,dm^{-3}$. Calculate K_c at $448\,°C$.

- Stage 1. Tabulate the known initial and equilibrium concentrations of all species involved in the equilibrium.

H_2 (g) + I_2(g) $\rightleftharpoons$ 2HI(g)

I (initial)	$1.00 \times 10^{-3}\,mol\,dm^{-3}$	$2.00 \times 10^{-3}\,mol\,dm^{-3}$	$0.00\,mol\,dm^{-3}$
C (change)			
E (equilibrium)			$1.87 \times 10^{-3}\,mol\,dm^{-3}$

- Stage 2. For those species where both the initial and equilibrium concentrations are known, calculate the change in concentration that takes place as the system reaches equilibrium. That change is that $1.87 \times 10^{-3}\,mol\,dm^{-3}$ of HI is formed = change in [HI(g)].
- Stage 3. Use the stoichiometry of the equation to calculate the changes in concentration for both the reactants. The ratio of HI to the other two species, H_2(g) and I_2(g), is 2:1 so their change will be

$1.87 \times 10^{-3}\,mol\,dm^{-3}/2 = 0.935 \times 10^{-3}\,mol\,dm^{-3}$

From the initial concentrations and the changes in concentration, then calculate the equilibrium concentrations.

I (initial)	$1.00 \times 10^{-3}\,mol\,dm^{-3}$	$2.00 \times 10^{-3}\,mol\,dm^{-3}$	$0\,mol\,dm^{-3}$
C (change)	$-0.935 \times 10^{-3}\,mol\,dm^{-3}$	$-0.935 \times 10^{-3}\,mol\,dm^{-3}$	$+1.87 \times 10^{-3}\,mol\,dm^{-3}$
E (equilibrium)	$0.065 \times 10^{-3}\,mol\,dm^{-3}$	$1.065 \times 10^{-3}\,mol\,dm^{-3}$	$1.87 \times 10^{-3}\,mol\,dm^{-3}$

- Stage 4. Then calculate K_c.

$$K_c = \frac{[HI]^2}{[H_2][I_2]} = \frac{(1.87 \times 10^{-3})^2}{(0.065 \times 10^{-3}) \times (1.065 \times 10^{-3})} = 51$$

Worked example

An organic compound X exists in equilibrium with its isomer, Y, in the liquid state at a particular temperature.

X(l) $\rightleftharpoons$ Y(l)

Calculate how many moles of Y are formed at equilibrium if 1 mole of X is allowed to reach equilibrium at this temperature, if K_c has a value of 0.020.

Let the amount of Y at equilibrium = x moles

X(l) $\rightleftharpoons$ Y(l)

(I) Starting amount (moles   1.00 0.00

(C) Change in amount (moles)   −x x

The process now is similar to that above.

From the equation, if y moles of the isomer Y are present then y moles of X must have reacted. Therefore, (1.00 − y) moles of X must remain at equilibrium. Also, if we call the volume of liquid V dm³, then we can complete the table as follows:

X(l) $\rightleftharpoons$ Y(l)

(I) Starting amount (moles)   1.00   0.00

(E) Equilibrium amount (moles) (1.00 − x)   x

Equilibrium concentration (mol dm⁻³) (1.00 − y)/V y/V

$$K_c = \frac{[Y]}{[X]}$$

$$0.020 = \frac{y/V}{(1.00 - y)/V} \quad \text{note the 'V' terms cancel}$$

$$0.020 = \frac{y}{(1.00 - y)}$$

$$0.020(1.00 - y) = y$$

$$0.020 - 0.020y = y$$

and so

$$1.020y = 0.020$$

therefore

$$y = \frac{0.02}{1.020} = 0.0196 = 0.020 \text{ moles}$$

■ QUICK CHECK QUESTIONS

4   A reaction was carried out in which 0.20 mol of $PCl_3(g)$ and 0.10 mol of $Cl_2(g)$ were mixed in a 1.0 dm³ flask at 350 °C. An equilibrium was established in which some $PCl_5(g)$ had been formed. This equilibrium mixture was found to contain 0.12 mol of $PCl_3(g)$. Calculate the K_c for this reaction at 350 °C.

5   The reversible reaction, $CO(g) + 2H_2(g) \rightleftharpoons CH_3OH(g)$ has a K_c value of 0.500 at 350 K. The equilibrium concentrations of carbon monoxide and hydrogen were found to be as follows:

[CO] = 0.200 mol dm⁻³

and

[H₂] = 0.155 mol dm⁻³

Calculate the equilibrium concentration of methanol, [$CH_3OH(g)$].

6   The thermal decomposition of water is difficult to achieve. The reaction has a very small value of K_c at 1000 °C. At this temperature
$K_c = 7.3 \times 10^{-18}$.

An equilibrium is set up at this temperature starting with an initial concentration of $H_2O(g)$ at 0.10 mol dm⁻³. Calculate the equilibrium hydrogen concentration, [$H_2(g)$].

■ Relationship between thermodynamics and equilibrium

Theoretically any chemical reaction is reversible and therefore has a dynamic equilibrium including very spontaneous reactions. A chemical equilibrium can happen only if both the forward and reverse reactions occur at the same time and are both spontaneous.

Gibbs free energy is a measure that indicates the spontaneity of a reaction. It also indicates the amount of useful energy (work) that can be released during the reaction. This means that for both the forward and reverse reactions to be spontaneous – which is what happens in a chemical equilibrium – a decrease in Gibbs free energy must occur in both the forward and reverse reactions; Gibbs free energy decreases in both directions. The free energy of the system decreases.

In the case of the forward reaction there must be a decrease in the free energy of the reactants while in the case of the reverse reaction there must be a decrease in the free energy of the products. The position of equilibrium (as measured by the equilibrium constant, K_c), corresponds to

■ a minimum in the value of the Gibbs free energy change (Figure 17.4) and
■ a maximum in the value of entropy

for a chemical reaction or physical change. When equilibrium is reached the Gibbs free energy change, ΔG, is zero.

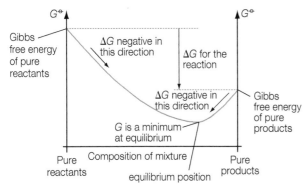

Figure 17.4 The variation in the Gibbs free energy for a reaction for which the overall ΔG is negative

The value and sign of $\Delta G^\ominus$ provide information about the position of equilibrium. If $\Delta G^\ominus$ is negative then the position of equilibrium will be closer to the products than the reactants (Figure 17.5a). The more negative the value of $\Delta G^\ominus$, the closer the position of equilibrium lies towards the products. If $\Delta G^\ominus$ is numerically very large and negative then the position of equilibrium lies very close to pure products (see Table 17.2) – a reaction for which $\Delta G^\ominus$ is negative proceeds spontaneously from reactants to products.

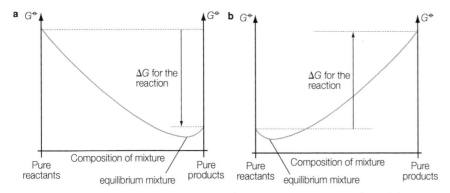

Figure 17.5 The influence of the value of $\Delta G^\ominus$ on the position of equilibrium:
a $\Delta G^\ominus$ is negative and the position of equilibrium lies closer to the products;
b $\Delta G^\ominus$ is positive and the position of equilibrium lies closer to the reactants

If $\Delta G^{\ominus}$ is positive then the position of equilibrium lies more towards the reactants (Figure 17.5b) – the more positive the value, the closer the position of equilibrium lies towards pure reactants.

In summary, $\Delta G^{\ominus}$ can be used to predict the spontaneity of a reaction and position of equilibrium as follows:

- A reaction that has a value of $\Delta G^{\ominus}$ that is both large and negative takes place spontaneously and reaches an equilibrium position to the right, favouring the products. The equilibrium mixture contains a large proportion of products.
- A reaction with a value of $\Delta G^{\ominus}$ that is large and positive does not take place spontaneously and reaches an equilibrium position to the left favouring the reactants. The equilibrium mixture contains predominantly reactants, with only a limited amounts of the products formed.

The Gibbs free energy change and the equilibrium constant are related by the following thermodynamic relationship:

$$\Delta G^{\ominus} = -RT \ln K$$

where R represents the molar gas constant and T represents the absolute temperature in kelvin. For reactions in solution or the liquid phase, *the value of K derived is that of* K_c.

The relationship between K_c and $\Delta G^{\ominus}$ obtained from this expression is summarized in Table 17.2 and Figure 17.5.

Table 17.2 A summary of the relationship between ΔG^1 and K_c

$\Delta G^{\ominus}$	$\ln K_c$	K_c	Position of equilibrium
Negative	Positive	>1	To the right – products favoured
Zero	Zero	= 1	Equal proportions of reactants and products
Positive	Negative	<1	To the left – reactants favoured

Figure 17.6 illustrates how the balance of spontaneity between forward and reverse reactions influences whether a reaction establishes a dynamic equilibrium or not. In a reversible reaction neither the forward nor the reverse reaction goes to completion as both become non-spontaneous at some point. This is the point where there cannot be a further decrease of free energy in either direction – each reaction (forward and reverse) has reached its minimum free energy and $\Delta G^{\ominus} = 0$. The difference in Gibbs free energy between reactants and products and the direction of that difference determine the position of equilibrium of the reaction and therefore the composition of the equilibrium mixture (i.e. more reactant or more product).

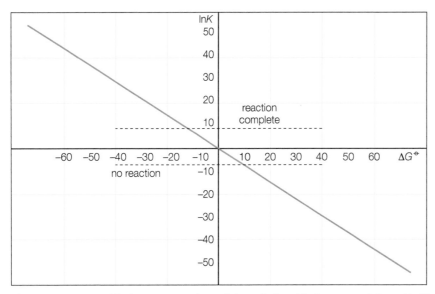

Figure 17.6 A graphical representation of the relationship between $\Delta G^{\ominus}$ and $\ln K$

The following broad relationship exists between $\Delta G^\ominus$ and the equilibrium constant, K_c:

- If $\Delta G^\ominus$ is negative, K_c is greater than 1 and the products predominate in the equilibrium mixture.
- Alternatively, if $\Delta G^\ominus$ is positive, K_c will be less than 1 and the reactants will predominate in the equilibrium mixture.

Expert tip

When using the relationship $\Delta G^\ominus = -RT \ln K_c$, the value of $\Delta G^\ominus$ must be converted from $kJ\,mol^{-1}$ to $J\,mol^{-1}$ to be consistent with the units of R ($J\,K^{-1}\,mol^{-1}$).

Worked example

The esterification reaction between ethanol and ethanoic acid has a free energy change ($\Delta G^\ominus$) of $-4.38\,kJ\,mol^{-1}$. Calculate the value of K_c for this reaction at 25 °C (298 K).

($R = 8.31\,J\,K^{-1}\,mol^{-1}$)

$CH_3COOH(l) + C_2H_5OH(l) \rightleftharpoons CH_3COOC_2H_5(l) + H_2O(l)$

$\Delta G^\ominus = -RT \ln K_c$

$-4.38 \times 1000 = -(8.31 \times 298 \times \ln K_c)$

Note that the value for $\Delta G^\ominus$ is converted to $J\,mol^{-1}$ to be consistent with the units for R.

$\ln K_c = 4380/2478 = 1.77$

Therefore

$K_c = e^{1.77} = 5.9$

QUICK CHECK QUESTIONS

7 The reaction between nitrogen and hydrogen to form ammonia is an equilibrium reaction. The value of K_c for the reaction is 1.45×10^{-6} at 298 K.

Determine the value of $\Delta G^\ominus$ for this reaction at 298 K.

8 a What is the value of $\Delta G^\ominus$ for a reaction when $K = 1$?

b Nitrogen(II) oxide, NO, is oxidized in air to nitrogen(IV) oxide, NO_2:

$2NO(g) + O_2(g) \rightleftharpoons 2NO_2(g)$

The value of K_c for this reaction at 298 K is 1.7×10^{12}. Calculate the value of $\Delta G^\ominus$ for the reaction.

9 The gas phase reaction between hydrogen and iodine to produce hydrogen iodide has a value of $\Delta G^\ominus$ of $+1.38\,kJ\,mol^{-1}$ at 298 *K*.

$H_2(g) + I_2(g) \rightleftharpoons 2HI(g)$

a Calculate the value of the equilibrium constant at this temperature given the relationship $\Delta G^\ominus = -RT \ln K$ for a gas phase reaction.

b From your calculated value for the equilibrium constant predict whether the position of the equilibrium achieved is closer to the reactants or products side of the equation.

18.1 Lewis acids and bases

Essential idea: The acid–base concept can be extended to reactions that do not involve proton transfer.

Lewis theory

- A Lewis acid is a lone pair acceptor and a Lewis base is a lone pair donor.
- When a Lewis base reacts with a Lewis acid a coordinate bond is formed.
- A nucleophile is a Lewis base and an electrophile is a Lewis acid.

Lewis theory

The Lewis concept of acids and bases (Figure 18.1) emphasizes the shared electron pair (in the coordinate bond formed) rather than the proton (H^+) transfer. A Lewis acid is an electron pair acceptor, and a Lewis base is an electron pair donor.

Lewis acid (electron pair acceptor)   Lewis base (electron pair donor)   coordinate covalent bond

Figure 18.1 Lewis theory of acids and bases

Any Lewis acid must contain at least one empty orbital in the valence shell of one of its atoms to accept an electron pair from a Lewis base (Figure 18.2).

> **Expert tip**
>
> The Lewis concept is more general than the Brønsted–Lowry concept because it can apply to reactions in which there is no proton transfer.

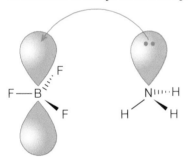

Figure 18.2 The reaction between ammonia and boron trifluoride to form ammonium boron trifluoride (Lewis adduct)

■ **QUICK CHECK QUESTIONS**

1. Compare and contrast the acid–base properties of water molecules in the two reactions below.

   $H_2O(l) + HBr(aq) \rightarrow H_3O^+(aq) + Br^-(aq)$

   $6H_2O(l) + Fe^{3+}(aq) \rightarrow [Fe(H_2O)_6]^{3+}(aq)$

2. Propylamine, $C_3H_7NH_2$, can behave as both a Brønsted–Lowry base and a Lewis base. Define the terms Brønsted–Lowry base and Lewis base.

Figure 18.3 shows a range of examples of Lewis acid–base reactions involving inorganic species. The electron pair donor is the Lewis base and the electron pair acceptor is the Lewis acid.

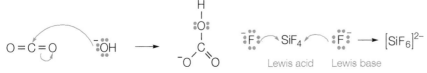

Figure 18.3 A selection of Lewis acid–base reactions

The formation of a transition metal complex is an example of a Lewis acid–base reaction, where the central transition metal ion is a Lewis acid and the ligands are Lewis bases (Figure 18.4).

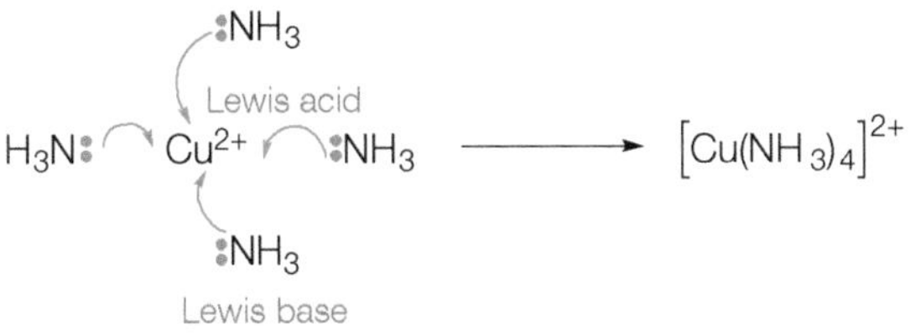

Figure 18.4 Formation of a complex ion

■ **QUICK CHECK QUESTION**

3 Boric acid, H_3BO_3, reacts with water molecules to form protons and tetrahydroxyborate ions, $B(OH)_4^-(aq)$:

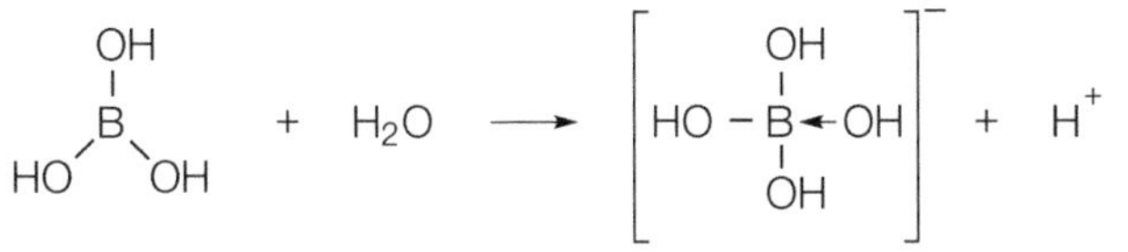

Discuss boric acid in terms of the Lewis and Brønsted–Lowry theories of acids and bases.

■ Nucleophiles and electrophiles

The Lewis acid–base concept is used in organic chemistry to identify reacting species and in the use of 'curly arrows' to show electron pair movement during organic reaction mechanisms.

The term nucleophile ('nucleus loving') indicates a species that is electron-rich (Lewis base) and attracted to the site of an organic molecule that is electron deficient. The electron-deficient site (which may be positive) is termed an electrophile (electron loving). Electrophiles are Lewis acids. Organic chemistry often involves nucleophilic reactions that occur at a carbon atom.

For example, the addition of a hydrogen bromide molecule across the carbon–carbon double bond of an alkene proceeds via an electrophilic addition mechanism. In the first slow step the hydrogen atom in the hydrogen bromide molecule accepts a pair of pi electrons from the alkene. Hence, the electrophile is a Lewis acid and the alkene is a Lewis base.

A second fast step occurs when the bromide ion (acting as nucleophile and Lewis base) donates a pair of electrons to the electron-deficient carbon in the carbocation intermediate to form the addition product.

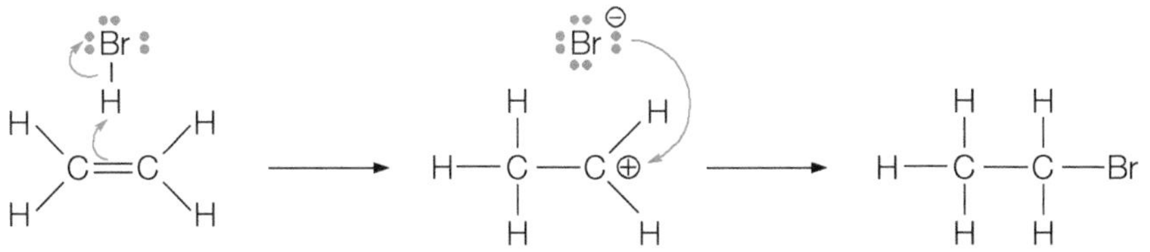

Figure 18.5 The electrophilic addition of hydrogen bromide to ethene showing formation of carbocation intermediate and movement of electron pairs by curly arrows

■ Classes of Lewis acids and bases

The following species can act as Lewis acids:
- molecules in which the central atom is electron deficient, for example, BCl_3
- molecules in which the central atom has empty d orbitals (to accept lone pairs), for example, $SiCl_4$
- simple cations, especially those that have a high charge density, for example, Al^{3+}
- molecules which have multiple bonds between atoms of different electronegativity, for example, SO_2.

Expert tip

The reaction between a Lewis base and a Lewis acid is analogous to a redox reaction where an oxidizing agent is defined as an electron acceptor and a reducing agent is a donor of electrons.

■ **QUICK CHECK QUESTION**

4 Classify the following species as nucleophiles or electrophiles:

$B(CH_3)_3$, OH^-, $HOCH_3$, CN^-, $CH_3CH_2O^-$, $C_6H_5-CH_2^+$, HI

Lewis bases are neutral species with at least one lone pair of electrons, for example, NH_3, or anions with available lone pairs, for example, OH^-.

NATURE OF SCIENCE

Table 18.1 summarizes the various theories of acids and bases, including Lewis theory. The Lewis theory is more general than the Brønsted–Lowry theory.

Table 18.1 Summary of the various acid and base theories

Theory	Basic principle
Traditional approach	Acid: a substance that has certain properties (for example, sour taste, turns litmus red)
Arrhenius	Acid: H^+ present in aqueous solution
	Base: OH^- present in aqueous solution
	At neutrality: $[H^+] = [OH^-]$
Brønsted–Lowry	Acid: H^+ donor
	Base: H^+ acceptor
	Conjugate acid–base pairs
	No concept of neutrality
Lewis	Acid: an electron pair acceptor
	Base: an electron pair donor
Usanovich	Acid: a substance that donates a cation, or accepts an anion or an electron
	Base: a substance that donates an anion or an electron, or accepts a cation

■ **QUICK CHECK QUESTION**

5   Identify the Lewis bases in the following reactions:

a $H_3CCOO^- + 2HF \rightarrow H_3CCOOH + HF_2^-$

b $FeCl_3 + Cl^- \rightarrow FeCl_4^-$

c $Ag^+ + 2NH_3 \rightarrow [Ag(NH_3)_2]^+$

18.2 Calculations involving acids and bases

Revised ☐

Essential idea: The equilibrium law can be applied to acid–base reactions. Numerical problems can be simplified by making assumptions about the relative concentrations of the species involved. The use of logarithms is also significant here.

Calculations involving acids and bases

Revised ☐

■ The expression for the dissociation constant of a weak acid (K_a) and a weak base (K_b).
■ For a conjugate acid–base pair, $K_a \times K_b = K_w$.
■ The relationship between K_a and pK_a is ($pK_a = -\log K_a$), and between K_b and pK_b is ($pK_b = -\log K_b$).

■ Acid dissociation constant

Acid dissociation constants are equilibrium constants that show the extent to which a weak acid dissociates into ions in aqueous solution. For a weak monoprotic acid represented by the general formula HA:

$$HA(aq) + H_2O(l) \rightleftharpoons H_3O^+(aq) + A^-(aq)$$

According to the equilibrium law:

$$K_a = \frac{[H_3O^+][A^-]}{[HA(aq)]}$$

where K_a is an equilibrium constant known as the acid dissociation constant.

Expert tip

Water undergoes auto-ionization or self-dissociation:

$$2H_2O(l) \rightleftharpoons H_3O^+(aq) + OH^-(aq)$$

which is often simplified to

$$H_2O(l) \rightleftharpoons H^+(aq) + OH^-(aq)$$

The equilibrium constant is known as the ionic product constant of water, K_w.

Common mistake

The expressions for an acid dissociation constant, K_a, or base dissociation constant, K_b, do not include $[H_2O]$ because this is considered a constant (because water is a pure liquid and pure liquids do not appear in equilibrium expressions). This expression can be used to calculate the pH of an aqueous solution of a weak acid. Two approximations simplify the calculation:

$[H_3O^+(aq)] = [A^-(aq)]$

so the expression can be simplified to

$K_a = [H_3O^+(aq)]^2/[HA(aq)]$

This assumption ignores the very low concentration of hydrogen ions from the dissociation of water molecules (which is suppressed by the protons from the acid). It is also assumed that so little of the weak acid dissociates that the concentration of the weak acid is equal to the concentration of the undissociated acid.

The pH of a solution of an acid does not indicate whether or not the acid is strong or weak – it is simply a measure of concentration (Figure 18.6). A solution of an acid with a pH of 4.00 might be a very dilute solution of a strong acid or a concentrated solution of a weak acid.

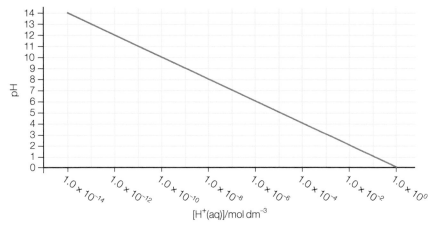

Figure 18.6 The variation of pH with $[H^+(aq)]$

Worked example

A solution consists of an aqueous solution of hydrazoic acid acid (weak and monoprotic), HN_3 at 25 °C. Calculate the pH of the solution. The pK_a of HN_3 is 4.63. State two assumptions used in the equilibrium calculation.

Since $pK_a = -\log_{10}K_a = 4.63$; $K_a = 10^{-4.63} = 2.34 \times 10^{-5}$. The reaction table is:

	$HN_3(aq)$	$\rightleftharpoons$	$H^+(aq)$	$N_3^-(aq)$
Initial	0.25		0	0
Change	$-x$		$+x$	$+x$
Equilibrium	$0.25 - x$		x	x

Hence

$$K_a = \frac{[H^+(aq)] \times [N_3^-(aq)]}{[HN_3(aq)]}$$

$$= \frac{x^2}{(0.25 - x)}$$

$$= 2.34 \times 10^{-5}$$

As the K_a is a very small value, very little HN_3 dissociates in solution and x is negligible so $(0.25 - x)$ is approximately 0.25. The contribution of protons from the dissociation of water molecules is also negligible and ignored.

Hence

$$\frac{x^2}{(0.25)} = 2.34 \times 10^{-5}$$

or

$$[H^+(aq)] = 2.42 \times 10^{-3}\,mol\,dm^{-3}$$

$$pH = -\log_{10}(2.42 \times 10^{-3}) = 2.62$$

■ Base dissociation constant

Base dissociation constants are equilibrium constants that show the extent to which a weak base dissociates into ions in aqueous solution.

For a weak monoprotic base represented by the general formula BH:

$$BH(aq) + H_2O(l) \rightleftharpoons OH^-(aq) + B^+(aq)$$

According to the equilibrium law

$$K_b = \frac{[OH^-(aq)]\,[B^+(aq)]}{[BH(aq)]}$$

where K_b is an equilibrium constant known as the base dissociation constant. Considering the reverse reaction of BH^+ acting as an acid (proton donor) to form the weak base, B, and H^+, then:

$$K_a = \frac{[B(aq)] \times [H^+(aq)]}{[BH^+(aq)]}$$

Then

$$K_a \times K_b = \frac{[B(aq)] \times [H^+(aq)]}{[BH^+(aq)]} \times \frac{[BH^+(aq)] \times [OH^-(aq)]}{[B(aq)]}$$

$$= [H^+(aq)] \times [OH^-(aq)]$$

Since $pK_a = -\log_{10}K_a$; $pK_b = -\log_{10}K_b$ and $pK_w = -\log_{10}K_w = 14$, this can be expressed as $pK_a + pK_b = 14$.

pK_a and pK_b are measures of acid and base strength. They are in an inverse relationship with acid and base dissociation constants, K_a and K_b (Figure 18.7).

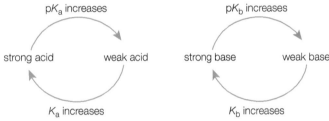

Figure 18.7 Measures of Brønsted–Lowry acid and base strength

NATURE OF SCIENCE

The strengths of acids and bases are ranked by the extent to which they ionize or dissociate in aqueous solution. Strength is based directly on the equilibrium constant (thermodynamics) or its negative logarithm. Any trends in acidity or basicity can then be explained by reference to molecular structure, including concepts such as bond enthalpy, bond polarity and resonance.

Expert tip

The relationships between K_a, K_b, K_w and pK_a, pK_b and pK_w are valid only for a conjugate acid–base pair.

■ QUICK CHECK QUESTIONS

6 Determine the pH of an ammonia solution ($pK_b = 4.75$) with a concentration of $0.121\,mol\,dm^{-3}$.

7 K_b for the nitrite ion is 2.22×10^{-11}. Deduce K_a for its conjugate acid.

8 An ethanoic acid solution contains $[CH_3COOH(aq)] = 0.0787\,mol\,dm^{-3}$ and $[H_3O^+(aq)] = [CH_3COO^-(aq)]$. Calculate the value of K_a for ethanoic acid.

9 Determine the hydrogen ion concentration and pH of a $0.534\,mol\,dm^{-3}$ solution of methanoic acid ($K_a = 1.8 \times 10^{-4}$).

10 Determine the hydroxide ion concentration of a $0.250\,mol\,dm^{-3}$ solution of trimethylamine ($K_b = 7.4 \times 10^{-5}$).

11 The table below shows data for the K_a and pK_b values for some acids and bases at 25 °C.

Acid	K_a	Base	pK_b
Chloric(I) acid, HClO	2.9×10^{-8}	Ammonia, NH_3	4.75
Phenylethanoic acid, $C_6H_5CH_2COOH$	4.9×10^{-5}	Phenylamine, $C_6H_5NH_2$	9.13

State which two formulas represent the weakest acid and the weakest base in the table.

12 Using the information provided in the table below, arrange the acids P, Q, R and S in order of decreasing acid strength.

Acid	pKa	Ka
P	2.58	
Q		1.05×10^{-2}
R	0.65	
S		1.05×10^{-3}

13 Calculate the concentration of $H^+(aq)$ in a 0.2 mol dm^{-3} HCN(aq) solution. (pKa for HCN is 9.3.)

14 The pH of 0.15 mol dm^{-3} butylamine solution, $C_4H_9NH_2$(aq), is 12.0. Determine the base dissociation constant.

Ionic product of water

There are low concentrations of hydrogen and hydroxide ions in pure water because of the transfer of hydrogen ions (protons) between water molecules. There is an equilibrium system, but the extent of ionization is very small:

$$H_2O(l) + H_2O(l) \rightleftharpoons H_3O^+(aq) + OH^-(aq)$$

which can be written more simply as:

$$H_2O(l) \rightleftharpoons H^+(aq) + OH^-(aq)$$

The equilibrium constant, $K_w = [H^+(aq)] \times [OH^-(aq)]$ since pure liquids are not included in equilibrium expressions. The value of K_w at 25 °C is 1.00×10^{-14}. K_w is the ionic product constant for water.

The dissociation of water molecules into protons and hydroxide ions is an endothermic process because it involves breaking an O–H bond in the water molecule. Because the forward reaction is endothermic the value of K_w will increase with temperature (Figure 18.8) which means that neutral $[H^+(aq)] = [OH^-(aq)]$ is only pH = 7 at 298 K.

> **Expert tip**
>
> The electrical conductivity of even the purest form of water never falls to zero suggesting the presence of ions.

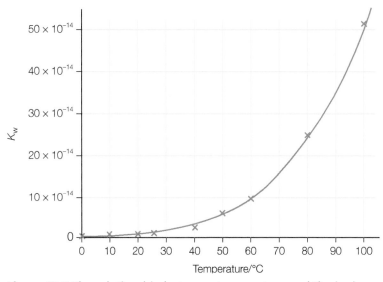

Figure 18.8 The relationship between temperature and the ionic product constant of water

QUICK CHECK QUESTIONS

15 The ionic product constant of water, K_w, at 100 °C has a numerical value of 5.60×10^{-13}. Determine the pH of pure water at 100 °C.

16 At 100 °C, a solution of sodium hydroxide has a hydroxide ion concentration of 0.100 mol dm^{-3}. Determine its pH at 100 °C.

17 Explain why the ionic product constant for water, K_w, is higher at 50 °C than at 25 °C. Explain whether water at 50 °C is acidic, alkaline or neutral.

■ Calculations involving pH, pOH for strong acids and bases

For strong monoprotic acids, such as HCl(aq), the hydrogen ion concentration will be equal to the concentration of the acid and will be twice the value of the acid concentration for the strong diprotic sulfuric acid:

$$H_2SO_4(aq) \rightarrow 2H^+(aq) + SO_4^{2-}(aq)$$

The use of the logarithmic scale (to the base 10) can be extended to other values, e.g. pOH and pK_w.

$$pOH = -\log_{10}[OH^-(aq)]$$

and

$$pK_w = -\log_{10}K_w; \quad K_w = [H^+(aq)] \times [OH^-(aq)]$$

If logarithms to the base ten of both sides of the expression are taken:

$$-\log_{10}K_w = -\log_{10}[H^+(aq)] - \log_{10}[OH^-(aq)]$$

then $pK_w = pH + pOH$.

At 25 °C, $K_w = 1.00 \times 10^{-14}$ and pH + pOH = 14.

This expression gives another way of calculating the pH of a strong base since the value of pOH can be determined directly from the hydroxide ion concentration then subtracted from 14.

■ QUICK CHECK QUESTIONS

18 Determine the pH of the solution when 50.00 cm³ of 0.50 mol dm⁻³ KOH solution is mixed with 200.00 cm³ of 0.10 mol dm⁻³ HCl solution.

19 Determine the pH and pOH of a 0.0125 mol dm⁻³ solution of KOH.

20 The hydrogen ion concentration of ethanoic acid is 4×10^{-3} mol dm⁻³. Determine the pH and pOH.

21 Calculate the pH of a solution prepared by mixing 100.00 cm³ of 0.500 mol dm⁻³ ethylamine solution, $CH_3CH_2NH_2(aq)$, and 40.00 cm³ of 0.500 mol dm⁻³ hydrochloric acid, HCl(aq). (pK_b of ethylamine = 3.35).

■ Buffers

A buffer solution is a mixture of molecules and ions in solution that maintains an approximately constant pH upon the addition of small amounts of acid or base. Buffers are equilibrium systems with dynamic equilibria that illustrate the practical importance of Le Châtelier's principle.

An acidic buffer mixture consists of a solution of a weak acid and one of its salts, for example, a mixture of ethanoic acid and sodium ethanoate (Figure 18.9). For good buffering capacity there should be a high concentration of the acid and its salt (conjugate base).

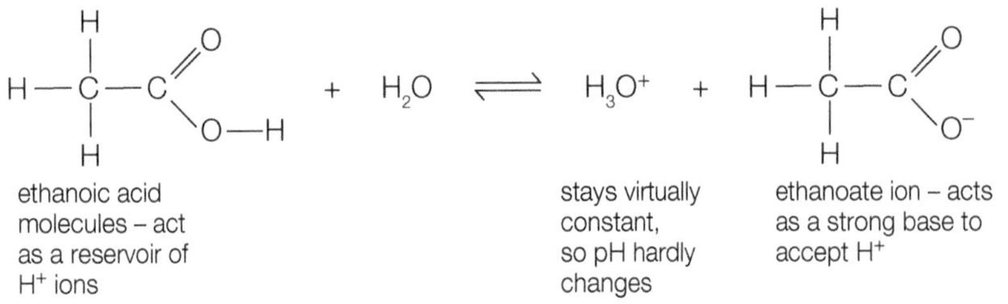

ethanoic acid molecules – act as a reservoir of H⁺ ions

stays virtually constant, so pH hardly changes

ethanoate ion – acts as a strong base to accept H⁺

Figure 18.9 The action of an ethanoic acid/ethanoate ion buffer. Note that the sodium ions from the salt simply acts as a spectator ions

By selecting the appropriate weak acid it is possible to prepare buffers at any pH value across the pH scale. If the concentrations of the weak acid and its salt are the same then the pH of the buffer is equal to pK_a.

An alkaline buffer consists of a weak base and its salt, for example, ammonia solution and ammonium chloride (Figure 18.10). The chloride ion is a spectator ion and for good buffering capacity there should be a high concentration of the base and its salt (conjugate acid).

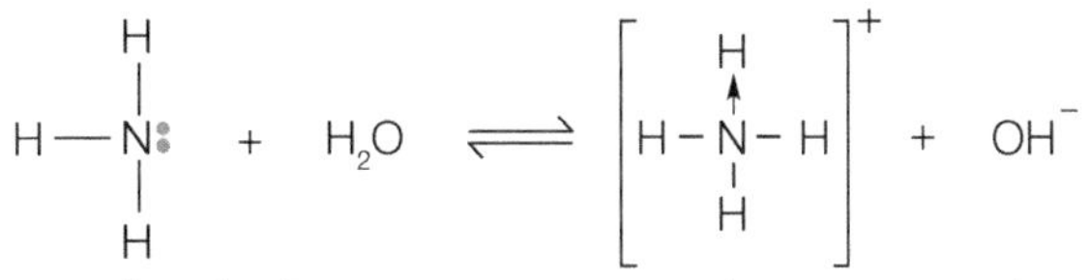

Figure 18.10 The action of an ammonia/ammonium ion buffer. Note that the chloride ions from the salt simply act as spectator ions

The mixture cannot be a mixture of a strong acid and a strong alkali, or the two will react with each other (e.g. a mixture of NaOH and HCl would react with each other and thus not behave as an effective buffer). If the acid and alkali in the buffer are too weak, however, they will not react effectively with the acid or alkali that are added. A suitable mixture is one which contains a mixture of acid and alkali strong enough to react with H_3O^+ and OH^-, but weak enough not to react with each other.

For the acid more generally the pH of a buffer mixture can be calculated with the logarithmic form of the equilibrium law (see Options B4 and D7).

$$pH = pK_a + \log_{10} \frac{[\text{salt}]}{[\text{acid}]}$$

This expression is known as the Henderson–Hasselbalch equation. The buffer will be most effective when the concentration of the weak acid is equal to the concentration of the salt of the weak acid. Diluting a buffer solution with water does not change the concentration ratio of salt to acid and so the pH does not change. On dilution, both the weak acid and the weak base can dissociate more to compensate for the dilution:

$$CH_3COOH(aq) + H_2O(l) \rightarrow CH_3COO^-(aq) + H_3O^+(aq)$$

$$CH_3COO^-(aq) + H_2O(l) \rightarrow CH_3COOH(aq) + OH^-(aq)$$

$$NH_4^+(aq) + H_2O(l) \rightarrow NH_3(aq) + H_3O^+(aq)$$

$$NH_3(aq) + H_2O(l) \rightarrow NH_4^+(aq) + OH^-(aq)$$

A buffer does not have to be a mixture of a weak acid and its conjugate base; any mixture of a weak acid and a weak base will have the same effect. Substances that can behave as both weak acids and weak bases can also behave as buffers.

One important example is sodium hydrogen carbonate; the hydrogen carbonate ion, HCO_3^-, can behave as either an acid or a base. This ion is partly responsible for the buffering of blood.

$$HCO_3^-(aq) + H_3O^+(aq) \rightarrow CO_2(g) + 2H_2O(l)$$

$$HCO_3^-(aq) + OH^-(aq) \rightarrow CO_3^{2-}(aq) + H_2O(l)$$

Amino acids can also behave as buffers. They are the main buffer in tears.

For example, $CH_3CH(NH_2)COOH$, 2-aminopropanoic acid

$$CH_3CH(NH_2)COOH(aq) + OH^-(aq) \rightarrow CH_3CH(NH_2)COO^-(aq) + H_2O(l)$$

$$CH_3CH(NH_2)COOH(aq) + H_3O^+(aq) \rightarrow CH_3CH(NH_3^+)COOH(aq) + H_2O(l)$$

There are two ways to prepare a buffer solution. A salt of a weak acid (CH_3COONa) can be mixed directly with a solution of weak acid (CH_3COOH); or salt of a weak base (NH_4Cl) to a solution of weak base (NH_3). Another way is to neutralize excess weak acid with a strong alkali; or excess weak base with a strong acid.

Worked example

State and explain which of the following mixtures, in an aqueous solution, will produce a buffer solution.

1 $50\,cm^3$ of $0.1\,mol\,dm^{-3}$ CH_3COOK and $50\,cm^3$ of $0.1\,mol\,dm^{-3}$ CH_3COOH

2 $50\,cm^3$ of $0.1\,mol\,dm^{-3}$ NH_3 and $50\,cm^3$ of $0.1\,mol\,dm^{-3}$ NH_4Cl

3 $50\,cm^3$ of $0.2\,mol\,dm^{-3}$ KOH and $50\,cm^3$ of $0.2\,mol\,dm^{-3}$ CH_3COOH

1 This is an acidic buffer: the salt of a weak acid and the weak acid are present.

2 This is an alkaline buffer: a weak base and its salt are present.

3 This will not be a buffer: after neutralization, only a salt (potassium ethanoate) and water are present. Table 18.2 summarizes acidic and basic buffers.

Table 18.2 Summary of acidic and basic buffers

	Acidic buffer	Alkaline buffer
Components	A weak acid (HA) and its salt (conjugate base, A⁻).	Weak base (B) and its salt (conjugate acid, BH⁺)
Equilibrium system	$HA + H_2O \rightleftharpoons H_3O^+ + A^-$ (or $HA \rightleftharpoons H^+ + A^-$)	$B + H_2O \rightleftharpoons BH^+ + OH^-$
Concentrations of components at equilibrium	High [HA] equilibrium ≈ [HA] initial; high [A⁻] equilibrium ≈ [salt]; low [H⁺]	High [B] equilibrium ≈ [B] initial; high [BH⁺] equilibrium ≈ [salt]; low [OH⁻]
pH of a buffer solution	$K_a = \dfrac{[H^+]\,[salt]}{[acid]}$	$K_b = \dfrac{[OH^-]\,[salt]}{[base]}$
When a small amount of acid (H⁺) is added...	the conjugate base reacts with and removes added H⁺: $A^- + H^+ \rightarrow HA$	the weak base reacts with, and removes added H⁺: $B + H^+ \rightarrow BH^+$
When a small amount of alkali (OH⁻) is added...	the weak acid reacts with, and removes added OH⁻: $HA + OH^- \rightarrow A^- + H_2O$	the conjugate acid reacts with, and removes added OH⁻: $BH^+ + OH^- \rightarrow B + H_2O$
Overall effect on pH	[H⁺], and hence pH, virtually unchanged	[OH⁻], and hence pOH (and therefore pH) virtually unchanged
Solution behaves as a buffer in region...	pH = pK_a ±1, where [salt]:[acid] ranges from 1:10 to 10:1	pOH = pK_b ±1, where [salt]:[base] ranges from 1:10 to 10:1
Maximum buffer capacity	When [salt] = [acid] and pH = pK_a.	When [salt] = [base] and pOH = pK_b.

■ **QUICK CHECK QUESTIONS**

22 $53.50\,g$ of ammonium chloride, NH_4Cl, is dissolved in $400\,cm^3$ of $14.00\,mol\,dm^{-3}$ ammonia and the mixture is diluted to $1.0\,dm^3$. Calculate the concentration, in $mol\,dm^{-3}$, in the prepared buffer solution of $NH_4^+(aq)$ and $NH_3(aq)$.

23 Define the term buffer solution. Explain, using appropriate equation/s how a mixture of excess ethanoic acid and aqueous NaOH can resist changes in pH when small amounts of aqueous HCl or KOH are added.

18.3 pH curves

Revised ▢

Essential idea: pH curves can be investigated experimentally but are mathematically determined by the dissociation constants of the acid and base. An indicator with an appropriate end point can be used to determine the equivalence point of the reaction.

pH curves

Revised ▢

■ The characteristics of the pH curves produced by the different combinations of strong and weak acids and bases.

■ An acid–base indicator is a weak acid or a weak base where the components of the conjugate acid–base pair have different colours.

■ The relationship between the pH range of an acid–base indicator, which is a weak acid, and its pK_a value.

■ The buffer region on the pH curve represents the region where small additions of acid or base result in little or no change in pH.

■ The composition and action of a buffer solution.

pH curves

Plotting a graph of pH against volume of alkali added during an acid–base titration gives a curve that is determined by the nature of the acid and base. The indicator chosen to detect the end point must change colour completely in the pH range of the near vertical section of the pH (titration) curve. The equivalence point is when stoichiometric amounts of acid and base have been added. Only a salt and water are present. With a suitable indicator the end point will correspond to the equivalence point.

Table 18.3 Acid–base titration types

Titration type	pH at equivalence point	Suitable indicator
Strong acid versus strong base	pH = 7	Any
Weak acid versus strong base	pH > 7	Phenolphthalein
Weak base versus strong acid	pH < 7	Methyl orange
Weak acid versus weak base	Unpredictable	None

> **Expert tip**
>
> The pH is not always neutral at the equivalence point due to salt hydrolysis. One or more ions from the salt can react with water molecules to release either protons or hydroxide ions.

Calculating pH during titrations:

1 At the mid-point of the titration (half-neutralization), pH = pK_a (for weak acids) or pK_b (for weak bases).

2 For incomplete titrations in general, the pH can be calculated by treating it as a buffer solution.

3 For complete titration, only the salt remains at the equivalence point and the pH of the solution may be found by considering the hydrolysis of the salt.

Worked examples

1 A 20.00 cm³ solution of nitric(III) acid, HNO_2, (pK_a = 3.15) was titrated to its equivalence point with 24.80 cm³ of 0.020 mol dm⁻³ NaOH. Calculate the concentration and pH of the nitric(III) acid solution.

Amount of NaOH added at the equivalence point = 0.020 mol dm⁻³ × 0.0248 dm³ = 0.00050 mol; HNO_2 + NaOH → $NaNO_2$ + H_2O; concentration of HNO_2 = (0.00050 mol)/(0.020 dm³) = 0.025 mol dm⁻³. K_a = [HNO_2^-] [H^+]/[HNO_2]; $10^{-3.15}$ = x^2/0.025; x = 0.0042 mol dm⁻³; pH = $-\log_{10}$[H^+] = 2.38.

2 Determine the pH after 12.40 cm³ and 24.80 cm³ of NaOH.

12.40 cm³ represents the half equivalence point. When this much OH^- is added, the amount of HNO_2 is reduced to half its initial value and an equal amount of NO_2^- is produced. With [HNO_2(aq)] = [NO_2^-(aq)], the Henderson–Hasselbalch equation gives the pH as:

pH = pK_a + $\log_{10}$[base]/[acid] = 3.15 + $\log_{10}$(1) = 3.15

24.80 cm³ represents the equivalence point. When this much OH^- is added, the amount of HNO_2 is reduced zero and all of the initial HNO_2 is now present as NO_2^-. The amount of NO_2^- is therefore 0.00050 mol. The total volume is now (20.00 + 24.80) cm³ = 44.8 cm³ so:

[NO_2^-(aq)] = (0.00050 mol)/(0.0448 dm³)

= 0.0112 mol dm⁻³

K_b = [HNO_2] [OH^-]/[NO_2^-] = y^2/0.0112; pK_a + pK_b = 14.00; pK_b = 14.00 − 3.15 = 10.85; K_b = $10^{-10.85}$; y = 3.97 × 10^{-7} mol dm⁻³ = [OH^-]; pOH = $-\log_{10}$(3.97 × 10^{-7}) = 6.40; pH + pOH = 14.00, so pH = 7.60.

3 Qualitatively, state how each of these three pH values would be affected if 5.00 cm³ of water were added to the 20.00 cm³ of nitric(III) acid before performing the titration: the initial pH; the pH at half equivalence; the final pH.

The initial pH would increase slightly as the nitric(III) acid solution would be more dilute.

The pH at half-equivalence point would not change (as pH = pK_a).

The final pH would decrease slightly as the NO_2^- solution produced would also be more dilute.

◼ Titration of a strong acid against a strong base

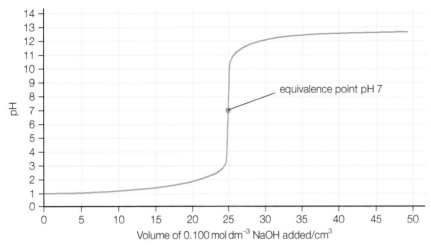

Figure 18.11 The titration curve for the titration of 25.00 cm³ of 0.100 mol dm⁻³ hydrochoric acid with 0.100 mol dm⁻³ sodium hydroxide

The titration curve starts at 1 which indicates the presence of strong acid. The sharp increase in pH indicates the equivalence point of the titration. The pH of the resulting salt solution at the equivalence point is 7 as the salt formed does not undergo hydrolysis. There is no buffer region as no weak acid/base is used. When excess base is added, the resulting solution has a pH close to 13 which is consistent with the addition of strong base.

The shape of the curve is purely because a logarithmic scale is being used. To get from pH = 1 to pH = 2 you have to remove 90% of the hydrogen ions so acid concentration decreases from 0.1 to 0.01 mol dm⁻³. To get from pH 2 to pH 3 you have to remove 90% of what hydrogen ions are left – far less $H^+(aq)$ to decrease the concentration from 0.01 mol dm⁻³ to 0.001 mol dm⁻³.

> **Expert tip**
>
> The initial and end values of pH would vary with the concentration of the acid and alkali.

NATURE OF SCIENCE

The modern pH meter was developed in 1934 and measures the difference in electrical potential between a pH electrode and a reference electrode. It usually has a glass electrode plus a reference electrode. The pH meter measures the difference in electrical potential between the glass electrode and a reference electrode.

◼ Titration of a strong acid against a weak base

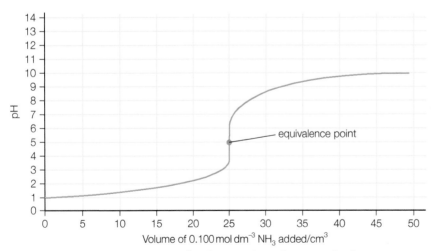

Figure 18.12 The titration curve for the titration of 25.00 cm³ of 0.100 mol dm⁻³ hydrochoric acid with 0.100 mol dm⁻³ ammonia solution

The titration curves starts at 1, which indicates the presence of strong acid. The sharp increase in pH indicates the equivalence point of the titration. At the equivalence point the salt formed undergoes hydrolysis. In general:

$$HB^+(aq) \rightleftharpoons B(aq) + H^+(aq)$$

The $H^+(aq)$ ions produced result in a pH < 7 at the equivalence point. A buffer region is formed when there is a large reservoir of weak acid/base and the pH changes more slowly. When excess acid or base is added the resulting solution has a pH close to 10 which reflects the addition of a weak base.

Titration of a weak acid against a strong base

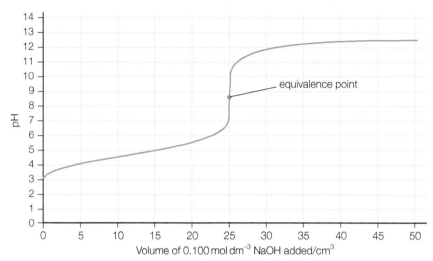

Figure 18.13 The titration curve for the titration of 25.00 cm³ of 0.100 mol dm⁻³ ethanoic acid with 0.100 mol dm⁻³ sodium hydroxide

The titration curve starts at 3 which indicates the presence of weak acid. The sharp increase in pH indicates the equivalence point of titration. At the equivalence point, the salt formed undergoes hydrolysis. In general:

$$A^-(aq) + H_2O(l) \rightleftharpoons HA(aq) + OH^-(aq)$$

The $OH^-(aq)$ ions produced result in a pH > 7 at the equivalence point. A buffer region is formed in which there is a large reservoir of weak acid/base and the pH changes more slowly. When excess base is added, the resulting solution has a pH close to 13 which reflects the addition of strong base.

Worked example

50.00 cm³ of an aqueous solution of propanoic acid (CH_3CH_2COOH) was titrated with 0.100 mol dm⁻³ sodium hydroxide.

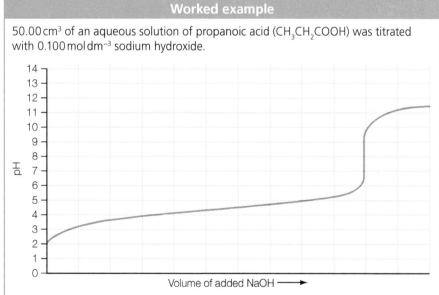

Using information only from the graph calculate an approximate value of the acid dissociation constant, K_a, of propanoic acid and the initial concentrations of $H^+(aq)$ ions and propanoic acid. Suggest a suitable indicator.

At half-equivalence point, pH = pK_a; pH = 4.5; K_a = $10^{-4.5}$ = 3.2 × 10⁻⁵; pH = 2; [$H^+(aq)$] = 10^{-2} = 0.01 mol dm⁻³; [$H^+(aq)$] = $\sqrt{K_a \times c}$; c = $(0.01)^2/3.2 × 10^{-5}$ = 3.125 mol dm⁻³.

Phenolphthalein.

■ Titration of a weak acid against a weak base

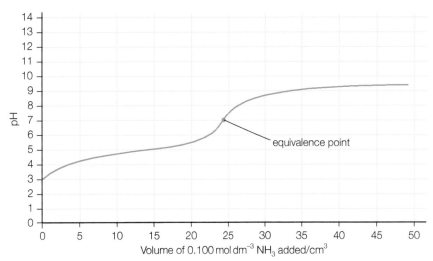

Figure 18.14 The titration curve for the titration of 25.00 cm³ of 0.100 mol dm⁻³ ethanoic acid with 0.100 mol dm⁻³ ammonia solution

The titration curve starts at 3 which reflects the presence of weak base. The pH changes gradually and there is no sharp increase in pH. At the equivalence point. the salt formed undergoes hydrolysis. In general:

$$HB^+(aq) \rightleftharpoons B(aq) + H^+(aq)$$

$$A^-(aq) + H_2O(l) \rightleftharpoons HA(aq) + OH^-(aq)$$

The amount of $H^+(aq)$ and $OH^-(aq)$ ions formed in hydrolysis is approximately equal so the pH at the equivalence point is approximately 7. No buffer regions are formed. When excess base is added the resulting solutions has a pH close to 10 which is consistent with the addition of a weak base.

■ QUICK CHECK QUESTION

24 An unknown solution is either chloroethanoic acid or ethanoic acid. 20.00 cm³ of the acid is titrated with aqueous potassium hydroxide.

  a Using the pH curve, state the volume of aqueous potassium hydroxide needed to neutralize 20.00 cm³ of the acid.

  b Using information from Section 21 of the IB Chemistry *data booklet* and the titration curve, identify the acid in the unknown solution. Explain your answer.

  c Suggest why chloroethanoic acid is a stronger acid than ethanoic acid.

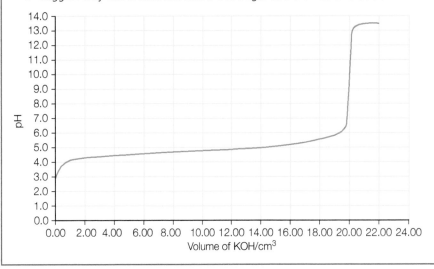

■ Salt hydrolysis

The acid–base properties of salts can be explained by the behaviour of their respective cations and anions. The reaction of ions with water molecules, with a resultant change in pH, is called hydrolysis. The equilibrium constant, K_a, has to be maintained in any solution which contains the acid–base pair, so if you add A^- anions in the form of salt then HA has to be formed until the equilibrium constant is restored.

Table 18.4 Summary of salt hydrolysis

Reaction between acid and base	Nature of salt solution
Strong acid and strong base	No hydrolysis – anion and cation do not react with water molecules. Solution is neutral.
Strong acid and weak base	Hydrolysis between cation and water molecules to release hydronium ions (Figure 18.15). Solution is acidic.
Weak acid and strong base	Hydrolysis between anion and water molecules to release hydroxide ions (Figure 18.16). Solution is alkaline.
Weak acid and weak base	Hydrolysis occurs between anions and water molecules and between cations and water molecules. It is assumed that the hydrogen and hydroxide ions neutralize each other. The solution is assumed to be neutral.

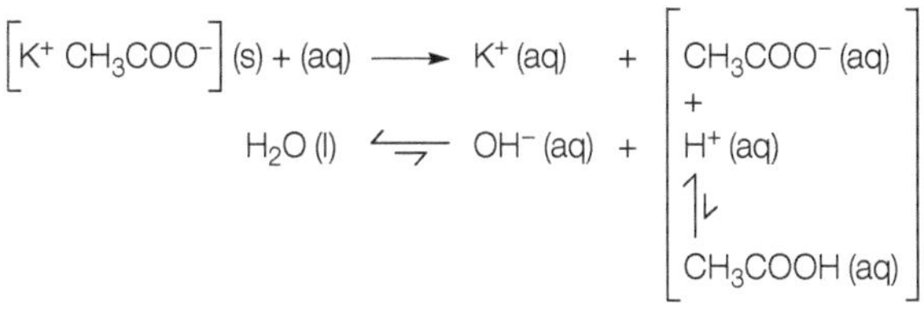

Figure 18.15 Hydrolysis reaction between potassium ethanoate and water – ethanoate ions act as a weak base and combine with protons to form water molecules leaving excess hydroxide ions in solution

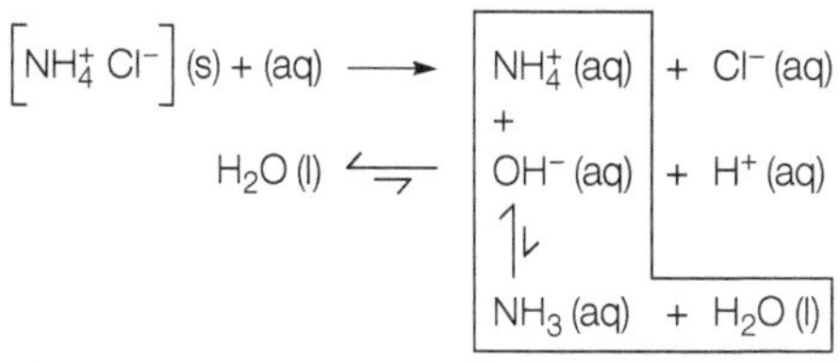

Figure 18.16 Hydrolysis reaction between ammonium chloride and water – ammonium ions act as an acid and donate protons to hydroxide ions to form water molecules leaving excess protons in solution

---

■ **QUICK CHECK QUESTIONS**

25 State whether aqueous solutions of the following salts are acidic, basic or neutral:

K_2CO_3, $CaCl_2$, KH_2PO_4, $(NH_4)_2CO_3$, $AlBr_3$, $CsNO_3$.

26 Suggest why potassium ions do not undergo hydrolysis in aqueous solution.

27 Predict and explain, using equations where appropriate, if each of the following aqueous solutions is acidic, alkaline or neutral:

$1.0\,mol\,dm^{-3}$ $Cu(NO_3)_2$, $1.0\,mol\,dm^{-3}$ $Ba(NO_3)_2$, $1.0\,mol\,dm^{-3}$ $NaHCO_3(aq)$

■ Indicators

Acid–base indicators reveal changes in the pH of aqueous solutions via colour changes. Indicators are usually weak acids that change colour when their molecules lose or gain hydrogen ions (protons) (Figure 18.17). The two forms will be in a dynamic equilibrium.

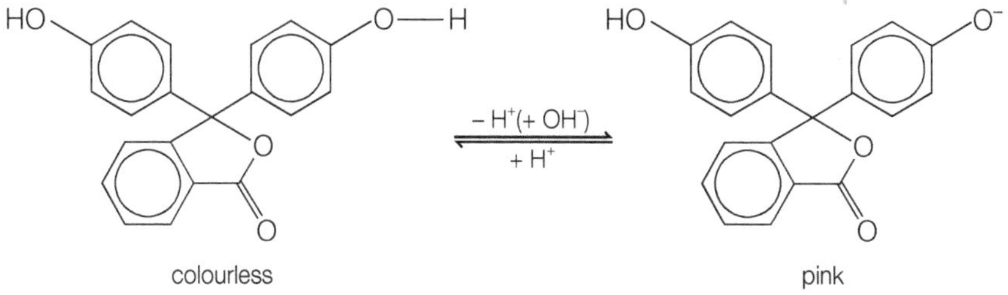

Figure 18.17 Phenolphthalein is colourless in acid, but pink in alkaline conditions

It is conventional to represent a weak acid indicator as HIn, where In is a shorthand form for the rest of the molecule. In water the following equilibrium is established:

$$HIn(aq) \rightleftharpoons H^+(aq) + In^-(aq)$$
colour in acidic solution   colour in alkaline solution

The indicator equilibrium shifts according to the pH of the aqueous solution in which it is present. In acidic solution where the concentration of hydronium ions, H_3O^+, is high the equilibrium shifts to the left and the HIn colour predominates. In an alkaline solution, the concentration of hydronium ions, H_3O^+, is low and the equilibrium shifts to the right and the In^- colour predominates.

The acid dissociation constant for an indicator is known as the indicator constant, K_{in}.

$$K_{in} = \frac{[H^+(aq)] \, [In^-(aq)]}{[HIn(aq)]}$$

rearranging

$$\frac{K_{in} \, [HIn(aq)]}{[In^-(aq)]} = [H^+(aq)]$$

When $[H^+(aq)] = K_{in}$, pH = pK_{in} and the concentrations of ionized and unionized forms will be equal; and so the indicator will appears as a mixture of the two coloured forms. The smallest change in the ratios of the two forms that can be detected by the human eye is a factor of 10 (Figure 18.18).

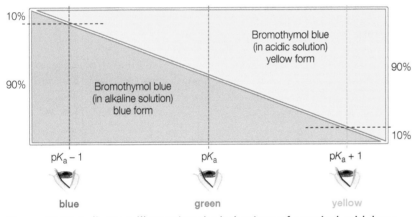

Figure 18.18 A diagram illustrating the behaviour of a typical acid–base indicator: bromothymol blue is placed into a transparent plastic box diagonally divided into halves

Indicators change colour over a range of pH values (Table 18.5). The pH range over which an indicator changes colour is determined by the strength of the weak acid. Typically the range is given by $pK_a \pm 1$.

Table 18.5 The colour changes, and pH range over which the given change takes place, for several commonly used indicators

Indicator	'Acid colour'	'Alkaline colour'	pH range and pK_a
Methyl orange	Red	Yellow	3.1–4.4 and 3.7
Bromothymol blue	Yellow	Blue	6.0–7.6 and 7.0
Bromophenol blue	Yellow	Blue	3.0–4.6 and 4.2
Phenolphthalein	Colourless	Pink	8.3–10.0 and 9.6
Thymol blue	Red	Yellow	1.2–2.8 and 1.6
	Yellow	Blue	8.0–9.6 and 8.9
Methyl red	Red	Yellow	4.4–6.2 and 5.1
Litmus	Red	Blue	5.0–8.0 and 6.5

In general, if $\frac{[HIn]}{[In^-]} > 10$, then colour 1 will dominate. If $\frac{[In^-]}{[HIn]} > 10$, then colour 2 will dominate. The pH at which these transitions will occur depends on the K_{In} of the indicator.

Worked example

The K_a of an indicator (that is, a weak acid) is 1.0×10^{-5}.

If $[H_3O^+] > 1 \times 10^{-4}$ (pH < 4), then $\frac{[HIn]}{[In^-]} > 10$ and colour 1 dominates.

If $[H_3O^+] < 1 \times 10^{-6}$ (pH > 6), then $\frac{[HIn]}{[In^-]} < 0.1$ and colour 2 dominates.

If the pH is between pH 4 and 6, then neither colour dominates.

A few indicators, such as methyl orange, are weak bases and can be described by the following equilibrium:

$$BOH(aq) \rightleftharpoons \quad B^+(aq) + OH^-(aq)$$

base   conjugate acid

Application of Le Châtelier's principle predicts that in alkaline solution the base will predominate and in acidic solution the conjugate acid will predominate.

During an acid–base titration, the pH changes very rapidly around the equivalence point and hence any indicator that changes colour close to the near vertical region of the pH or titration curve will be appropriate to use experimentally.

For a strong acid/strong base titration, any of the indicators could be used since all of them change colour within the vertical portion of the titration curve between about pH 4 and pH 11. In other words, the pK_a values of suitable indicators must lie between 4 and 11, and be centred around 7.

Two common indicators are methyl orange and phenolphthalein. Methyl orange changes colour over the pH range 3.1–4.4 and phenolphthalein changes over the pH range 8.3–10.0. Both indicators are suitable for titrations involving a strong acid and a strong base (Figure 18.19). The same principle applies to the other three possible titrations, except for a weak acid/weak base titration where the almost vertical region is relatively small.

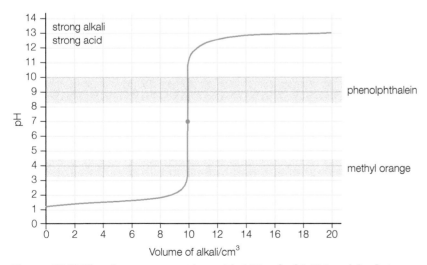

Figure 18.19 Titration curve starting with $100\,cm^3$ of $0.100\,mol\,dm^{-3}$ strong acid and adding $1.0\,mol\,dm^{-3}$ strong alkali with phenolphthalein and methyl orange as indicators

19.1 Electrochemical cells

Essential idea: Energy conversions between electrical and chemical energy lie at the core of electrochemical cells.

Electrochemical cells

- A voltaic cell generates an electromotive force (EMF) resulting in the movement of electrons from the anode (negative electrode) to the cathode (positive electrode) via the external circuit. The EMF is termed the cell potential ($E^{\ominus}$).
- The standard hydrogen electrode (SHE) consists of an inert platinum electrode in contact with 1 mol dm^{-3} hydrogen ion and hydrogen gas at 100 kPa and 298 K. The standard electrode potential ($E^{\ominus}$) is the potential (voltage) of the reduction half-equation under standard conditions measured relative to the SHE. Solute concentration is 1 mol dm^{-3} or 100 kPa for gases. $E^{\ominus}$ of the SHE is 0 V.
- When aqueous solutions are electrolysed, water can be oxidized to oxygen at the anode and reduced to hydrogen at the cathode.
- $\Delta G^{\ominus} = -nFE^{\ominus}$: When $E^{\ominus}$ is positive, $\Delta G^{\ominus}$ is negative, indicative of a spontaneous process. When $E^{\ominus}$ is negative, $\Delta G^{\ominus}$ is positive, indicative of a non-spontaneous process. When $E^{\ominus}$ is 0, then $\Delta G^{\ominus}$ is 0.
- Current, duration of electrolysis and charge on the ion affect the amount of product formed at the electrodes during electrolysis.

■ Electrochemistry

Electrochemistry is the study of the interconversion of chemical and electrical energy. There are two types of cells: voltaic and electrolytic cells. A voltaic cell involves the change of chemical energy to electrical energy. A spontaneous reaction occurs and a voltage and current are produced. Electrolytic cells involve the change of electrical energy to chemical energy.

■ Concept of standard electrode potential

The change in electron density around atoms, ions or molecules during reduction and oxidation means there is an electrical difference between the oxidized and reduced forms of the element (Figure 19.1).

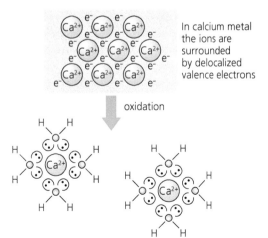

In calcium metal the ions are surrounded by delocalized valence electrons

oxidation

Figure 19.1 The change in electron density between calcium ions in a metal lattice and hydrated calcium ion

In aqueous solution the calcium ions are surrounded by water molecules forming dative bonds with the lone pairs on the oxygen atoms.

$E^{\ominus}$ is the potential difference (voltage difference) between the two chemical species. The measurement of this potential difference under standard conditions gives a measure of the oxidizing or reducing power of a substance (in aqueous solution).

Standard conditions are 298 K, 100 kPa (for any gases) and 1 mol dm^{-3} concentration for all ions in aqueous solution. A change in any of these conditions will cause a change in electrode potential.

The measured potential (voltage) is called the standard electrode potential. It has the symbol $E^{\ominus}$ and is measured in volts by a voltmeter.

■ Measuring standard electrode potentials

Standard electrode potentials are *relative* values. They are always compared to the standard hydrogen electrode (SHE). The standard electrode potential ($E^{\ominus}$) is defined as the potential difference between a standard half-cell and the SHE.

The SHE (Figure 19.2) consists of hydrogen gas bubbling over an inert platinum electrode immersed in a 1.00 mol dm^{-3} solution of H$^+$(aq) (e.g. 1.00 mol dm^{-3} HCl(aq)). Under standard conditions, the hydrogen gas must be at a pressure of 100 kPa (approximately 1 atm).

On the surface of the platinum, an equilibrium is set up between the adsorbed layer of hydrogen molecules and the hydrogen ions in the aqueous solution at 25 °C, i.e. H$_2$(g) $\rightleftharpoons$ 2H$^+$(aq) + 2e$^-$; $E^{\ominus}$ = 0.00 V. The platinised electrode acts as an inert conductor and catalyses the setting up of this equilibrium. The SHE potential is taken to be 0.00 V by definition.

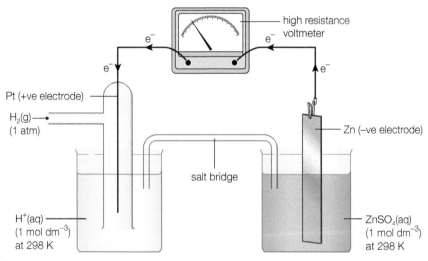

Figure 19.2 The standard hydrogen electrode (SHE) being used to measure the standard electrode potential of the zinc half cell

NATURE OF SCIENCE

Any redox half-cell can in theory be used as a reference electrode. The choice of the SHE and assigning a zero voltage is like choosing sea level to act as a zero reference for height. The deepest part of the ocean could be used as a reference point and assigned a value of zero metres. This is similar to the choice of where to place zero degrees on the Celsius temperature scale.

■ QUICK CHECK QUESTION

1 Describe the materials and conditions used in the standard hydrogen electrode (SHE).

Any redox equilibrium can be arranged into a half-cell including where the oxidizing and reducing agent (conjugate redox pair) are both ions (Figure 19.3). In this case the redox equilibrium occurs on the surface of a platinum wire.

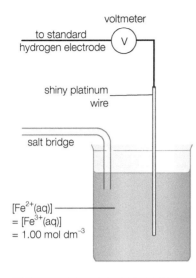

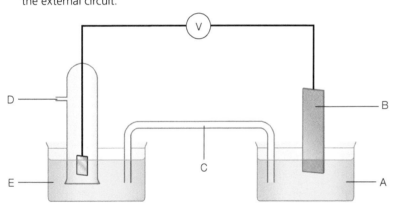

Figure 19.3 Measuring the standard electrode potential of an iron(II)/iron(III) redox couple

[Fe²⁺(aq)] = [Fe³⁺(aq)] = 1.00 mol dm⁻³

■ QUICK CHECK QUESTION

2 An ability to interconvert between two oxidation states is a characteristic of transition elements, including rhenium:

$$ReO_4^-(aq) + 8H^+(aq) + 7e^- \rightleftharpoons Re(s) + 4H_2O(l) \qquad E^\ominus = +0.34\,V$$

The apparatus shown is used to determine the standard electrode potential for the above reaction involving rhenium. Identify clearly the parts labelled from **A** to **E** in the diagram. Include all concentrations and operating conditions whenever applicable. Describe the direction of electron flow in the external circuit.

▨ Predicting redox reactions

The value of the standard electrode potential, $E^\ominus$, for a standard half-cell gives a measure of the relative stability of the oxidizing and reducing agents in a redox equilibrium.

By convention the reactions in half-cells are written as reduction potentials. The reduced form is on the left and the oxidized form on the right with the electrons on the left as a reactant.

A positive value for $E^\ominus$ indicates that the equilibrium position for the half-cell reaction (Figure 19.4) lies to the right compared to the hydrogen half-equation ($K_c > 1$). It is positive because electrons (negatively charged) are a reactant and being used up.

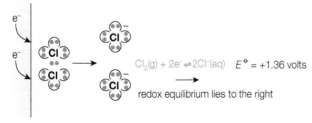

$$Cl_2(g) + 2e^- \rightleftharpoons 2Cl^-(aq) \quad E^\ominus = +1.36 \text{ volts}$$

redox equilibrium lies to the right

Figure 19.4 Reduction at the cathode

The zero value of $E^{\ominus}$ for the standard hydrogen electrode indicates that the redox equilibrium is at 'perfect' equilibrium ($K_c = 1$)

$$2H^+(aq) + 2e^- \rightleftharpoons H_2(g)$$

A negative value for $E^{\ominus}$ indicates that the equilibrium position for the half-cell reaction (Figure 19.5) lies to the left compared with the hydrogen half-equation ($K_c < 1$). It is negative because electrons are being produced on the surface of the electrode.

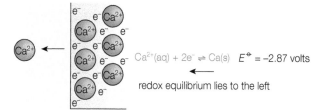

$$Ca^{2+}(aq) + 2e^- \rightleftharpoons Ca(s) \quad E^{\ominus} = -2.87 \text{ volts}$$

redox equilibrium lies to the left

Figure 19.5 Oxidation at the anode

Table 19.1 shows the standard electrode potentials for some redox half-reactions. They are written as reduction half potentials. The table is extracted from the IB Chemistry *data booklet*.

Table 19.1 Standard reduction electrode potentials

Oxidized species	$\rightleftharpoons$	Reduced species	$E^{\ominus}$/V
$Li^+(aq) + e^-$	$\rightleftharpoons$	$Li(s)$	−3.04
$K^+(aq) + e^-$	$\rightleftharpoons$	$K(s)$	−2.93
$Ca^{2+}(aq) + 2e^-$	$\rightleftharpoons$	$Ca(s)$	−2.87
$Na^+(aq) + e^-$	$\rightleftharpoons$	$Na(s)$	−2.71
$Mg^{2+}(aq) + 2e^-$	$\rightleftharpoons$	$Mg(s)$	−2.37
$Al^{3+}(aq) + 3e^-$	$\rightleftharpoons$	$Al(s)$	−1.66
$Mn^{2+}(aq) + 2e^-$	$\rightleftharpoons$	$Mn(s)$	−1.19
$H_2O(l) + e^-$	$\rightleftharpoons$	$\frac{1}{2}H_2(g) + OH^-(aq)$	−0.83
$Zn^{2+}(aq) + 2e^-$	$\rightleftharpoons$	$Zn(s)$	−0.76
$Fe^{2+}(aq) + 2e^-$	$\rightleftharpoons$	$Fe(s)$	−0.45
$Ni^{2+}(aq) + 2e^-$	$\rightleftharpoons$	$Ni(s)$	−0.26
$Sn^{2+}(aq) + 2e^-$	$\rightleftharpoons$	$Sn(s)$	−0.14
$Pb^{2+}(aq) + 2e^-$	$\rightleftharpoons$	$Pb(s)$	−0.13
$H^+(aq) + e^-$	$\rightleftharpoons$	$\frac{1}{2}H_2(g)$	0.00
$Cu^{2+}(aq) + e^-$	$\rightleftharpoons$	$Cu^+(aq)$	+0.15
$SO_4^{2-}(aq) + 4H^+(aq) + 2e^-$	$\rightleftharpoons$	$H_2SO_3(aq) + H_2O(l)$	+0.17
$Cu^{2+}(aq) + 2e^-$	$\rightleftharpoons$	$Cu(s)$	+0.34
$\frac{1}{2}O_2(g) + H_2O + 2e^-$	$\rightleftharpoons$	$2OH^-(aq)$	+0.40
$Cu^+(aq) + e^-$	$\rightleftharpoons$	$Cu(s)$	+0.52
$\frac{1}{2}I_2(s) + e^-$	$\rightleftharpoons$	$I^-(aq)$	+0.54
$Fe^{3+}(aq) + e^-$	$\rightleftharpoons$	$Fe^{2+}(aq)$	+0.77
$Ag^+(aq) + e^-$	$\rightleftharpoons$	$Ag(s)$	+0.80
$\frac{1}{2}Br_2(l) + e^-$	$\rightleftharpoons$	$Br^-(aq)$	+1.07
$\frac{1}{2}O_2(g) + 2H^+(aq) + 2e^-$	$\rightleftharpoons$	$H_2O(l)$	+1.23
$Cr_2O_7^{2-}(aq) + 14H^+(aq) + 6e^-$	$\rightleftharpoons$	$2Cr^{3+}(aq) + 7H_2O(l)$	+1.33
$\frac{1}{2}Cl_2(g) + e^-$	$\rightleftharpoons$	$Cl^-(aq)$	+1.36
$MnO_4^-(aq) + 8H^+(aq) + 5e^-$	$\rightleftharpoons$	$Mn^{2+}(aq) + 4H_2O(l)$	+1.51
$\frac{1}{2}F_2(g) + e^-$	$\rightleftharpoons$	$F^-(aq)$	+2.87

■ **QUICK CHECK QUESTIONS**

3 Aluminium and copper have standard electrode potentials of −1.66 V and +0.34 V, respectively. Explain the meaning of these values.

4 Consider the following values of electrode potential.

Electrode	$E^{\ominus}$/V
$Fe^{2+}(aq)/Fe(s)$	−0.45
$Na^+(aq)/Na(s)$	−2.71
$Cu^{2+}(aq)/Cu(s)$	+0.34
$Ag^+(aq)/Ag(s)$	+0.80

Determine which species is the strongest oxidizing agent.

Calculating cell potentials

The standard cell potential, $E^{\ominus}_{cell}$, is a measure of the tendency of electrons to flow through the external circuit (wire and voltmeter) of a voltaic cell under standard conditions of 25 °C, 100 kPa (1 atmosphere pressure) and 1.00 mol dm^{-3} ion concentrations. It is the maximum potential difference (voltage) between the electrodes. This is known as the electromotive force (EMF) and is measured using a high resistance voltmeter so negligible current flows.

The cell potential can be determined from the cell diagram of the voltaic cell. (This is the formal approach adopted by IUPAC.)

$$E^{\ominus}_{cell} = E^{\ominus}_{C} - E^{\ominus}_{A}$$

where $E^{\ominus}_{C} = E^{\ominus}$ of the cathode (reduction)

$E^{\ominus}_{A} = E^{\ominus}$ of the anode (oxidation)

$E^{\ominus}$ values used are 'as given' in the IB Chemistry *data booklet*. Do *not* change the sign. The electrode with the more positive $E^{\ominus}$ value is more oxidizing and so will undergo reduction, i.e. the half-cell is on the right.

Worked example

A voltaic cell consisting of half-cells: copper in $CuSO_4(aq)$ and silver in $AgNO_3(aq)$.

The cell diagram:

$Cu^{2+} \xrightarrow{\text{electron flow}} Ag(s)$

$(-)Cu(s)/Cu^{2+}(aq) \quad Ag^{+}(aq)/Ag(s) \ (+)$

$\quad E^{\ominus}_{A} = +0.34\,V \quad E^{\ominus}_{C} = +0.80\,V$

$\quad E^{\ominus}_{cell} = E^{\ominus}_{C} - E^{\ominus}_{A}$

$\qquad = +0.80\,V - (+0.34)\,V$

$\qquad = +0.46\,V$

The half-equations for the reactions at the electrodes are:

$Cu(s) \rightarrow Cu^{2+}(aq) + 2e^{-}$ (oxidation)

$2Ag^{+}(aq) + 2e^{-} \rightarrow 2Ag(s)$ (reduction)

The ionic equation for the overall cell reaction is:

$Cu(s) + 2Ag^{+}(aq) \rightarrow Cu^{2+}(aq) + 2Ag(s)$

This can be written as a cell diagram using the following convention:

phase boundary   salt bridge   phase boundary

electrode(s) | electrolyte(aq) || electrolyte(aq) | electrode(s)

(anode half-cell)   (cathode half-cell)

oxidation   reduction

electrons flow this way →

Another approach is to find the two half-equations in the IB Chemistry *data booklet*. They are always written as *reduction* potentials.

For example, to determine the cell potential of the copper/silver voltaic cell.

$Cu^{2+}(aq) + 2e^{-} \rightarrow Cu(s) \quad E^{\ominus} = +0.34\,V$

$Ag^{+}(aq) + e^{-} \rightarrow Ag(s) \qquad E^{\ominus} = +0.80\,V$

Reverse the half-equation with the *more negative* (or *less positive*) electrode potential and add it to the other half-equation. Since the half-equation is reversed the sign must be reversed (cf. Hess's law in Topic 5 Energetics/thermochemistry).

$Cu(s) \rightarrow Cu^{2+}(aq) + 2e^{-} \quad E^{\ominus} = -0.34\,V$

$Ag^{+}(aq) + e^{-} \rightarrow Ag(s) \qquad E^{\ominus} = +0.80\,V$

Adjust the number of electrons for the silver half-cell so it can be added to the copper half-cell. Show the signs when adding the two standard electrode potentials.

$Cu(s) \rightarrow Cu^{2+}(aq) + 2e^- \quad E^\ominus = -0.34\,V$

$2Ag^+(aq) + 2e^- \rightarrow Ag(s) \quad E^\ominus = +0.80\,V$

$E^\ominus = (-0.34\,V) + (0.80\,V) = +0.46\,V$

Common mistake

A common misconception is that the electrode potential of a half-cell is changed if the stoichiometry is changed; it is not. The voltage is a measure of the energy of an electron. Current is a measure of the number of electrons that flow past a point. The voltage of a simple voltaic cell does not depend on the amounts of chemicals used.

■ QUICK CHECK QUESTIONS

5   State the half-cells involved in the redox reactions below.

$Zn(s) + Pb^{2+}(aq) \rightarrow Zn^{2+}(aq) + Pb(s)$

$2H^+(aq) + Ca(s) \rightarrow Ca^{2+}(aq) + H_2(g)$

6   A voltaic cell is constructed by connecting an aluminium half-cell to a silver half-cell under standard conditions. Write the cell diagram for the cell. Deduce the ionic equation and its cell potential.

7   Manganate(VII) ions oxidize iron(II) ions to iron(II) ions and they reduced to manganese(II) ions in the presence of aqueous acid. Deduce the ionic equation that occurs during the reaction and write a cell diagram for the electrochemical cell that can be constructed for the reaction.

▧ Uses of standard electrode potential ($E^\ominus$) and standard cell potential, $E^\ominus_{cell}$

1   To determine the strengths of both oxidizing and reducing agents.
   ■   Strong oxidizing agents have high positive $E^\ominus$ values and undergo reduction easily by gaining electrons.
   ■   Strong reducing agents have high negative $E^\ominus$ values and undergo oxidation easily by losing electrons.
   ■   Therefore, the $E^\ominus$ values can be used to compare the relative tendencies of half-cells to undergo oxidation or reduction.

2   To predict the direction of electron flow in a simple voltaic cell.
   ■   In the external circuit, electrons flow from the negative electrode (oxidation) to the positive electrode (reduction).
   ■   Therefore, electrons flow from the half-cell with the more negative $E^\ominus$ value to the half-cell with the more positive $E^\ominus$ value.

3   To predict the feasibility of a reaction.
   ■   Reactions with a positive $E^\ominus_{cell}$ are thermodynamically feasible or spontaneous. It is important to understand the relationship between $E^\ominus_{cell}$ and $\Delta G^\ominus$. A *positive* $E^\ominus_{cell}$ value implies a *negative* $\Delta G^\ominus$; hence it will be a thermodynamically spontaneous reaction (under standard conditions).
   ■   This relationship follows from the expression: $\Delta G^\ominus = -nFE^\ominus$, where n is the number of electrons transferred during the redox reaction and F is the Faraday constant (the charge carried by one mole of electrons).
   ■   Thermodynamics indicates nothing about rates, so a spontaneous redox reaction may appear not to happen. The reaction will be occurring at a very low rate due to a large activation energy.
   ■   The predictions are only accurate for standard conditions. If the temperatures and/or concentrations are non-standard then the predicted direction for the redox equilibrium may not be accurate.

Expert tip

The standard electrode potentials for reactive metals, such as calcium and sodium are not measured experimentally since they undergo a direct reaction with water to release hydrogen. The values of their electrode potentials are calculated indirectly using Hess's law.

---

8 An Fe^{3+}/Fe^{2+} half-cell was connected to a Ni^{2+}/Ni cell under standard conditions. Determine the cell potential; deduce the ionic equation and state the observation at each electrode when the cell is discharging.

9 Determine $\Delta G^{\ominus}$ for the ionic reaction:

$Zn(s) + 2Ag^+(aq) \rightarrow Zn^{2+}(aq) + 2Ag(s)$.

10 The standard reduction potentials in volts, for Ag^+ to Ag and Fe^{3+} to Fe^{2+} are +0.80 and +0.77 V, respectively. Calculate K_c, the equilibrium constant, for the reaction under standard conditions.

11 The standard electrode potentials of four half-cells are given below.

Half-equation		$E^{\ominus}$/V
$Zn^{2+}(aq) + 2e^-$	$\rightleftharpoons Zn(s)$	−0.76 V
$V^{3+}(aq) + e^-$	$\rightleftharpoons V^{2+}(aq)$	−0.26 V
$VO^{2+}(aq) + 2H^+(aq) + e^-$	$\rightleftharpoons V^{3+}(aq) + H_2O(l)$	+0.34 V
$VO_2^+(aq) + 2H^+(aq) + e^-$	$\rightleftharpoons VO^{2+}(aq) + H_2O(l)$	+1.00 V

Identify which species above are the strongest oxidizing and reducing agents, giving a reason in terms of electrons for your answer. Deduce the balanced equation for the ionic reaction with the most positive cell potential. State and explain the sign of $\Delta G^{\ominus}$ for the reaction.

Factors affecting electrode potential

Table 19.2 summarizes the effect of various factors affecting the electrode potential of the following generalized redox equilibria:

$M^{n+}(aq) + ne^- \rightleftharpoons M(s)$

$L_2(g) + 2ne^- \rightleftharpoons 2L^{n-}(aq)$

Table 19.2 Factors affecting electrode potential

Factor	Concentration of M^{n+}	Temperature	Pressure of L_2
Change in factor	Increase	Increase	Increase
Change in equilibrium (application of Le Châtelier's principle)	Equilibrium shifts to the right	Equilibrium shifts to the left	Equilibrium shifts to the right
	Favours reduction to remove M^{n+} ions	More metal* dissolves at a higher temperature	Favours reduction to remove L_2 molecules
		Favours oxidation to produce more M^{n+}	
Outcome	$E^{\ominus}$ value is more positive; more metal, M, is deposited	$E^{\ominus}$ value is more negative; more M^{n+} ions released	$E^{\ominus}$ value is more positive; more L^{n-} ions released

*Applies to active metals (above hydrogen in the activity series)

12 Consider the cell: $Pt(s) \mid H_2(g); H^+(aq) \parallel Fe^{3+}(aq); Fe^{2+}(aq) \mid Pt(s)$. Calculate the cell potential, show the direction of electron flow and predict how the cell potential will change if the concentration of $Fe^{3+}(aq)$ is increased.

Electrolysis of aqueous solutions

The electrolysis of aqueous solutions of ionic compounds (soluble salts, alkalis and acids) is more complicated than the electrolysis of molten ionic compounds. This is because ions from the water molecules present in the solution of the electrolyte will also undergo electrolysis. This occurs because of the self-ionization of water:

$H_2O(l) \rightleftharpoons H^+(aq) + OH^-(aq)$

The hydrogen and hydroxide ions migrate with the ions in the ionic compound and compete with them to accept electrons (reduction) or release electrons (oxidation) at the cathode and anode:

$2e^- + 2H^+(aq) \rightarrow H_2(g)$

and

$4OH^-(aq) \rightarrow 4e^- + 2H_2O(l) + O_2(g)$

Although the ion concentrations are very low, they are maintained since the equilibrium will shift to the right to replace any ions removed by redox reactions.

An alternative (but *equivalent*) explanation is to suggest that water molecules undergo reduction and oxidation at the cathode and anode:

$2H_2O(l) + 2e^- \rightarrow H_2(g) + OH^-(aq)$

and

$2H_2O(l) \rightarrow O_2(g) + 4H^+(aq) + 4e^-$

It is important to understand the factors affecting the discharge of ions during the electrolysis of aqueous solutions. The factors that determine the ions discharged at the electrodes are:

- the standard electrode potentials of the ions, $E^\ominus$ values
- the relative concentration of the competing ions and
- the chemical nature of electrodes used.

The standard electrode potentials of the ions, $E^\ominus$ values

During electrolysis of aqueous solutions, cations are discharged at the cathode, $M^{n+}(aq) + ne^- \rightarrow M(s)$. For example, if hydrogen ions (H^+) are discharged, molecules of hydrogen gas, H_2, will be obtained: $2H^+(aq) + 2e^- \rightarrow H_2(g)$. Since the discharge of ions at the cathode involves reduction, ions that accept electrons readily will be reduced first.

Therefore, strong oxidizing agents (electron acceptors) with more positive $E^\ominus$ values are preferentially discharged.

For example, it is easier to discharge $Cu^{2+}(aq)$ than $Fe^{2+}(aq)$ at the cathode because of their respective $E^\ominus$ values: $Fe^{2+}(aq) + 2e^- \rightarrow Fe(s)$; $E^\ominus = -0.45\,V$ and $Cu^{2+}(aq) + 2e^- \rightarrow Cu(s)$; $E^\ominus = +0.34\,V$.

Hence, if iron(II) and copper(II) ions (of equal concentration) migrate to a platinum electrode, the copper(II) ions will be preferentially discharged, i.e. $Cu^{2+}(aq) + 2e^- \rightarrow Cu(s)$.

During electrolysis, anions (often halide ions) are discharged at the anode: $2X^{n-}(aq) \rightarrow X_2(g) + 2ne^-$ Since this is an oxidation reaction, ions that lose electrons readily will be oxidized first.

Therefore, strong reducing agents (electron donors) with more negative $E^\ominus$ values are preferentially discharged.

For example, it is easier to discharge iodide ions than chloride ions (of equal concentration) at the anode because of their respective $E^\ominus$ values:

$I_2(s) + 2e^- \rightarrow 2I^-(aq) \qquad E^\ominus = +0.54\,V$

$Cl_2(aq) + 2e^- \rightarrow 2Cl^-(aq) \quad E^\ominus = +1.36\,V$

Hence, if chloride and iodide ions migrate to a platinum electrode, the iodide ions will be preferentially discharged, i.e. $2I^-(aq) \rightarrow I_2(s) + 2e^-$.

The nature of the electrodes used

Usually, inert electrodes such as graphite or platinum are used for electrolysis. These electrodes do not interfere with the chemical reactions occurring at the surface of the electrode. They simply act as a connector between the electrical circuit and the solution.

If metal electrodes are used in metal ion solutions, they can get involved in the reactions by dissolving as ions, leaving their electrons behind. (This can only happen when the metal takes the place of the anode, the positive electrode.)

■ QUICK CHECK QUESTION

13 Using the data below and other data from Table 19.1, predict and explain which metal, nickel or manganese, may be obtained by electrolysis of separate aqueous solutions of $Ni^{2+}(aq)$ ions and $Mn^{2+}(aq)$ ions (under standard conditions).

$Ni^{2+}(aq) + 2e^- \rightarrow Ni(s)$;
$$E^\ominus = -0.25\,V$$

$Mn^{2+}(aq) + 2e^- \rightarrow Mn(s)$;
$$E^\ominus = -1.18\,V$$

▓ Relative concentration of cations or anions

If the difference between the standard electrode potential values of two competing ions is not large, the ion of a higher concentration will be selectively discharged.

For example, if an aqueous electrolyte contains both hydroxide ions, $OH^-(aq)$ and chloride ions, $Cl^-(aq)$, both anions will migrate to an inert platinum or graphite anode. Either the $OH^-(aq)$ or the $Cl^-(aq)$ ions may be discharged at the anode. The relevant $E^{\ominus}$ values are listed below:

$$Cl_2(g) + 2e^- \rightarrow 2Cl^-(aq) \qquad\qquad E^{\ominus} = +1.36\,V$$

$$O_2(g) + 2H_2O(l) + 4e^- \rightarrow 4OH^-(aq) \qquad E^{\ominus} = +0.40\,V$$

Based on standard electrode potential values, $OH^-(aq)$ will be selected for discharge, since its value is less positive and hence its discharge is more thermodynamically favourable.

However, if the concentration of $Cl^-(aq)$ is very high then chloride ions will be discharged at the anode instead, i.e. $2Cl^-(aq) \rightarrow Cl_2(g) + 2e^-$. (This is a kinetic effect and in practice chlorine mixed with a trace of oxygen is produced.)

In a redox equilibrium, the $E^{\ominus}$ of a half-cell or redox reaction changes with the concentration of the ions in the cell. The influence is directly predicted by Le Châtelier's principle. If the position of equilibrium is shifted to the right, the $E^{\ominus}$ becomes more positive and vice versa.

However, if the difference between the $E^{\ominus}$ values of the competing ions is very large, concentration is no longer an important factor. Consider the electrolysis of a solution containing $H^+(aq)$ and $Na^+(aq)$ ions (with equal concentrations).

$$Na^+(aq) + e^- \rightarrow Na(s) \qquad E^{\ominus} = -2.71\,V$$

$$2H^+(aq) + 2e^- \rightarrow H_2(g) \qquad E^{\ominus} = 0.00\,V$$

Since the $E^{\ominus}$ of $H^+(aq)$ is much more positive than $E^{\ominus}$ of $Na^+(aq)$, $H^+(aq)$ will be discharged at the cathode to produce hydrogen gas i.e. $2H^+(aq) + 2e^- \rightarrow H_2(g)$.

Table 19.3 summarizes some common electrolytic processes:

Table 19.3 Some common electrolytic processes

Electrolyte	Electrode	Anode reaction	Cathode reaction	Remarks
Molten NaCl	C	$2Cl^- \rightarrow Cl_2 + 2e^-$	$2Na^+ + 2e^- \rightarrow 2Na$	–
Dilute aqueous NaCl	C	$4OH^- \rightarrow O_2 + 2H_2O + 4e^-$ or $(H_2O \rightarrow \frac{1}{2}O_2 + 2H^+ + 2e^-)$	$4H^+ + 4e^- \rightarrow 2H_2$ or $(H_2O + e^- \rightarrow \frac{1}{2}H_2 + OH^-)$	If phenolphthalein is added to the solution around the cathode it turns pink
Concentrated aqueous NaCl	C	$2Cl^- \rightarrow Cl_2 + 2e^-$	$2H^+ + 2e^- \rightarrow H_2$ or $(H_2O + e^- \rightarrow \frac{1}{2}H_2 + OH^-)$	Cl_2 selectively liberated instead of O_2 due to high concentration of chloride ions
$H_2SO_4(aq)$	Pt	$4OH^- \rightarrow O_2 + 2H_2O + 4e^-$ or $H_2O \rightarrow \frac{1}{2}O_2 + 2H^+ + 2e^-$	$4H^+ + 4e^- \rightarrow 2H_2$	$H_2SO_4(aq)$ slowly becomes more concentrated
$H_2SO_4(aq)$	C	$4OH^- \rightarrow O_2 + 2H_2O + 4e^-$ or $H_2O \rightarrow \frac{1}{2}O_2 + 2H^+ + 2e^-$	$4H^+ + 4e^- \rightarrow 2H_2$	$H_2SO_4(aq)$ slowly becomes more concentrated
$CuSO_4(aq)$	Pt	$4OH^- \rightarrow O_2 + 2H_2O + 4e^-$ or $H_2O \rightarrow \frac{1}{2}O_2 + 2H^+ + 2e^-$	$2Cu^{2+} + 4e^- \rightarrow 2Cu$	Solution slowly becomes more concentrated in H^+ and SO_4^{2-} ions
$CuSO_4(aq)$	*Cu	*$Cu \rightarrow Cu^{2+} + 2e^-$	$Cu^{2+} + 2e^- \rightarrow Cu$	Transfer of copper from anode to cathode. No change to $CuSO_4(aq)$
Molten $PbBr_2$	C	$2Br^- \rightarrow Br_2 + 2e^-$	$Pb^{2+} + 2e^- \rightarrow Pb$	–
Molten PbO	C	$2O^{2-} \rightarrow O_2 + 4e^-$	$2Pb^{2+} + 4e^- \rightarrow 2Pb$	O_2 liberated
$Na_2SO_4(aq)$	Pt	$4OH^- \rightarrow O_2 + 2H_2O + 4e^-$ or $H_2O \rightarrow \frac{1}{2}O_2 + 2H^+ + 2e^-$	$4H^+ + 4e^- \rightarrow 2H_2$ or $H_2O + e^- \rightarrow \frac{1}{2}H_2 + OH^-$	Solution slowly becomes more concentrated

■ QUICK CHECK QUESTIONS

14 Deduce the equations for the formation of the major product at the positive electrode (anode) when dilute sodium chloride and concentrated sodium chloride solutions are electrolysed.

15 Explain how the pH of an aqueous solution of copper(II) sulfate changes when it is electrolysed with inert platinum electrodes.

16 Write equations describing the anode and cathode reactions when silver(I) nitrate solution is electrolysed using silver electrodes.

17 Explain the difference between the following two electrolytic processes: electrolysis of molten sodium chloride gives sodium at the cathode while aqueous sodium chloride gives hydrogen gas at the cathode.

■ Electroplating

Metals are electroplated to improve their appearance or to prevent corrosion (oxidation). During electroplating the object to be electroplated is made the cathode in an electrolytic cell. The anode is pure plating metal and the electrolyte contains the ions of the plating metal (Figure 19.6).

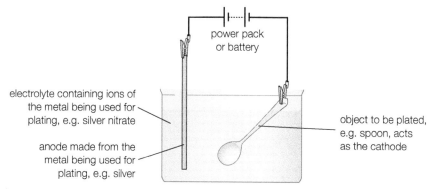

Figure 19.6 Electroplating apparatus: silver-plating

■ QUICK CHECK QUESTIONS

18 Draw a diagram to show the migration of ions in a nickel-plating cell containing nickel sulfate solution.

19 The apparatus used for chromium-plating is shown below.

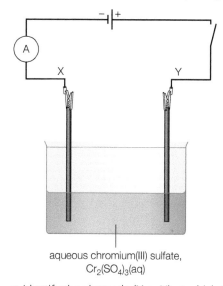

a Identify the electrode (X or Y) at which the object to be chromium-plated should be placed.

b Write the half-equations, including state symbols, for the reactions taking place at the two electrodes X and Y.

c Explain why aluminium-plating cannot be carried out if the electrolyte is replaced with aqueous aluminium nitrate.

Electrolysis calculations

We want to know how much of a specific product we obtain with electrolysis, especially electroplating. Consider the reduction of copper(II) ions to copper atoms:

$$Cu^{2+}(aq) + 2e^- \rightarrow Cu(s)$$

Two moles of electrons will plate 1 mole of copper atoms. The charge of 1 mole of electrons is $96\,500\,C$ and is known as the Faraday constant (1 Faraday = $96\,500\,C\,mol^{-1}$). A coulomb is the quantity of electrons passing a point in 1 second when the current is 1 ampere. The amount of copper can be calculated from the current in amps (I) and time in seconds (t) required to plate: $Q = It$.

Faraday's laws of electrolysis summarize the formation of products:

Expert tip

The Faraday constant is the charge (in coulombs) carried by 1 mole of electrons. It is good practice to include units in any electrochemical calculations. The Faraday constant allows a charge to be converted to amount of electrons.

Faraday's first law of electrolysis

The mass of any substance liberated or deposited at an electrode during electrolysis is directly proportional to the quantity of electricity that passed through the electrolyte. This means that the larger the current used or the longer the solution is electrolysed, the greater the amount of products that will be obtained.

Worked example

A current of 8.04 A is passed for 100 min through an aqueous solution of sulfuric acid using inert graphite electrodes. Determine the volume of gas produced at STP.

$Q = It$; $Q = 8.04\,A \times (100 \times 60\,s) = 48\,250\,C$; amount of electrons = $48\,250\,C/96\,500\,C\,mol^{-1} = 0.5\,mol$; $4H^+ + 4e^- \rightarrow 2H_2$; $4OH^- \rightarrow O_2 + 2H_2O + 4e^-$; $1/4\,mol\ H_2$ ($5.68\,dm^3$) and $1/8\,mol\ O_2$ ($2.84\,dm^3$); total volume = $8.52\,dm^3$.

Worked example

When a current of 3.00 A was passed through molten lead(II) bromide for 30 min lead of mass 5.60 g was obtained. The charge of the electron (e) is $1.6 \times 10^{-19}\,C$. Determine an experimental value for the Avogadro constant (L).

$Q = It$; $Q = 3.00\,A \times (30 \times 60\,s) = 5400\,C$; anode: $2Br^- \rightarrow Br_2 + 2e^-$; cathode: $Pb^{2+} + 2e^- \rightarrow Pb$; to deposit 5.60 g of lead 5400 C of charge has to be passed; to deposit 1 mole of lead, $(207.20/5.60) \times 5400\,C = 199\,800\,C$; 2 moles of electrons are required to deposit 1 mole of lead; 2 moles of electrons ($2L$) electrons will carry a charge of $2L \times e^- = 2Le^-$ coulombs, therefore $2Le^- = 199\,800\,C$, substituting e for $1.6 \times 10^{-19}\,C$ and solving for L: $(199\,800\,C)/(2 \times 1.6 \times 10^{-19}\,C) = 6.24 \times 10^{23}$.

Faraday's second law of electrolysis

The amount of electrons (in moles) required to discharge 1 mole of an ion at an electrode is equal to the number of charges on the ion. Hence, the mass of elements liberated during electrolysis depends on: the time of passing of a steady current, the size of the steady current used and the charge on the ion of the element. It is independent of the electrolyte concentration.

Faraday's laws of electrolysis can be illustrated by three electrolytic cells with aqueous solutions (cells 1, 2 and 3) joined in series. This arrangement means all three cells receive the same current (number of electrons per second).

Expert tip

The higher the charge on a cation, the lower the amount of the metal that will be formed (for the same current and same time).

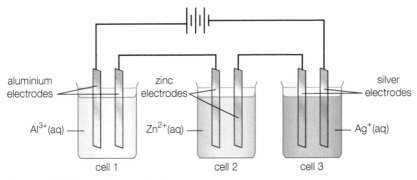

Figure 19.7 Electrolytic cells in series

When 3 mol of electrons ($96\,500\,C\,mol^{-1} \times 3\,mol = 289\,500\,C$) or $289\,500\,C$ of electricity are passed through the three electrolytic cells:

Cell I	Cell II	Cell III
Anode: $Al(s) \rightarrow Al^{3+}(aq) + 3e^-$	Anode: $Zn(s) \rightarrow Zn^{2+}(aq) + 2e^-$	Anode: $Ag(s) \rightarrow Ag^+(aq) + e^-$
Cathode: $Al^{3+}(aq) + 3e^- \rightarrow Al(s)$	Cathode: $Zn^{2+}(aq) + 2e^- \rightarrow Zn(s)$	Cathode: $Ag^+(aq) + e^- \rightarrow Ag(s)$
3 mol of electrons will deposit 1 mol of Al	2 mol of electrons will deposit 1 mol of Zn	1 mol of electrons will deposit 1 mol of Ag
3 mol of electrons will deposit 1 mol of aluminium atoms, Al	3 mol of electrons will deposit 1.5 mol of zinc atoms, Zn	3 mol of electrons will deposit 3 mol of silver atoms, Ag

■ Summary of the steps involved in calculating amounts of substances reduced or oxidized in electrolysis

The starting point is the balanced redox equation for the electrode process, and then:

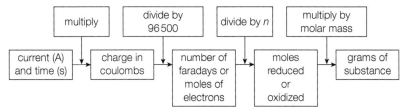

n = number of moles of electrons in balanced redox equation.

■ QUICK CHECK QUESTIONS

20 A current of 2.40 A is passed through an aqueous solution of copper(II) sulfate for 25 minutes using copper electrodes. Determine the change in mass of the electrodes.

21 Electrolytic cells containing aqueous sodium hydroxide, concentrated aqueous sodium chloride and aqueous silver(I) nitrate solution are connected in series. Each cell has carbon (graphite) electrodes.

Determine the ratio of the amount of hydrogen gas in the first cell, chlorine gas in the second cell and oxygen gas in the third cell when an electric current flows through the three cells.

22 A current is passed through two cells connected in series. The first cell contains $XSO_4(aq)$ while the second cell contains $Y_2SO_4(aq)$.

Given that the relative atomic masses of X and Y are in the ratio 1:2, deduce the ratio of mass of X liberated: mass of Y liberated.

23 An unknown amount of steady current was allowed to pass through a beaker containing silver nitrate solution for 1200 s using inert platinum electrodes. A mass of 0.200 g of silver was deposited at the cathode.

Calculate the current in mA that was used in the electrolysis process.
[1 faraday = $96\,500\ C\ mol^{-1}$.]

20.1 Types of organic reactions

Essential idea: Key organic reaction types include nucleophilic substitution, electrophilic addition, electrophilic substitution and redox reactions. Reaction mechanisms vary and help in understanding the different types of reaction taking place.

Nucleophilic substitution

- S_N1 represents a nucleophilic unimolecular substitution reaction and S_N2 represents a nucleophilic bimolecular substitution reaction. S_N1 involves a carbocation intermediate. S_N2 involves a concerted reaction with a transition state.
- For tertiary halogenoalkanes the predominant mechanism is S_N1 and for primary halogenoalkanes it is S_N2. Both mechanisms occur for secondary halogenoalkanes.
- The rate-determining step (slow step) in an S_N1 reaction depends only on the concentration of the halogenoalkane, rate = k[halogenoalkane]. For S_N2, rate = k[halogenoalkane][nucleophile]. S_N2 is stereospecific with an inversion of configuration at the carbon.
- S_N2 reactions are best conducted using aprotic, non-polar solvents and S_N1 reactions are best conducted using protic, polar solvents.

Halogenoalkanes undergo nucleophilic substitution reactions (Figure 20.1) with nucleophiles, most notably, OH⁻ and H_2O. The former is a better nucleophile due to its charge. One of the lone pairs on the oxygen attacks the electrophilic (partially positive) carbon atom attached to the halogen atom. A halide ion (leaving group) is removed along with the pair of electrons which made up the original C–X bond.

Figure 20.1 Diagrammatic representation of nucleophilic substitution

The mechanism of this nucleophilic substitution reaction varies with the structure of the halogenoalkane. Tertiary halogenoalkanes follow a two-step (S_N1) mechanism in which the halide ion leaves (in a slow step) before the nucleophilic attack to form a low energy carbocation intermediate. The reaction is completed (in a fast step) by nucleophilic attack (Figure 20.2).

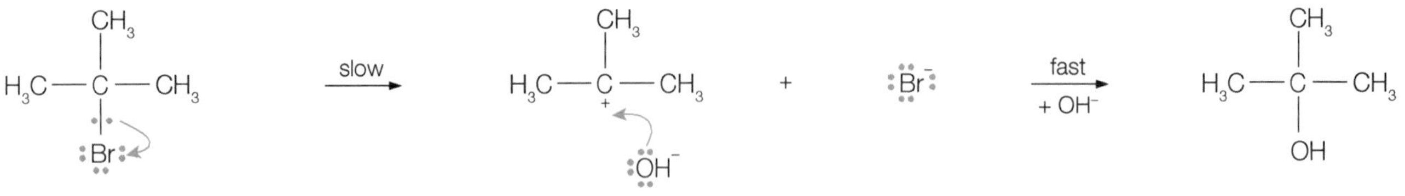

Figure 20.2 The S_N1 mechanism for the hydrolysis of 2-bromo-2- methylpropane

Primary halogenoalkanes (Figure 20.3) follow a one-step mechanism (S_N2) in which one pair of electrons from the nucleophile begins forming the new bond to the carbon atom at the same time as the halide ion is leaving from the other side of the molecule. This is known as a concerted reaction and a high-energy transition state is formed.

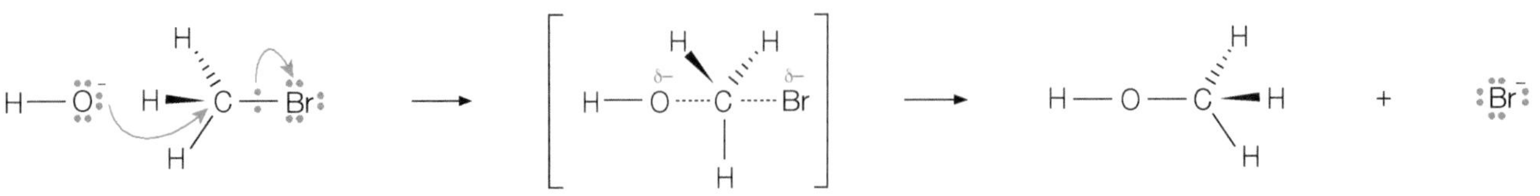

Figure 20.3 The single-step mechanism envisaged for an S_N2 reaction

The reasons for the different mechanisms are related to the difference in the energy required to reach the transition state in the two mechanisms. This depends on the degree of steric hindrance and the inductive effect, both maximized in tertiary halogenoalkanes.

Iodoalkanes are the most reactive halogenoalkanes, whereas fluoroalkanes are the least reactive. This is explained by the trend in the bond strength of the carbon–halogen bonds. The C–F bond is the most polar, and the shortest and strongest bond; the C–I bond is the least polar, and the longest and weakest bond. Hence, the C–I bond is the most easily broken during its substitution reactions.

Evidence for the S_N2 mechanism includes the observation of a second order rate expression. The S_N2 process is bimolecular, because two particles react in the rate-determining step (slow step).

When the carbon atom bearing the halogen is a chiral centre, the reaction proceeds with inversion of configuration (Figure 20.4). The requirement for back-side attack during the S_N2 mechanism is the need for effective overlap of orbitals. S_N2 reactions are said to be stereospecific because the configuration of the product is determined by the configuration of the reacting chiral halogenoalkane.

Evidence for the S_N1 mechanism includes the observation of a first order rate expression. The S_N1 process is unimolecular, because one particle reacts in the rate-determining step (slow step).

The carbocation intermediate (trigonal planar) can be attacked from either side by the nucleophile leading to racemization (Figure 20.5)

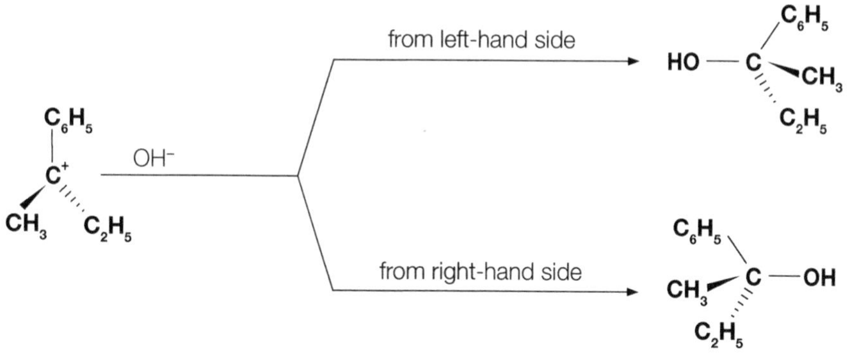

Figure 20.5 The racemization of an optically active halogenoalkane during nucleophilic substitution (S_N1)

Solvent effects in nucleophilic substitution

Protic solvents

Protic solvents contain at least one hydrogen atom connected directly to an electronegative atom. Examples include water, methanol, ethanol, ethanoic acid and ammonia.

Protic solvents stabilize cations and anions via the formation of ion–dipole interactions. This process is known as solvation. Cations are stabilized by lone pairs from the solvent molecules, while anions are stabilized by the formation of hydrogen bonds with the solvent molecules. As a result, cations and anions are both solvated and surrounded by a solvent shell (Figure 20.6).

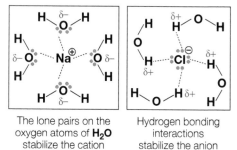

The lone pairs on the oxygen atoms of **H₂O** stabilize the cation

Hydrogen bonding interactions stabilize the anion

Figure 20.6 Cations and anions are both solvated by protic solvents such as water

Protic solvents favour the S_N1 mechanism by stabilizing polar intermediates and transition states.

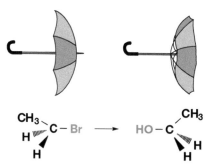

Figure 20.4 The inversion of the structure of groups around the carbon atom during an S_N2 reaction has been likened to an umbrella being blown inside-out in the wind

■ Aprotic solvents

Polar aprotic solvents contain no hydrogen atoms bonded directly to an electronegative atom. Examples include dimethylsulfoxide (DMSO), ethane nitrile and dimethylmethanamide.

Polar aprotic solvents stabilize cations, but not anions. Cations are stabilized by lone pairs from the solvent molecules, while anions are not stabilized by the solvent molecules.

Cations are solvated and surrounded by a solvent shell, but anions are not (Figure 20.7). As a result, nucleophiles are higher in energy when placed in a polar aprotic solvent.

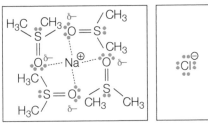

The lone pairs on the oxygen atoms of DMSO stabilize the cation

The anion is not stabilized by the solvent

Figure 20.7 Cations can be solvated by polar aprotic solvents such as DMSO; but not anions

Aprotic solvents favour the S_N2 mechanism by raising the energy of the nucleophile, resulting in a smaller activation energy barrier and hence a faster rate of reaction.

Table 20.1 summarizes the key differences between the two substitution mechanisms.

Table 20.1 A comparison between the two substitution mechanisms

	S_N1	S_N2
Kinetics	First order (two-step mechanism)	Second order (one-step mechanism)
Nature of mechanism	Two-step reaction via a carbocation intermediate	Concerted one-step mechanism with unstable transition state
Substrate (R–L), where R represents an alkyl group and L represents a leaving group	Tertiary halogenoalkane, as R–L that forms the most stable carbocation favoured	Primary halogenoalkane: methyl > primary > secondary (there should not be any bulky alkyl groups at the carbon centre)
Nucleophile	Not important	Non-bulky strong nucleophile favoured
Solvent	Polar, protic solvent favoured	Polar, aprotic solvent favoured
Stereochemistry	Racemization	Inversion
Leaving group (L)	Good leaving group favoured	Good leaving group favoured
Reaction profile		

Electrophilic addition reactions

Revised ☐

- An electrophile is an electron-deficient species that can accept electron pairs from a nucleophile. Electrophiles are Lewis acids.
- Markovnikov's rule can be applied to predict the major product in electrophilic addition reactions of unsymmetrical alkenes with hydrogen halides and interhalogens. The formation of the major product can be explained in terms of the relative stability of possible carbocations in the reaction mechanism.

The addition of halogens, such as bromine, proceeds via polarization of the halogen molecule (electrophile) resulting in the formation of an electron-deficient carbocation intermediate, which is attacked by the halide ion to form the product (Figure 20.8).

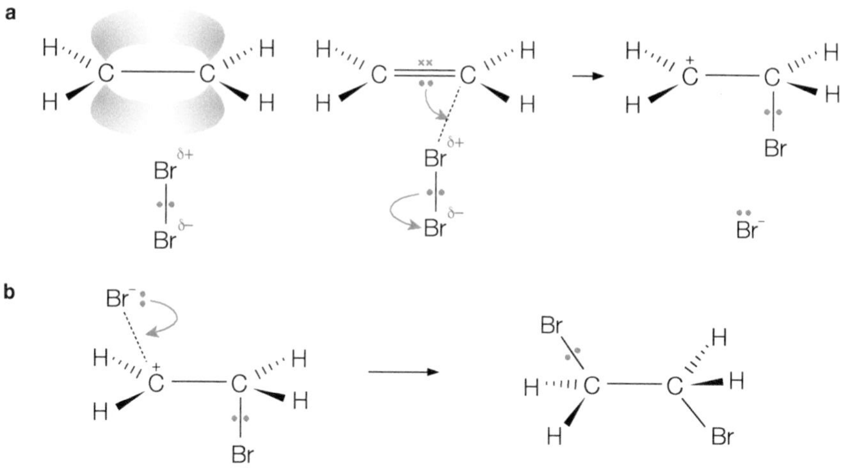

Figure 20.8 The mechanism of electrophilic addition of bromine

Addition to non-symmetrical alkenes

Electrophilic addition to alkenes, such as the addition of hydrogen halides (and interhalogens), follows Markovnikov's rule, which states that in the addition of a hydrogen halide, H–X, to an alkene, the hydrogen atom (or less electronegative atom) becomes bonded to the less-substituted carbon. The major product (Figure 20.9) is formed via the more stable carbocation: the inductive effect correlates with the number of alkyl groups.

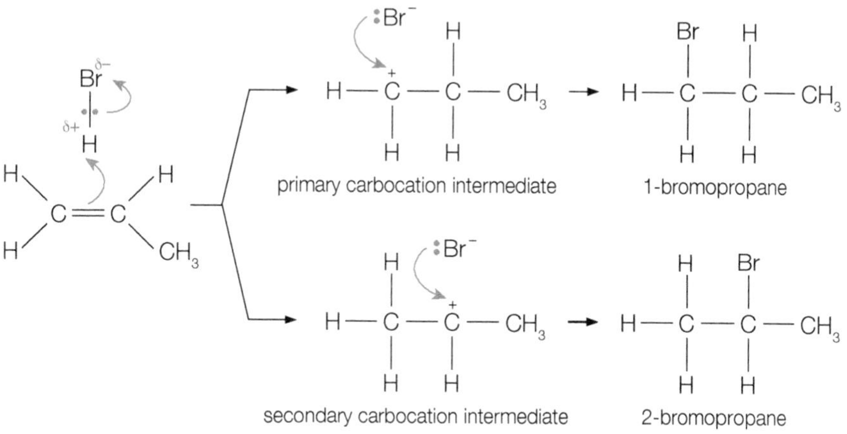

Figure 20.9 The possible reaction products when propene (an unsymmetrical alkene) reacts with hydrogen bromide

Expert tip

The reactivity of H–X increases when the position of X is lower down group 17. The increasing size of the X atom down the group lengthens and weakens the H–X bond. Order of reactivity: HF < HCl < HBr < HI.

NATURE OF SCIENCE

Markovnikov's rule (like Le Châtelier's principle), is a very useful predictive tool. However, it is not an explanation of why a particular product is formed; that is related to the stability of the carbocation formed.

Interhalogens are diatomic molecules containing two different halogen atoms bonded together. They are polar molecules and behave as electrophiles. Markovnikov's rule states that in the addition of an interhalogen, X–Y, to an alkene, the less electronegative halogen atom becomes bonded to the less-substituted carbon (Figure 20.10).

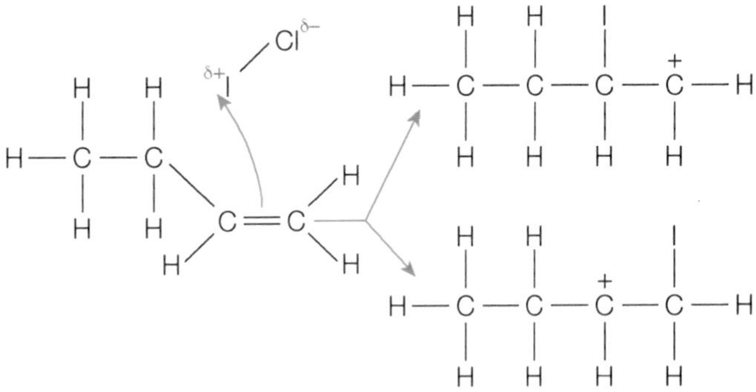

Figure 20.10 The possible reaction products when but-1-ene (an unsymmetrical alkene) reacts with iodine monochloride, ICl

NATURE OF SCIENCE

PRESENCE OF COMPETING SPECIES

Water molecules and chloride ions will compete with bromide ions for nucleophilic attack on the carbocation intermediate of ethene. Water is present in aqueous bromine, and chloride ions can be introduced to the reaction mixture by the addition of sodium chloride. The formation of a mixture of products (Figure 20.11) is supporting evidence for the electrophilic addition mechanism of alkenes.

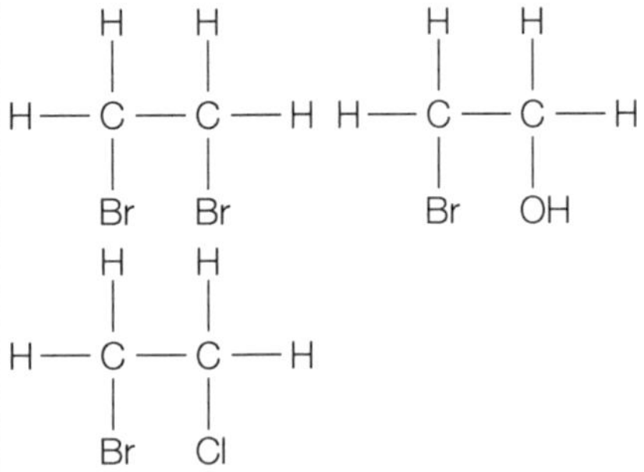

Figure 20.11 Formation of a mixture of products when ethene is reacted with bromine water containing sodium chloride

Electrophilic substitution reactions

Revised ☐

- Benzene is the simplest aromatic hydrocarbon compound (or arene) and has a delocalized structure of π bonds around its ring. Each carbon to carbon bond has a bond order of 1.5. Benzene is susceptible to attack by electrophiles.

Arenes

The electrophilic substitution of benzene involves two steps: a slow step involving attack by the pi electrons of the benzene ring to form a carbocation intermediate, followed by a fast deprotonation step which restores the aromatic benzene ring (Figure 20.12)

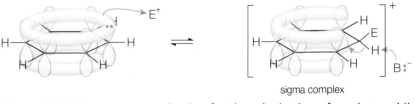
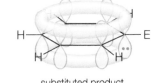

sigma complex

substituted product

Figure 20.12 The common mechanism for the substitution of an electrophile (E⁺) into the benzene ring, and the subsequent restoration of the stable ring structure involving a base

A mixture of concentrated sulfuric and nitric acids produces (via an acid–base reaction) the nitronium cation, NO_2^+, a powerful electrophile. Benzene reacts with the nitronium ion to produce nitrobenzene in a process called nitration (Figure 20.13).

$$2H_2SO_4 + HNO_3 \rightarrow NO_2^+ + 2HSO_4^- + H_3O^+$$

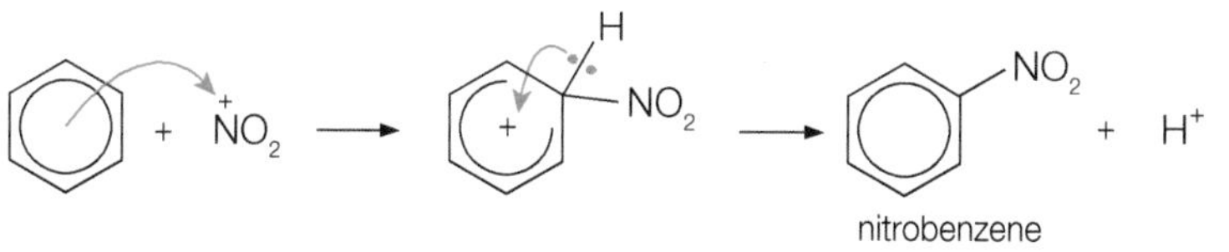

Figure 20.13 The formation of the nitronium cation and the electrophilic substitution of benzene to form nitrobenzene

Reduction reactions

■ Carboxylic acids can be reduced to primary alcohols (via the aldehyde). Ketones can be reduced to secondary alcohols. Typical reducing agents are lithium aluminium hydride (used to reduce carboxylic acids) and sodium borohydride.

Reduction

Reduction is the loss of oxygen, gain of hydrogen, gain of electrons and decrease in oxidation state. In organic chemistry, the first definitions are used more frequently, because it is often difficult to determine the change in oxidation states of individual carbon atoms in an organic molecule.

A summary of the various oxidation state levels is provided in Table 20.2 for common functional groups. The oxidation state of each carbon atom is directly related to the number of heteroatoms (atoms other than carbon and hydrogen) bonded to that carbon atom.

Table 20.2 A summary of the various oxidation levels for several common functional groups

Alkane level	Alcohol level	Aldehyde level	Carboxylic acid level	Carbon dioxide level
Bonded to zero heteroatoms	Bonded to one heteroatom	Bonded to two heteroatoms	Bonded to three heteroatoms	Bonded to four heteroatoms

Organic chemists often represent reduction (or oxidation) reactions of organic compounds using the symbols [H] for reductions and [O] for oxidations.

This classification helps in identifying the type of reaction required to interconvert functional groups. Carbon atoms in the same oxidation level are interconverted without oxidation or reduction. Carbon atoms at different oxidation levels are interconverted by oxidation or reduction.

Reduction of carbonyl compounds and carboxylic acids

The reduction of carbonyl compounds can be achieved by using lithium aluminium hydride, $LiAlH_4$, and sodium borohydride, $NaBH_4$. Both are ionic compounds and act as a source of hydride ions, H^-. This is a powerful nucleophile and attacks the carbon atom of a carbonyl group (Figure 20.14). Sodium borohydride is dissolved in aqueous or alcoholic solution, but the more powerful lithium aluminium hydride must be used in dry ethoxyethane ('ether').

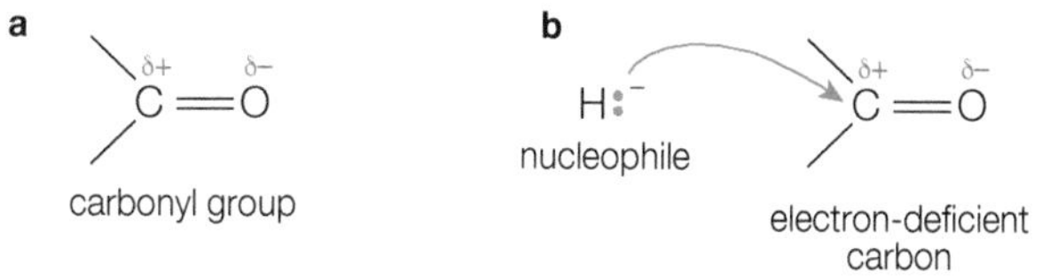

Figure 20.14 a The carbonyl group contains an electron-deficient carbon atom. **b** Reduction begins with the nucleophilic attack of an H^- ion

Expert tip

Lithium aluminium hydride hydrolyses in water, so it must be used under dry conditions. Water is added at the end of the reaction to complete the conversion.

LiAlH$_4$ and NaBH$_4$ can be used to reduce aldehydes to primary alcohols and ketones to secondary alcohols. LiAlH$_4$ is used to reduce carboxylic acids to primary alcohols.

■ Reduction of nitrobenzene

Phenylamine is prepared by reducing nitrobenzene using tin and concentrated hydrochloric acid. Tin atoms act as a reducing agent and supply electrons to form the phenylammonium ion. The addition of sodium hydroxide releases the free amine via an acid–base reaction (Figure 20.15).

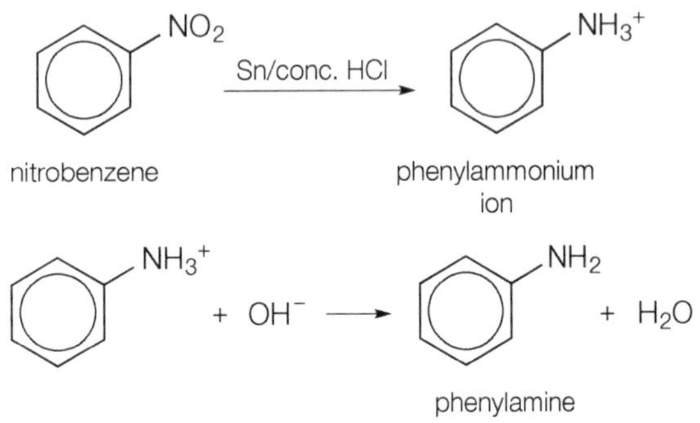

Figure 20.15 The reactions involved in the reduction of nitrobenzene by tin and concentrated hydrochloric acid

---

■ QUICK CHECK QUESTIONS

1. **a** Suggest why polar aprotic solvents are more suitable for S$_N$2 reactions whereas polar protic solvents favour S$_N$1 reactions.

   b Deduce, with a reason, if water or DMF (*N,N*-dimethylformamide), HCON(CH$_3$)$_2$) is a better solvent for the reaction between 1-bromoethane and sodium hydroxide.

2. **a** Predict the major products of the reaction between hydrogen bromide and but-1-ene.

   b Describe the polarization of the interhalogen compound Br–Cl. Give the condensed structural formula of the product of the reaction between Br-Cl and 2-methylpent-2-ene.

3. Give the names of the products of the following reactions:

   a CH$_3$CH$_2$CH$_2$CHO and NaBH$_4$

   b (CH$_3$)$_2$CHCOCH$_3$ and LiAlH$_4$

20.2 Synthetic routes

Revised ☐

Essential idea: Organic synthesis is the systematic preparation of a compound from a widely available starting material or the synthesis of a compound via a synthetic route that often can involve a series of different steps.

Synthetic routes

Revised ☐

- The synthesis of an organic compound stems from a readily available starting material via a series of discrete steps. Functional group interconversions are the basis of such synthetic routes.
- Retro-synthesis of organic compounds.

Recognizing the functional groups in an organic molecule enables chemists to predict its reactions. Knowing the reactions of the different functional groups in organic reactions enables chemists to devise synthetic routes to prepare organic compounds (Figure 20.16).

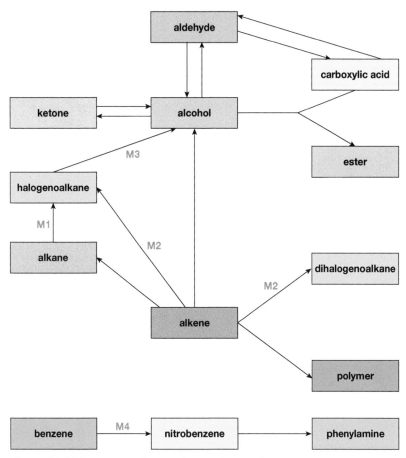

Figure 20.16 Summary of organic reaction pathways

The reaction mechanisms for these conversions that you are required to know are also shown on Figure 20.16; they are

■ M1, free radical substitution
■ M2, electrophilic addition
■ M3, nucleophilic substitution and
■ M4, electrophilic substitution.

Table 20.3 shows the reactions that you need to know and can therefore be used to deduce the synthetic routes.

Table 20.3 Useful reactions in organic synthesis

Functional group	Reaction	Products
Alkane	Free radical substitution	Halogenoalkane
Alkenyl	Electrophilic addition	Halogenoalkanes, dihalogenoalkanes, alcohols, polymers
Hydroxyl	• Esterification/condensation	• Ester
	• Oxidation	• Aldehydes, ketones, carboxylic acids
Halogeno	Nucleophilic substitution	Alcohols
Carboxyl	Reduction	Aldehydes, ketones and alcohols
Phenyl	Electrophilic substitution	
Carbonyl	Reduction	Alcohol

■ Retrosynthesis

In devising new synthetic routes, chemists use a strategy involving retrosynthesis. Starting with knowledge of the *target molecule* (the TM), they think in reverse to determine possible pathways to produce it. In retrosynthetic analysis, the last step of the synthesis is first established and the remaining steps are determined, working backwards from the product.

The strategy then involves progressively breaking down the target molecule into fragments, known as precursors. Each precursor then itself becomes the target of further analysis, eventually yielding a familiar and accessible starting molecule from which to begin the synthetic route:

target molecule (TM) → precursor 1 → precursor 2 → starting material

Retrosynthesis essentially involves looking at a target molecule and considering the bonds that can reasonably be made, and then generating a sub-structure and so working backwards. The principles and approach of retrosynthesis have proved immensely successful, with important features of the approach being:

- the starting materials should be readily available
- the techniques involved should be as straightforward as possible
- the number of stages required should be as few as possible to enable as high a yield as feasible.

Worked examples

Devise two-step synthetic routes to produce the following products from the given starting compound. Give the necessary conditions and equations for the reactions involved.

1 propanone from propene
2 pentan-1-ol from pentane
3 ethyl ethanoate from ethanol

1 propene $\xrightarrow{\text{conc. H}_2\text{SO}_4\text{/heat/H}_2\text{O}}$ propan-2-ol $\xrightarrow{\text{Na}_2\text{Cr}_2\text{O}_7\text{/H}^+\text{/reflux}}$ propanone

$CH_3CH=CH_2 + H_2O \rightarrow CH_3CH(OH)CH_3$

$CH_3CH(OH)CH_3 + [O] \rightarrow CH_3COCH_3 + H_2O$

2 pentane $\xrightarrow{\text{Cl}_2\text{/UV radiation}}$ 1-chloropentane $\xrightarrow{\text{NaOH(aq)/reflux}}$ pentan-1-ol

$CH_3(CH_2)_3CH_3 + Cl_2 \rightarrow CH_3(CH_2)_3CH_2Cl + HCl$

$CH_3(CH_2)_3CH_2Cl + NaOH \rightarrow CH_3(CH_2)_3CH_2OH + NaCl$

3 ethanol $\xrightarrow{\text{Na}_2\text{Cr}_2\text{O}_7\text{/H}^+\text{/reflux}}$ ethanoic acid $\xrightarrow{\text{more C}_2\text{H}_5\text{OH/H}^+\text{/reflux}}$ ethyl ethanoate

$C_2H_5OH + 2[O] \rightarrow CH_3COOH + H_2O$

$CH_3COOH + C_2H_5OH \rightarrow CH_3COOC_2H_5 + H_2O$

■ QUICK CHECK QUESTIONS

4 State broadly the essential conditions and reagents needed to carry out the following conversions:

a a carboxylic acid to a primary alcohol

b a ketone to a carboxylic acid

c an alkane to a bromoalkane

d an alkene to a dihalogenoalkane

e propene to propan-2-ol

f benzene to nitrobenzene

g nitrobenzene to phenylamine

h a primary alcohol to an aldehyde.

5 Outline the steps in a method for synthesizing the ketone, butanone, starting from an alkene. There may be more than one possible route; comment on these possibilities and on which route may give the best yield of the ketone.

6 How would you convert the halogenoalkane, 1-chlorobutane, into butanoic acid? Outline the steps used giving the reagents, conditions and equations for each.

20.3 Stereoisomerism

Essential idea: Stereoisomerism involves isomers which have different arrangements of atoms in space but do not differ in connectivity or bond multiplicity (i.e. whether single, double or triple) between the isomers themselves.

Stereoisomerism

- Stereoisomers are subdivided into two classes – conformational isomers, which interconvert by rotation about a sigma bond and configurational that interconvert only by breaking and reforming a bond. Configurational isomers are further subdivided into *cis–trans* and *E/Z* isomers and optical isomers.
- *Cis–trans* isomers can occur in alkenes or cycloalkanes (or heteroanalogues) and differ in the positions of atoms (or groups) relative to a reference plane. According to IUPAC, *E/Z* isomers refer to alkenes of the form $R_1R_2C=CR_3R_4$ (where $R_1 \neq R_2$, $R_3 \neq R_4$).
- A chiral carbon is a carbon joined to four different atoms or groups.
- An optically active compound can rotate the plane of polarized light as it passes through a solution of the compound. Optical isomers are enantiomers. Enantiomers are non-superimposeable mirror images of each other. Diastereomers are not mirror images of each other.
- A racemic mixture (or racemate) is a mixture of two enantiomers in equal amounts and is optically inactive.

Stereoisomers of a molecule contain the same atoms with the same bonding arrangement but differ in their three-dimensional (spatial) arrangement. Optical and *cis–trans* and *E/Z* isomers are examples of stereoisomers.

According to the IB syllabus, stereoisomers (Figure 20.17) are divided into two classes: conformational isomers, which interconvert by rotation about a sigma bond, and configurational isomers, which can only be interconverted by breaking and reforming a bond.

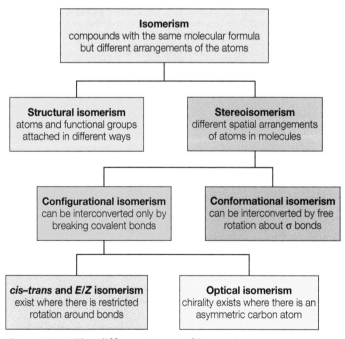

Figure 20.17 The different types of isomerism

Stereochemistry is the study of the spatial arrangements of atoms in molecules.

Cis–trans isomerism

Cis–trans isomers are found in alkenes where the carbon–carbon double bond, prevents free rotation, due to the presence of a pi bond (Figure 20.18). There is not sufficient heat energy to break the pi bond at room temperature and allow free rotation around the sigma bond.

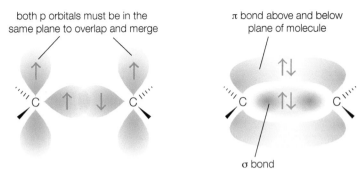

Figure 20.18 The nature of the carbon–carbon double bond in an alkene prevents rotation about the carbon–carbon axis. Single bonds are sigma (σ) only, and so rotation can take place without bond breaking (except in cyclic compounds)

The presence of two identical atoms or functional groups on opposite sides of the double bond can then give rise to *cis–trans* isomers (Figure 20.19). If two substituents are on the same side of the carbon–carbon double bond, the isomer is termed *cis*, whereas if they are on opposite sides the isomer is *trans*.

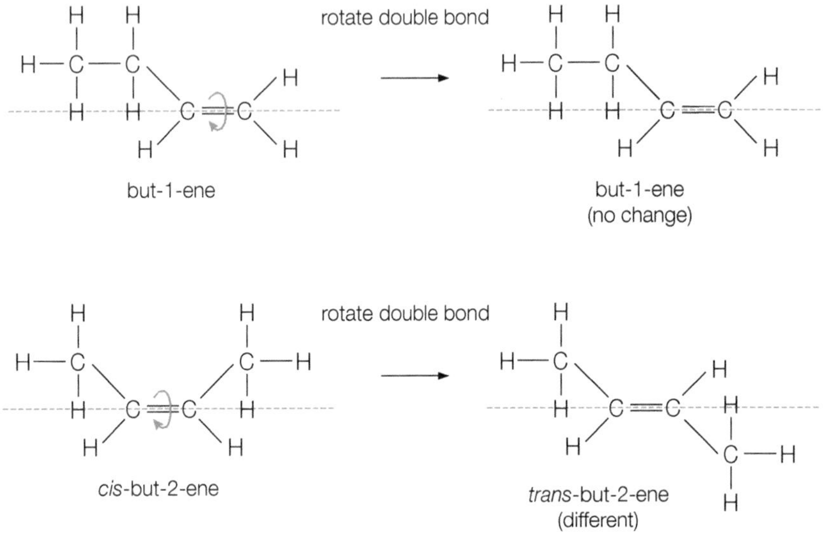

Figure 20.19 The different spatial arrangements possible for but-1-ene and but-2-ene

The priority sequencing rules of the Cahn–Ingold–Prelog convention are used to name di-, tri- and tetra-substituted alkenes (Figure 20.20). The higher priority substituent at carbon (1) is identified, as is that at carbon (2). If the two higher priority substituents are on opposite sides of the double bond, the alkene configuration is *E*; if they are on the same side, the configuration is *Z*. The convention is also applied to multiple carbon–carbon double bonds.

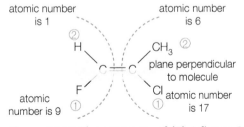

Figure 20.20 The structure of (*Z*)-1-fluoro-2-chloropropene. Note that the high priority groups (labelled 1) are both on the same side of the double bond

Cis–trans isomers of substituted alkenes generally have similar chemical properties but different physical properties. The *cis* isomer usually has a lower melting point than its *trans* isomer owing to its lower symmetry as it fits into a crystalline lattice more poorly.

However, the *cis* isomer usually has a higher boiling point than its *trans* isomer as the symmetrical nature of the *trans* isomer results in a net dipole moment of zero. Thus, the *trans* isomer is a non-polar compound. Since both isomers have the same molecular mass, the intermolecular forces (permanent dipole–permanent dipole) in the *cis* isomer is stronger than the London (dispersion forces) present in the *trans* isomer.

An irregularity in the melting point of *cis–trans* isomers occurs when there is a chemical interaction between the substituents. For example, *cis*-but-2-ene-1,4-dioic acid melts with decomposition at 130–131 °C. However, *trans*-but-2-ene,1,4-dioic acid does not melt until 286 °C.

In the *cis* isomer, the two carboxylic acid groups are closer together so intra-molecular hydrogen bonding occurs. In the *trans* isomer, the two carboxylic acid groups are too far apart to attract each other, resulting in stronger intermolecular forces of attraction between adjacent molecules and higher melting point. The *cis* isomer also reacts when heated to lose water and forms a cyclic acid anhydride. However, the *trans* isomer cannot undergo this reaction (Figure 20.21).

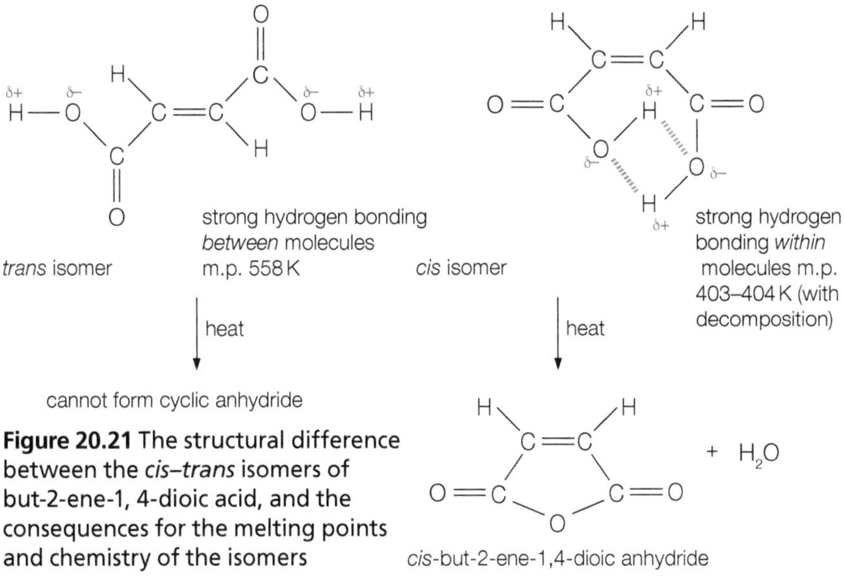

Figure 20.21 The structural difference between the *cis–trans* isomers of but-2-ene-1, 4-dioic acid, and the consequences for the melting points and chemistry of the isomers

Other double bond systems also show *cis–trans* isomerism (Figure 20.22). Oximes and imines contain C=N double bonds, so can exist in two forms.

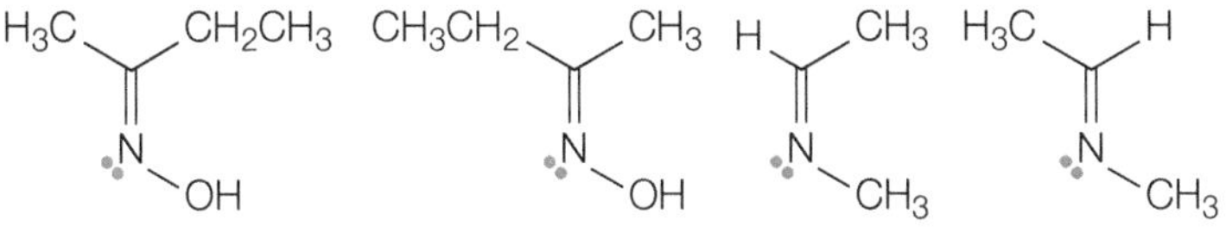

Figure 20.22 The two forms of butanone oxime and ethanal methylimine

Cis–trans isomerism is also possible in substituted cyclic alkanes, such as 1,2-dichlorocyclopropane because the rigid ring prevents free rotation (Figure 20.23) around the single bonds. In cyclic compounds, there must be at least two substituent groups bonded to different carbon atoms of the ring; the carbon atoms need not be adjacent and double bonds in a ring do not have geometric isomers as they can only exist in the *cis* configuration.

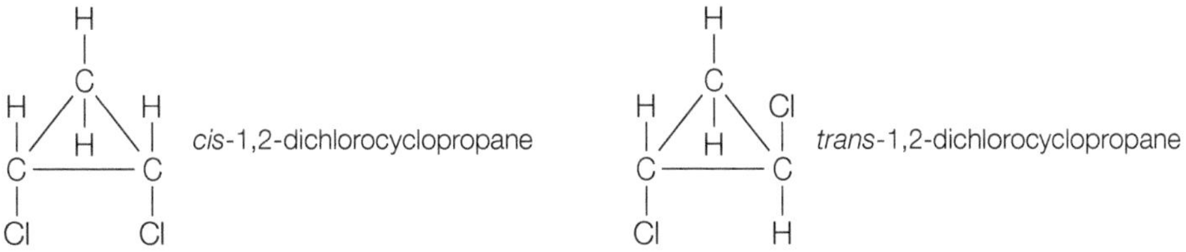

Figure 20.23 Simplified representations of the *cis* and *trans* isomers of 1,2-dichlorocyclopropane

■ Optical isomerism

Optical isomers are molecules that are non-superimposable mirror images of each other. Such molecules often contain a chiral centre. This is often a carbon atom bonded to four different atoms or functional groups (Figure 20.24).

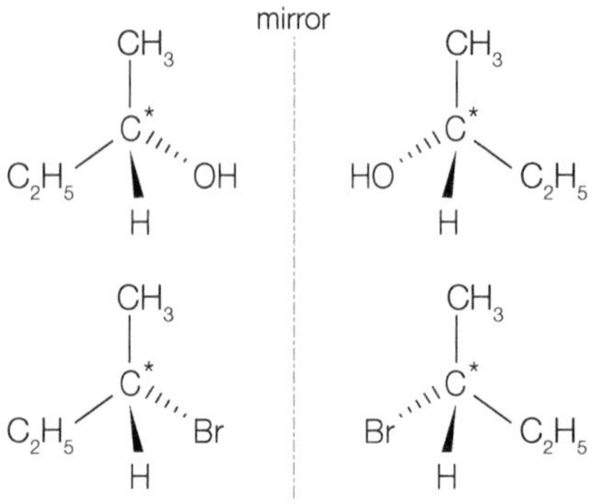

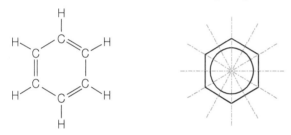

Figure 20.24 The enantiomers of butan-2-ol and 2-bromobutane

For a molecule to be chiral it should not possess either a plane of symmetry or centre of symmetry. A plane of symmetry ('mirror plane') is a plane which divides the molecule into two halves that are superimposable (Figure 20.25).

Expert tip

Chiral objects, for example, a pair of left and right hands, are not superimposable on their mirror images.

Figure 20.25 Benzene is a molecule with seven planes of symmetry; the six shown here and the plane of the molecule itself

A centre of symmetry is a point or an atom from which, if a straight line is drawn from any part of the molecule through that atom and extended to an equal distance by a straight line on the other side, it meets the same point (or an atom).

Fischer projections are drawings that show the configuration of chiral centres, without the use of wedges and dashes (Figure 20.26). All horizontal lines represent wedges and all vertical lines represent dashes.

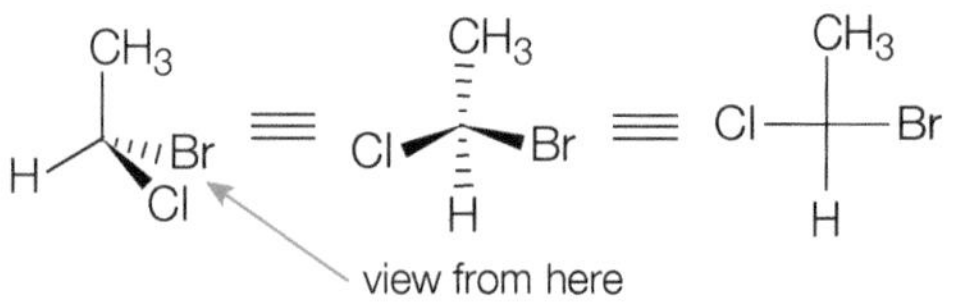

Figure 20.26 A Fischer projection for 1-bromo-1-chloro-ethane

Enantiomers have identical physical properties, except for their effect on plane-polarized light, and have identical chemical properties, except in their interactions with another chiral molecule (including enzymes).

Sugars, amino acids (except glycine) and most biological molecules are chiral. Usually, only one of the two isomers is synthesized in nature. Pharmaceutical drugs that are designed to interact with receptors in cells of organisms also tend to have chiral centres. The active site of enzymes usually react with only one of the enantiomers and not the other.

■ Optical activity

Ordinary visible light is an electromagnetic wave which has vibrations in all three planes. Plane-polarized light can be obtained by passing the light through a diffraction grating or calcite crystal (Figure 20.27). Polarized light has all the waves vibrating in the same plane.

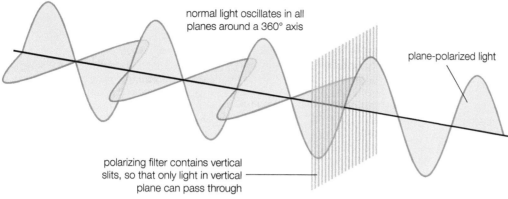

Figure 20.27 A representation of how a polarizing filter works

Compounds which rotate the plane of polarized light are called optically active compounds. Optical activity is characteristic of chiral compounds. A polarimeter (Figure 20.28) is an instrument used to measure the angle of rotation of plane-polarized light by optically active compounds.

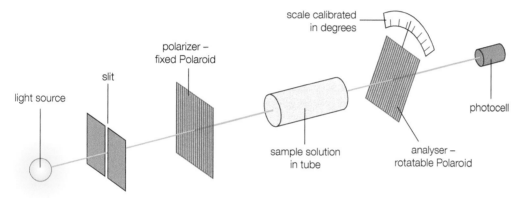

Figure 20.28 The various stages of a polarimeter for analysing optical isomers

A single enantiomer will rotate the plane of plane-polarized light and is referred to as optical activity. An enantiomer is given the prefix (+), if the rotation is clockwise and (−) if the rotation is anticlockwise.

A racemic mixture is a mixture of equal amounts of the enantiomers of the same chiral compound. The mixture does not rotate plane-polarized light because the two optical isomers (enantiomers) have equal and opposite effects, so they cancel each other out.

▧ Diastereoisomers

For a compound with multiple chiral carbon atoms, a family of stereoisomers exists (Figure 20.29). Each stereoisomer will have at most one enantiomer, with the remaining members of the family being diastereoisomers. The number of stereoisomers of a compound can be no larger than 2^n, where n is the number of chiral centres.

mirror plane

H
|
H_3C ⅶC^* CHClCH$_3$
Br

Compound A

H
|
H_3CClHC C^* ⅲCH$_3$
Br

Compound B

H
|
H_3C ⅶC^* CHBrCH$_3$
Cl

Compound C

H
|
H_3CBrHC C^* ⅲCH$_3$
Cl

Compound D

mirror plane

Figure 20.29 2-bromo-3-chlorobutane has two chiral centres, giving rise to $2^2 = 4$ stereoisomers

Compounds A and B or C and D exist as a pair of enantiomers. However, compound A and compound C or D exist as a pair of diastereoisomers.

Resolution is the separation of enantiomers from a racemic mixture and can be achieved by conversion of enantiomers into stereoisomers, which are usually salts, and then using solubility differences to achieve separation.

The flow chart in Figure 20.30 presents questions, the answers to which lead to the classification of either structural or stereoisomers.

Expert tip

Enantiomers are mirror images; diastereoisomers are not mirror images. Enantiomers and diastereoisomers must be drawn using solid and dashed wedges to show the three-dimensional shape.

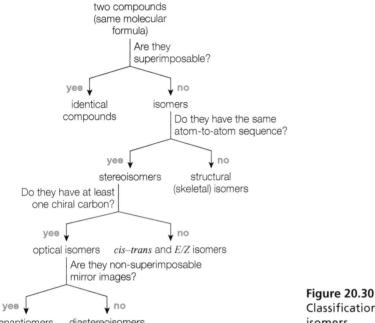

Figure 20.30
Classification of isomers

Conformations

Conformational isomers differ from one another in the arrangement of atoms around a single bond. They can be interconverted by rotation about a sigma bond without breaking any bonds. Different conformations can have different amounts of strain and so have different energies.

Expert tip

Conformations cannot be separated (even at low temperatures) because they rapidly interconvert.

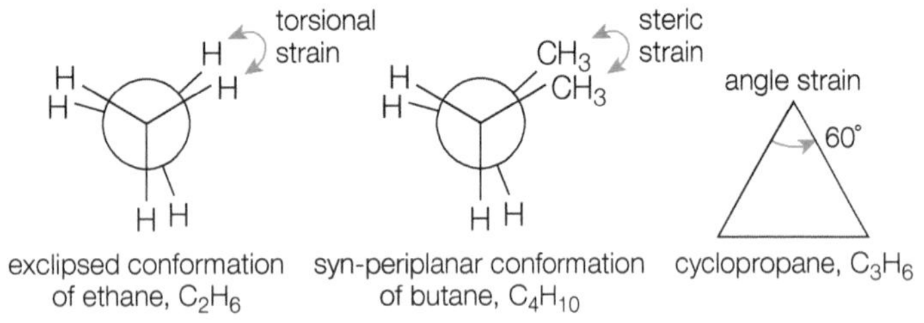

There are three types of molecular strain: torsional, steric and angle strain (Figure 20.31). Torsional strain is due to electron–electron repulsions in adjacent bonds. Steric strain is due to crowding of (particularly large-sized) groups in a certain conformation. Angle strain arises when the C–C–C bond angle (in non-aromatic cyclic compounds) is different from the normal tetrahedral bond angle (109.5).

torsional strain	steric strain	angle strain
exclipsed conformation of ethane, C_2H_6	syn-periplanar conformation of butane, C_4H_{10}	cyclopropane, C_3H_6

Figure 20.31 Types of molecular strain

Alkanes

Small three-dimensional molecules are commonly viewed by sawhorse or Newman projections (Figure 20.32).

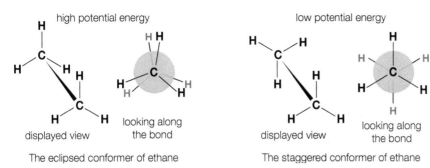

Figure 20.32 The eclipsed and staggered conformations of ethane as shown by sawhorse and Newman projection

Rotation around the axis of the C–C sigma bond means that one set of three hydrogen atoms rotates relative to the other causing the conformation of ethane to change. As the hydrogens pass each other (eclipsed conformation) the energy of the ethane molecule rises (Figure 20.33). The angle between the C–H bonds on the front and back carbon atoms is called the torsional (dihedral) angle.

At 60° away from this eclipsed conformation is the most stable conformation (three-dimensional shape), known as staggered; intermediate conformations are termed skewed.

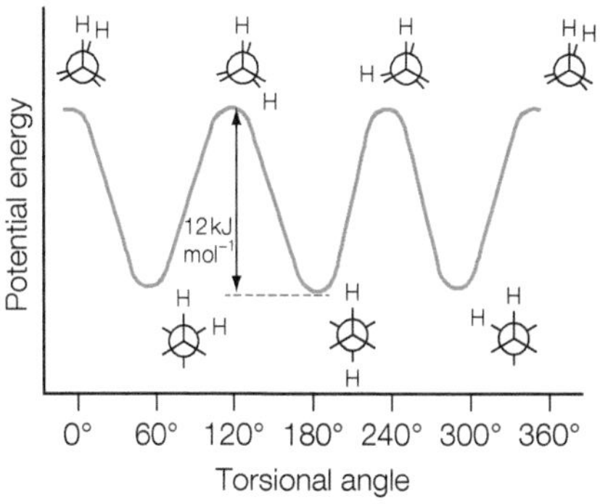

Cycloalkanes

The conformations of cycloalkanes are determined by torsional strain and steric strain, but they also have angle strain. Cyclopropane and cyclobutane are the least stable of the cycloalkanes because of high angle strain. Cyclopentane and cyclohexane are the most stable cycloalkanes (and therefore the easiest to prepare) because of low angle strain.

Except for cyclopropane, all cycloalkanes have non-planar rings. The non-planar conformations relieve the torsional strain that would result from eclipsed substituents and the angle strain that would be present in the planar form.

Cyclopropane

In cyclopropane (Figure 20.34), the three carbon atoms are in the same plane so that the C–C–C bond angle is only 60°. The three-membered ring is highly strained because of torsional strain, owing to eclipsed hydrogen atoms on adjacent carbon atoms, and also because of angle strain as the C–C–C bond angle is compressed (from 109.5° to 60°). This results in poor overlap between the sp³ hybridized orbitals, resulting in the bonds being bent (and therefore weak).

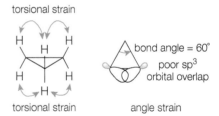

Figure 20.34 Cyclopropane

Figure 20.33 Energy diagram for rotation in ethane. Only two of the hydrogens are shown for clarity. For ethane, the rotational energy barrier is 12 kJ mol⁻¹

Cyclobutane

If all four carbon atoms of cyclobutane were in the same plane, the C–C–C bond angle would be 90°. The highly strained planar conformation can relieve the torsional strain by moving one of the carbon atoms out of the plane and forming a puckered ring (butterfly conformation) (Figure 20.35). The decrease in torsional strain outweighs the small increase in angle strain.

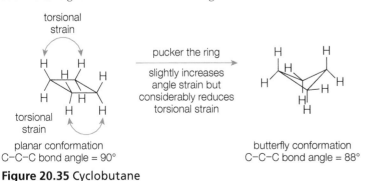

Figure 20.35 Cyclobutane

■ Cyclohexane

The planar conformation of cyclohexane has C–C–C bond angles of 120°, which means this conformation has severe angle strain in addition to the torsional strain caused by 12 eclipsing C–H bonds from the six carbons. If one carbon moves down, and the other up, the ring is puckered to form the chair conformation (Figure 20.36). This is the most stable conformation and has very little bond angle and torsional strain. Hydrogen atoms occupy equatorial and axial positions.

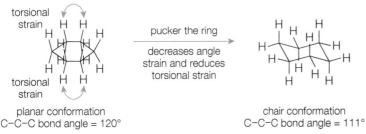

Figure 20.36 The chair conformation of cyclohexane

■ QUICK CHECK QUESTIONS

7 Which of the following molecules exhibit *cis–trans* isomerism? For those examples that do, draw their isomeric structures and label them as the *E* or *Z* form.

   a 1-chlorobut-1-ene

   b 2-methylbut-2-ene

   c 3-methylpent-2-ene

8 How many four-membered ring isomers are there of dichlorocyclobutane, $C_4H_6Cl_2$?

9 Which of the following compounds is optically active?

A	B
2,2-dimethylpropane	1-chlorobutane

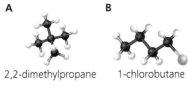

C	D
2-aminoethanoic acid	2-chlorobutane

10 The following molecules **A, B** and **C** all show stereoisomerism. State and explain whether each example is the *E*- or *Z*- form of the isomeric pair.

A

$$\begin{array}{c} H \qquad\quad CH_3 \\ \diagdown \qquad \diagup \\ C = C \\ \diagup \qquad\quad \diagdown \\ CH_3CH_2 \qquad H \end{array}$$

B

$$\begin{array}{c} HOCH_2 \qquad CH_2CH_2Cl \\ \diagdown \qquad\quad \diagup \\ C = C \\ \diagup \qquad\quad \diagdown \\ CH_3CH_2 \qquad CH_2Cl \end{array}$$

C

$$\begin{array}{c} \qquad\qquad\quad O \\ \qquad\qquad\quad \| \\ ClCH_2 \qquad\; C \\ \diagdown \quad\quad \diagup \diagdown \\ C = C \qquad OCH_3 \\ \diagup \quad\quad \diagdown \\ CH_3 \qquad CH_2OH \end{array}$$

21.1 Spectroscopic identification of organic compounds

Revised ☐

Essential idea: Although spectroscopic characterization techniques form the backbone of structural identification of compounds, typically no one technique results in a full structural identification of a molecule.

Spectroscopic identification of organic compounds

Revised ☐

- Structural identification of compounds involves several different analytical techniques including IR, ^{1}H NMR and MS.
- In a high-resolution ^{1}H NMR spectrum, single peaks present in low resolution can split into further clusters of peaks.
- The structural technique of single crystal X-ray crystallography can be used to identify the bond lengths and bond angles of crystalline compounds.

■ High-resolution NMR

In low-resolution NMR spectra, protons in a particular environment produce a single peak, but in high-resolution NMR spectra the single peaks may be split into a group of peaks because the nuclear spins of adjacent protons in molecules interact with one another. This is termed spin–spin splitting and occurs in the presence of a powerful and accurately controlled homogeneous magnetic field.

The exact splitting pattern of a peak depends on the number of hydrogen atoms on the adjacent carbon atom or atoms. The number of signals a given peak splits into equals $n + 1$, where n is the number of hydrogen atoms on the adjacent carbon atoms. The high-resolution ^{1}H NMR spectrum of ethanol (Figure 21.1) illustrates this $n + 1$ rule used to interpret splitting patterns.

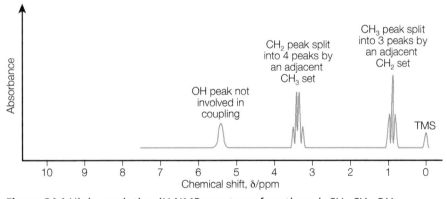

Figure 21.1 High-resolution ^{1}H NMR spectrum for ethanol, CH_3-CH_2-OH

The –CH_3 peak is split into three signals because there are two hydrogen atoms on the adjacent –CH_2– group. $n + 1 = 3$ (as $n = 2$); this is called a triplet.

The –CH_2– peak is split into four signals because there are three hydrogen atoms on the adjacent –CH_3 group. $n + 1 = 4$ (as $n = 3$); this is called a quartet.

The –OH peak is not usually split as its hydrogen atom is constantly being exchanged with the hydrogen atoms of other ethanol molecules and water molecules or acid present. This results in one averaged peak being observed.

Table 21.1 summarizes the descriptions of peaks found in high-resolution ^{1}H NMR spectra of organic molecules.

Table 21.1 Descriptions of simple peak clusters in a high- resolution ^1H NMR spectrum

Number of hydrogens on carbon adjacent to resonating hydrogen	Number of lines in cluster (multiplet)	Relative intensities
1	2	1:1
2	3	1:2:3
3	4	1:3:3:1
4	5	1:4:6:4:1

Expert tip

The relative intensities of the peaks in the cluster is given by Pascal's triangle.

Predictable patterns of split peaks due to the specific arrangements of adjacent hydrogen atoms are observed (Figure 21.2). The $n + 1$ rule predicts the splitting pattern: spin–spin splitting by an adjacent group of n hydrogen atoms will cause the signal to split into $n + 1$ peaks.

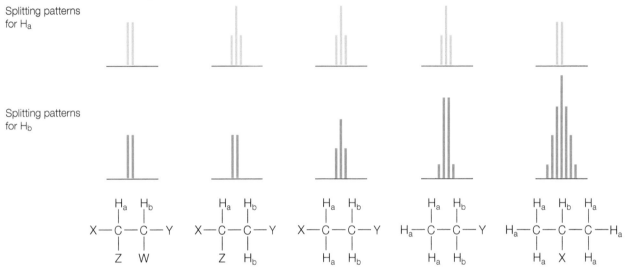

Figure 21.2 Commonly observed splitting patterns in high-resolution ^1H NMR spectra

---

Worked example

The deduction of the appearance of the high-resolution ^1H NMR spectrum of butan-1-ol (Figure 21.3) is discussed below, taking each individual carbon atom in turn.

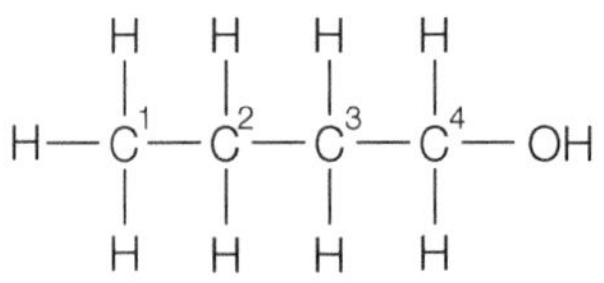

Figure 21.3 Butan-1-ol with numbered carbon atoms

CARBON ATOM 1

These three methyl hydrogen atoms are all equivalent and in the same chemical environment and will not couple or split with each other. However, they will couple with the two hydrogens attached to the adjacent carbon atom: $n + 1 = 2 + 1 = 3$ (a triplet).

CARBON ATOM 2

These two hydrogen atoms will couple with the three hydrogen atoms on the methyl group and the two carbon atoms on the adjacent methylene ($-CH_2-$) group: $n + 1 = 5 + 1 = 6$ (a sextet).

CARBON ATOM 3

The two hydrogen atoms will couple with both pairs of hydrogen atoms on the two adjacent methylene carbon atoms ($-CH_2-$): $n + 1 = 4 + 1 = 5$ (a quintet).

CARBON ATOM 4

The two hydrogen atoms will couple with the two hydrogen atoms attached to the methylene carbon (to the left in the diagram): $n + 1 = 2 + 1 = 3$ (triplet).

Expert tip

Note that a peak in a high-resolution ^1H NMR spectrum can be split *twice* by hydrogen atoms in two neighbouring atoms (Figure 21.4). For example, in the spectrum of propanal (Figure 21.5), the peak for the hydrogen atoms labelled II is split into four by the hydrogen atoms labelled I and then split again by the hydrogen atom labelled III. The peak for II is therefore split into a quartet of doublets.

■ QUICK CHECK QUESTION

1   Deduce the combination of peaks in the ^1H NMR spectrum of diethyl ether, $CH_3CH_2OCH_2CH_3$ and 2-methylpropane.

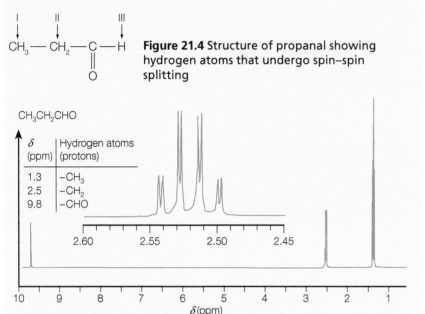

Figure 21.4 Structure of propanal showing hydrogen atoms that undergo spin–spin splitting

Figure 21.5 ^1H NMR spectrum of propanal showing a quartet of doublets for the $-CH_2-$ hydrogen atoms

■ QUICK CHECK QUESTIONS

2   Sketch the splitting pattern that will be observed in the high-resolution ^1H NMR spectra for 2-chloropropanoic acid, $CH_3CHClCOOH$, and 3-chloropropanoic acid, CH_2ClCH_2COOH.

3   The ^1H NMR spectrum of a compound with the formula $C_4H_8O_2$ exhibits three major peaks. The chemical shifts, areas and splitting patterns of the peaks are given below.

Chemical shift/ppm	Peak area	Splitting pattern
0.9	3	Triplet
2.0	2	Quartet
4.1	3	Singlet

Deduce the structure and explain your reasoning.

4   An unknown compound has the formula C_3H_7I and an IHD of zero. Its high resolution ^1H NMR spectrum consists of a doublet, δ 1.9 ppm, from six hydrogens, and a septet, δ 4.3 ppm. Deduce the structure of the compound. Explain your reasoning.

■ Tetramethylsilane

The different magnetic field strengths are measured relative to a reference compound (internal standard), which is given a value of zero. The standard compound chosen is tetramethylsilane (TMS). TMS is an inert, volatile liquid that mixes well with most organic compounds and is easily removed after analysis by evaporation.

Its formula is $Si(CH_3)_4$, so all its hydrogen atoms are chemically equivalent and in the same chemical environment. TMS gives one sharp peak which is at a higher frequency (upfield) than most other hydrogen atoms (protons). All other absorptions are measured by their shift relative (usually downfield) to the TMS line on the NMR spectrum (Figure 21.6).

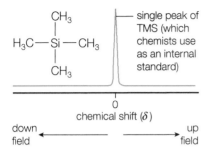

Figure 21.6 The standard TMS peak used as an internal reference on ¹H NMR spectra

---

■ **QUICK CHECK QUESTIONS**

5   Peaks in a ¹H NMR spectrum are measured relative to a reference standard. State the name of a substance used and identify **two** properties that makes it particularly suitable for this application.

6   In NMR tetrachloromethane is often used as a solvent to prepare samples for NMR analysis. Suggest why tetrachloromethane is used as solvent.

▦ Combined techniques

IR, NMR, UV-vis and mass spectra are all used in the analysis of organic compounds, but frequently it is not possible to establish an identity or structure from one technique alone. The compounds are most quickly and easily analysed by combining the four techniques (Figure 21.7). This is because the information provided by one spectrum may fill the gaps in information from another and the various spectra complement one another.

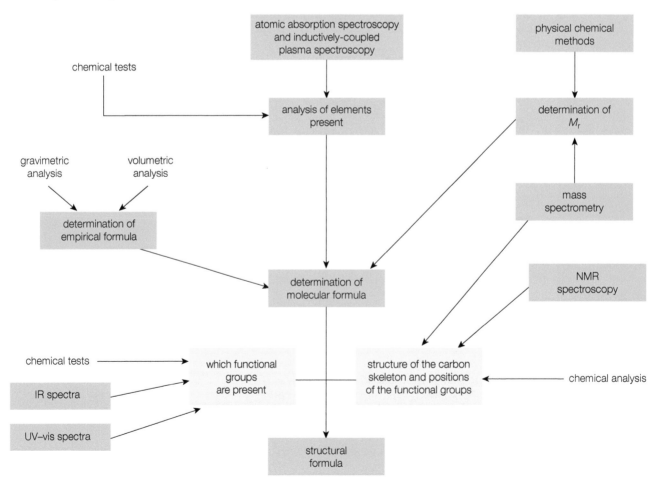

Figure 21.7 Summary of methods used to determine the structure of an organic compound

The contribution that each type of spectrum can contribute to the knowledge of an organic compound is shown in Table 21.2.

Table 21.2 Information provided by mass spectra, infrared spectra and ¹H NMR spectra

Type of spectrum	Information provided about the nature of the organic compound.
Mass	Accurate relative molecular mass from the molecular ion peak; possible structure from the fragmentation pattern and the presence of halogens (chlorine or bromine) from [M + 2].
Infrared	The presence of functional groups from wavenumbers of absorption peaks; the identity of the compound from the 'fingerprint' region.
Nuclear magnetic resonance	The identity of chemical groups containing hydrogen atoms from the chemical shift; the arrangement of hydrogen containing groups in the molecule from the high-resolution spin–spin splitting pattern.

Worked example

The infrared spectrum (Figure 21.8) has a sharp absorption band at 3000 cm⁻¹ indicating the presence of C–H bonds and the absorption band at 1200 cm⁻¹ suggests the presence of a C–O bond. The absorption band near 600 cm⁻¹ suggests the presence of a C–Br bond.

The high-resolution NMR spectrum shows two types of hydrogen atom in the ratio 6:1. The single hydrogen atom is the one bonded to the C–Br. It is split into a septet (a group of 7 peaks) by the six identical hydrogen atoms of the two –CH₃ groups. Its chemical shift value, δ, of 4.2 ppm shows that it relatively unshielded because of its proximity to the electronegative bromine atom.

The peak of area 6 is caused by the six hydrogen atoms of the two –CH₃ groups and is split into two by the single hydrogen atom of the C–Br group. These six hydrogen atoms are further from the bromine atom and are well shielded as shown by their chemical shift value, δ, of 1.8.

The mass spectrum shows an M⁺ and an [M + 2]⁺ peak at m/z 122 and 124, suggesting the presence of a halogen. The group of peaks at m/z 41 and 43 are due to the cations derived from the hydrocarbon part of the molecule without the halogen. The M⁺ peaks include the mass of the halogen, which is bromine: ⁷⁹Br and ⁸¹Br. The peak at m/z = 43 is due to the presence of [C₃H₇]⁺.

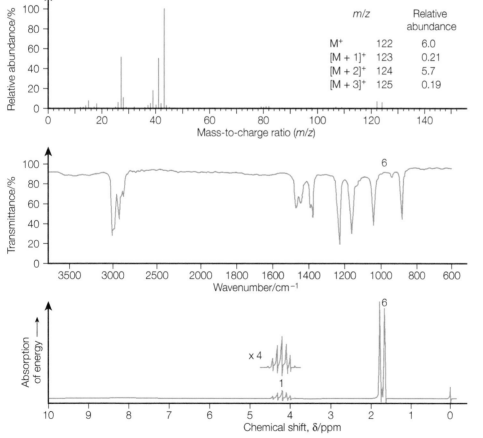

	m/z	Relative abundance
M⁺	122	6.0
[M + 1]⁺	123	0.21
[M + 2]⁺	124	5.7
[M + 3]⁺	125	0.19

Figure 21.8 The mass spectrum, infrared spectrum and high-resolution ¹H NMR spectrum of an unknown compound

■ **QUICK CHECK QUESTION**

7   Identify the compound from an analysis of its mass spectrum, infrared spectrum and NMR spectrum.

Elemental analysis reveals the following composition: carbon (64.86%), hydrogen (13.51%) and oxygen (21.62%). The molecule contains four carbon atoms.

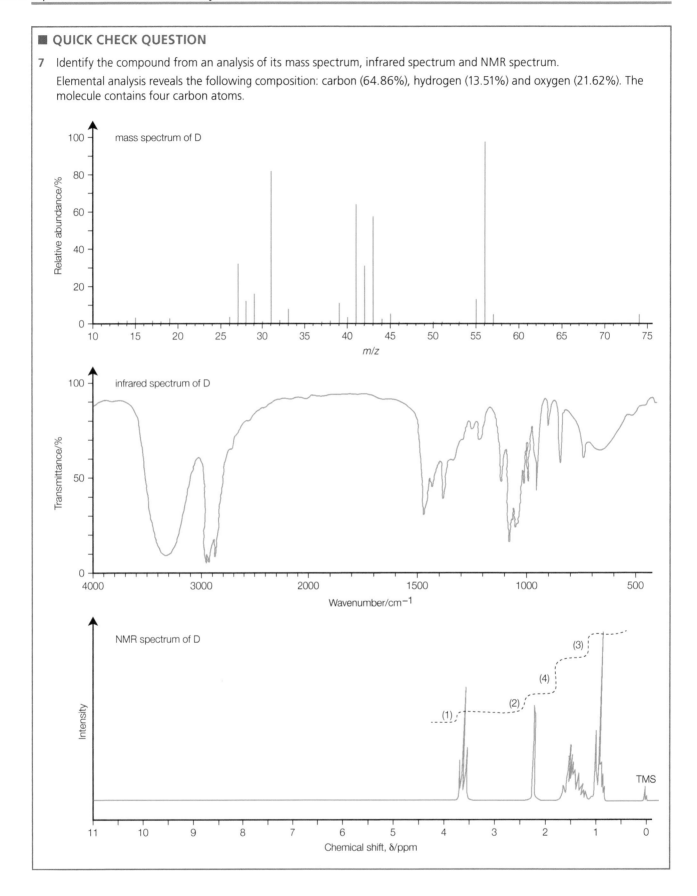

■ Magnetic resonance imaging (MRI)

Because the human body contains a high concentration of water within its tissues, this responds to nuclear magnetic resonance (NMR). Like NMR, MRI depends on low-energy nuclear spin processes occurring in water molecules. The energy absorbed from a short pulse of radio waves is released as a longer signal after the pulse has ended. This process is known as relaxation.

By scanning with pulses of radio waves of different energy, an image of the water distribution within the body can be built up, which is useful in the diagnosis of various illnesses, in particular brain disorders.

The hydrogen nucleus in a water molecule behaves like a bar magnet since it has a magnetic moment (due to the spinning charged proton). In a strong external magnetic field, the hydrogen nuclei either line up in the direction of the magnetic field (lower energy state), or in the opposite direction (higher energy state). Each nucleus (proton) behaves like a gyroscope and its magnetic moment precesses about the external magnetic field. A superconducting magnet is usually used to generate such a strong magnetic field in an MRI scanner (Figure 21.9).

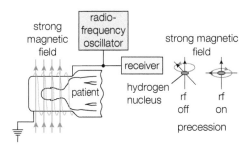

Figure 21.9 The MRI scanner and its effect on hydrogen nuclei (protons)

By applying magnetic field gradients that change strength along the x, y and z directions, the exact location where the radio signal is emitted from can be located in the body.

The magnetic field gradients are varied continuously so the point of location of the radio signal (relaxation signal) changes systematically. The strength of the signal is a measure of the concentration of water molecules, which varies between tissues and healthy and unhealthy tissue.

As the frequency of the radio waves is changed, hydrogen nuclei in the water molecules at different depths inside the body respond to the radio waves and resonate. The direction of the magnetic field gradient is rotated 360° which enables the computer to produce a two-dimensional image – a 'slice', which can distinguish between water in the grey or white tissue of the brain, or in cancerous (malignant) or normal cells. It is non-invasive and the radio waves are of very low energy compared to the X-rays used in computerized tomography (CT).

■ **QUICK CHECK QUESTION**

8   MRI and CT scans are important diagnostic tools in medicine. State **two** advantages of using MRI.

   Outline how NMR is used in an MRI scanner.

■ X-ray diffraction

Diffraction is a property of all waves and the separation distances of atoms in molecules (when in crystalline solids) is of the order to cause diffraction of X-rays (Option A Materials). The diffraction can be visualized as reflection from planes of atoms (Figure 21.10).

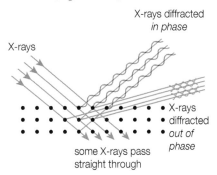

Figure 21.10 Diffraction of X-rays

The resultant diffraction pattern reveals information that enables the position of all atoms (except hydrogen) to be established in a complex molecule, such as a protein. X-ray diffraction can also give information about distances between adjacent nuclei and hence atomic radius, bond length and bond angle.

The analysis of X-ray diffraction patterns allows electron density maps (Figure 21.11) to be generated where atomic arrangements and separation distances are clearly shown. From the electron density map the precise location of each atom in the molecule can be determined and since heavier atoms have more electrons than lighter ones each atom in the molecule can be identified.

> **Expert tip**
>
> Since a hydrogen atom has a low electron density it is not easily detected by X-rays.

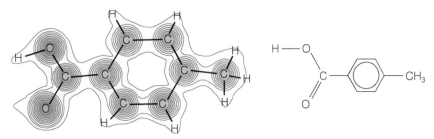

Figure 21.11 The structure of the 4-methyl benzoic acid molecule superimposed on its electron density map

---

■ QUICK CHECK QUESTION

9   The diagram below shows the electron density map of an aromatic compound with the molecular formula $C_6H_3I_3O_3$. Draw the structural formula for the compound and explain how your structure relates to the electron density map.

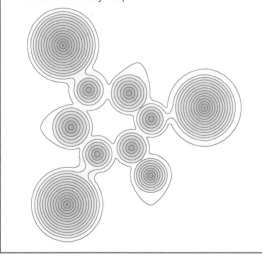

> **Expert tip**
>
> X-rays are not used in the same way as in classical spectroscopy, for example, IR. X-ray diffraction is not a spectroscopic technique.